Ecological Studies, Vol. 178

Analysis and Synthesis

Ecological Studies

Volumes published since 1992 are listed at the end of this book.

William K. Smith
Thomas C. Vogelmann
Christa Critchley
Editors

Photosynthetic Adaptation

Chloroplast to Landscape

With 94 illustrations

William K. Smith
Department of Biology
Wakeforest University
Winston-Salem, NC 27109
USA
smithwk@wfu.edu

Thomas C. Vogelmann
109 Carrigan Drive
Burlington, VT 05405-0086

Christa Critchley
Department of Botany
University of Queensland
Brisbane 4072
Australia
c.critchley@botany.uq.edu.au

Cover illustration: Schematic representation of the structural/spatial hierarchy of organizational complexity in the plant kingdom (see Figure 1.1 on page 5).

Library of Congress Cataloging-in-Publication Data
Photosynthetic adaptation : chloroplast to landscape / William K. Smith, Thomas C. Vogelmann, Christa Critchley.
p. cm. — (Ecological studies)
Includes bibliographical references and index.
ISBN 0-387-22079-8 (hc : alk. paper)
1. Chloroplasts. 2. Photosynthesis. 3. Plants—Adaptation. I. Smith, William K. (William Kirby), 1944- II. Vogelmann, Thomas Craig. III. Critchley, Christa. IV. Series.
QK725.P48 2004
571.6′592—dc22 2004049195

ISBN 0-387-22079-8 Printed on acid-free paper.

Printed in the United States of America. (MP/MVY)

9 8 7 6 5 4 3 2 1 SPIN 10977309

Springer is a part of Springer Science+Business Media
springeronline.com

Contents

Part 4. CO_2 Capture

Part 5. CO_2 Processing

Part 6. Environmental Constraints

Part 7. Overview

Preface

Across the broad structural and spatial hierarchy of organization found in the plant kingdom, chloroplasts are progressively packaged into more complex levels of organization, generating newly emerged limitations to photosynthesis at each level, as well as corresponding adaptive responses. Most likely, the adaptive advantage of size and height for enhanced sunlight competition also generated selective pressures for adaptations that maximized, for one example, the efficiency of light capture expressed per unit of biomass invested. The primary objective of this book is to identify and evaluate quantitatively fundamental adaptations at each level of the organizational hierarchy (i.e., chloroplast, cell, tissue, leaf, branch, canopy, stand, landscape). Thus, the present volume has been organized into two sections, the capture and processing of sunlight and CO_2 across two structural/spatial hierarchies (chloroplast-to-leaf and leaf-to-landscape). The most general objective was to identify which levels of the organizational hierarchy generates the greatest potential for influencing photosynthetic carbon gain in the plant kingdom. Within this framework, the potential effectiveness of metabolic (C_4 versus C_3) versus structural/developmental adaptations (e.g., sun/shade leaves) on the photosynthetic process is addressed specifically, as well as their integration. Such an understanding of the integrated mechanisms of plant form and physiology will be critical for making future decisions about the genomic manipulation of agricultural species to improve production efficiency and yield, as well as current policies in agroecology and land-use management. Similarly, incorporation of global change impacts on sources, sinks, and sequestration of photosynthetic carbon will also be critical for understanding future impacts on agricultural and native species.

Essential to answering the fundamental questions posed above is the improved capability for measuring photosynthetic light and CO_2 capture and processing across a broad range of structural and spatial scales, for example, from the chloroplast and cell (e.g., biochemistry, nanomicroscopy) to inside a leaf (e.g., fiber optic microprobes, fluorescence, laser beams, and photoacoustics), to the whole leaf, crown and whole plant (e.g., gas exchange chambers, fluorescence, stable isotope techniques), and to the community and landscape level (e.g., tower and aircraft eddy covariance, remote sensing). The above "footprints," or spatial resolution for measuring photosynthetic properties, range from nanometers to kilometers. Moreover, these measurement capabilities are now producing the comparative data necessary for evaluating comprehensively the adaptive significance of the emerging properties upscale that may require the concerted evolution of photosynthetic traits at lower levels of organization. These measurement capabilities also enable comparisons with adaptations in photosynthetic physiology across the broad scale of plant structural organization and complexity.

A workshop was held in conjunction with the Twelfth International Congress on Photosynthesis in Brisbane, Australia (August 18 to 23, 2001) to organize the content and contributors in this volume. Funding was provided by the National Science Foundation (Ecology and Evolutionary Physiology Program), the U.S. Department of Agriculture (Competitive Grants Program, Plant Biochemistry), and the U.S. Department of Energy (Global Change Program).

William K. Smith
Thomas C. Vogelmann
Christa Chritchley

Contributors

Neil R. Baker	Department of Biological Sciences, University of Essex, Colchester C04 3SQ, United Kingdom
Dennis D. Baldocchi	Ecosystem Science Division, Department of Environmental Science, Policy and Management, 151 Hilgard Hall, University of California, Berkeley, Berkeley, CA 94720, USA
Alessandro Cescatti	Centro di Ecologia Alpina, I-38040 Viote del Monte Bondone, Italy
Christa Critchley	Department of Botany, The University of Queensland, Brisbane, QLD 4072, Australia
David S. Ellsworth	School of Natural Resources and Environment, University of Michigan, Dana Building, 430 E. University, Ann Arbor, MI 48109-1115, USA

John R. Evans	Environmental Biology Group, Research School of Biological Sciences, Australian National University, GPO Box 475 Canberra, ACT 2601, Australia
Matthew J. Germino	Department of Biology, Idaho State University, Pocatello, ID 83209-8007, USA
Holly L. Gorton	Department of Biology, St. Marys College, St. Marys City, MD 20686-3001, USA
Yuko Hanba	Research Institute for Bioresources, Okayama University, Kurashiki, 710-0046, Japan
Jeremy Harbinson	Horticultural Production Chains (HPC) Group, Marijkeweg 22 (building number 527), Wageningen University, 6709 PG Wageningen, The Netherlands
Francesco Loreto	Consiglio Nazionale delle Ricerche (CNR), Istituto di Biochimica ed Ecofisiologia Vegetali (IBEV), Via Salaria Km. 29,300, 00016 Monterotondo Scalo (Roma), Italy
Elke Naumburg	MWH Energy and Infrastructure, Inc., 760 Whalers Way, Suite A-100, Fort Collins, CO 80525, USA
Ülo Niinemets	Department of Plant Physiology, University of Tartu, Riia 23,51011, Tartu, Estonia
Park S. Nobel	Department of Organismic Biology, Ecology and Evolution, 621 Young Drive South, University of California, Los Angeles, Los Angeles, California 90095-1606, USA
Donald R. Ort	Department of Plant Biology, University of Illinois, Urbana, IL 61801, USA

Peter B. Reich	Department of Forest Resources, University of Minnesota, 1530 N. Cleveland Avenue, St. Paul, MN 55108, USA
Thomas D. Sharkey	Department of Botany, 430 Lincoln Drive, University of Wisconsin, Madison, WI 53706, USA
Stanley D. Smith	Department of Biological Sciences, University of Nevada, Las Vegas, Las Vegas, NV 89154-4004, USA
William K. Smith	Department of Biology, Wake Forest University, Winston-Salem, NC 27109-7325, USA
Andrew J. Standish	University of Wisconsin-Madison, Department of Botany, 430 Lincoln Dr., Madison, WI 53706, USA
Ichiro Terashima	Department of Biology, Graduate School of Science, Osaka University, 1-16 Machikaneyama-cho, Toyonaka, Osaka, 560-0043, Japan
Thomas C. Vogelmann	Botany and Agricultural Biochemistry, University of Vermont, Burlington, VT 05405-0086, USA
Sean E. Weise	University of Wisconsin-Madison, Department of Botany, 430 Lincoln Dr., Madison, WI 53706, USA
John Whitmarsh	National Institutes of Health, National Institute of General Medical Sciences, 45 Center Drive, Room 2AS55F, Bethesda, Maryland 20892-6200, USA
Mathew Williams	Institute of Ecology and Resource Management, University of Edinburgh, Darwin Building, Mayfield Rd, Edinburgh, EH9 3JU, United Kingdom

William E. Williams — Department of Biology, St. Marys College, St. Marys City, MD 20686-3001, USA

Ian Woodward — Department of Animal and Plant Sciences, University of Sheffield, Sheffield, S10 2TN, United Kingdom

Part 1

Introduction

1 Background and Objectives

William K. Smith, Thomas C. Vogelmann, and Christa Critchley

Introduction

Photosynthesis is the fundamental process whereby plants capture and process sunlight and CO_2 as primary ecological resources supporting the growth and reproduction of an individual, and, ultimately, the survival of a species. Moreover, adaptations in resource acquisition and utilization are recognized as strong candidates for evolutionary selection in all species (Ackerly and Monson 2003a, Gutschick and BassiriRad 2003). Across the hierarchy of structural organization and spatial scale, extending from the chloroplast to the canopy and landscape, the efficiency of photosynthetic carbon assimilation declines substantially when assimilation is expressed per unit of plant biomass invested. For example, photosynthesis per unit mass is greatest for a single chloroplast, followed by a single cell, different cell layers, individual leaves, leaves arranged on a stem, and then leaves within crowns and canopies due, primarily, to increasing architectural constraints that require greater supportive biomass and generate increased mutual shading. Increasing size and accompanying structural complexity also place contraints on diffusional processes that require replacement by bulk transfer mechanisms. Thus, photosynthesis per unit biomass and ground area across any vegetative landscape is dramatically less than a hypothetical monolayer of all of the chloroplasts present, with each chloroplast photosynthesizing at its maximum capacity. Apparently, the adaptive advantages of size and greater structural/spatial complexity are most often related to a competition for sunlight that, apparently, outweighs any loss in photosynthetic efficiency expressed on an invested biomass basis. However, the accompanying evolution of mechanisms to enhance sunlight/CO_2 uptake and processing, in concert with sunlight exposure, is necessary (Smith et al. 1997, 1998). For example, it has been recognized for over half a century that leaves of numerous species adjust by increasing the mass per unit leaf area (specific leaf mass) as sunlight incidence increases (sun/shade leaves), generating greater photosynthesis per unit leaf area, without (or with relatively small) declines in photosynthesis expressed per unit biomass invested. Despite these structural changes at the internal leaf level, net assimilation rate expressed on an area or biomass basis continues to decline successively upscale at the branch, whole crown, and canopy levels. This decline in efficiency upscale is due most obviously to increased mutual shading among leaves, as well as the growth and maintenance costs of nonphotosynthetic organs necessary for optimizing display. Thus, evolutionary changes at the leaf level for increasing photosynthetic efficiency do not compensate for losses in efficiency due to a more crowded and costly leaf display. Independent measurements of photosynthesis at each structural/spatial level of organization have not been compared for any species, although substantial reduction in potential photosynthetic carbon gain is expected. Photosynthetic CO_2 gain declines sequentially when expressed per unit chlorophyll, chloroplast, tissue area, leaf area, or biomass invested, as well as per unit ground area occupied.

In response to the potential decline in light capture efficiency as structural scale increases, numerous venues for a variety of adaptive responses "emerge" at each successive level of organization and complexity. These adaptive, architectural properties may have evolved, fundamentally, in response to the increasing complexity in arranging and displaying chloroplasts for maximum sunlight and CO_2 capture and processing efficiency, while minimizing the cost/benefit ratio to the species (e.g., maximum photosynthesis per unit biomass invested). Despite decreases in photosynthetic efficiency at higher organizational levels, there are no published results showing specific declines among more than two, adjacent, organizational levels for any species. However, there is an increasing awareness that too much, as well as too little, sunlight exposure can result in photosynthetic impairment, for example, both low- and high-temperature photoinhibition (Baker and Bowyer 1994). Thus, the regulation of sun/shade exposure at higher levels of structural complexity appears to be a primary adaptive venue that includes adaptations in leaf structure and orientation, leaf aggregation on stems, crown shape, tree spacing, and so forth, as well as the concerted evolution of metabolic traits (Fig. 1.1).

Phylogeny of Photosynthesis: Structure versus Metabolism

Evolutionary adaptations, in response to this general scenario of increasing constraints on chloroplast display, exist not only in possible structural solutions, but also accompanying adaptations in metabolic biochemistry. Most likely, either or both of these adaptive venues have occurred throughout the evolutionary history of the plant kingdom, although little is known regarding the relative, quantitative contributions of either, or their potential adaptive role in future scenarios of global climate change (Ehleringer and Field 1993, Schulze and Caldwell 1995). It is thought provoking that the chloroplast genome, originally derived from an endosymbiotic association with cyanobacteria-like prokaryotes, appears to be highly conserved (Clegg et al. 1995) and that structural changes may have led to the appearance of relatively recent milestones in the evolution of plant adaptation via changes in the metabolic processes of photosynthesis, for example, the

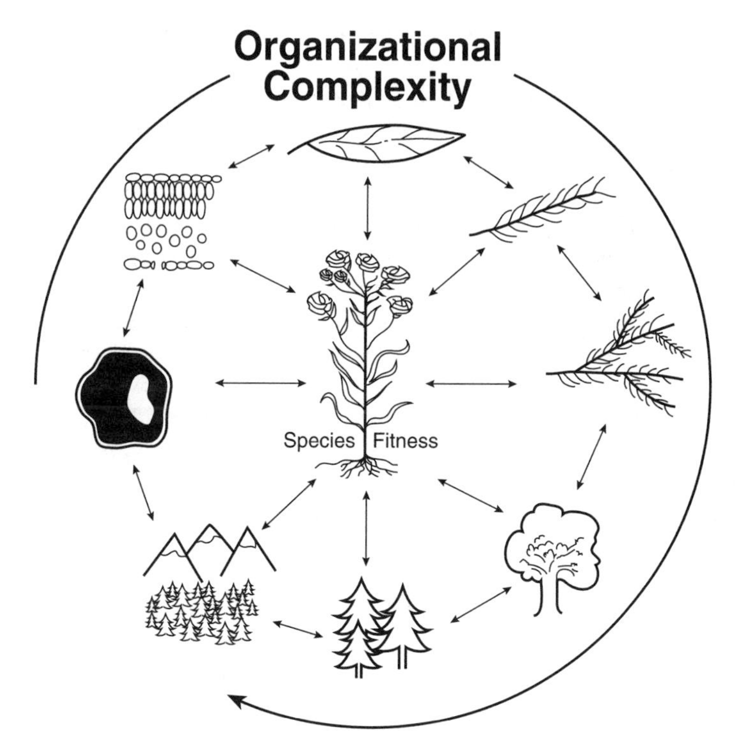

FIGURE 1.1. Schematic representation of the structural/spatial hierarchy of organizational complexity in the plant kingdom. Feed-forward and feedback between each level (↔) generates a concerted (circular) evolution of both structural/spatial and metabolic adaptations, possibly in response to the competitive advantage (sunlight capture) of being taller, and, thus, larger. Important adaptations at each level can feed back to preceding levels, and forward to higher levels (clockwise). Adaptations at all levels are generated, and persist, in response to enhanced organism fitness (inner arrows).

appearance of Kranz anatomy in the phylogenetic record before the appearance of the C_4 metabolic pathway (Sage 2004b).

In a recent review of the evolution of "functional" traits in plants, only 15 key papers or books were listed in the category of "selection analysis, adaptive value of individual traits" (Ackerly and Monson 2003a). Virtually all of these key studies evaluated the evolutionary selection of individual traits in extant species by utilizing historical perspectives derived from extinct species, phylogenetic and genetic comparisons of extant species, and/or correlations between traits or groups of traits and environmental variables (e.g., ecological convergence). However, studies that identify explicit increases in ecological fitness for vascular plants, associated with the selection and establishment of a specific trait in subsequent populations, are rare in the literature (Ackerly and Monson 2003b, Geber and Griffen 2003). The recently proposed dominance of extreme events in shaping the evolution of plants also needs further documentation (Gutschick and Bassiri-Rad 2003).

Importantly, phylogenetic studies have provided evidence for the evolution of angiosperms originating from shaded understory habitats vulnerable to disturbance characteristics that might also lead to variable sunlight regimes (Field et al. 2003). Paleobotanical data show trends toward the evolution of sun-leaf traits, such as palisade cell layers in response to increased sun exposure. Extant, basal angiosperms, such as found among the Chloranthaceae, also provide both structural and physiological data in support of this hypothesis, as well as in the family Nymphaeales, another angiosperm group that diverged early (Field et al. 2003). With regard to evolutionary "strategies" in photosynthetic traits, Field et al. (2003) concluded that photosynthesis in early (basal) angiosperms evolved in shade environments vulnerable to disturbance that ultimately led to the appearance of higher photosynthetic capability in more sun-exposed microsites, along with corresponding changes in leaf mesophyll structure from shade- to sun-type. Specifically, the leaves of floating lily pads (Nymphaeales) show both photosynthetic and leaf structural characteristics typical of early terrestrial angiosperms (Field et al. 2003). Many other leaf-level features of anatomy and morphology have been postulated as examples of an evolutionary trend toward the more efficient utilization of greater incident sunlight, culminating in such leaf types as the conifer needle (Smith et al. 1998, Field et al. 2003). Several other adaptive "strategies", including water repulsion from leaf surfaces, leaf guttation, and root pressure also appear to have evolved in concert with higher photosynthetic capacity. Also, the importance of extreme events to the evolution of plants deserves special consideration, and may be a potentially dominant influence in the evolution of physiological traits for resource acquisition and utilization, for example, photosynthesis (Gutschick and BassiriRad 2003).

A recent review of the genetic basis of adaptation in plants points to the occurrence of keystone mutational events in a relatively small number of regulatory loci and chromosomal rearrangements, ones that were largely responsible for the high diversity of life forms found in virtually all plant taxa (see Ackerly and Monson, 2003a for review). A better understanding of genetic mechanisms controlling functional traits of growth and development versus metabolic processes associated with physiological performance is needed (Ackerly et al. 2000, Monson 2003, Remington and Purugganan 2003, Sage 2004a). Although little is currently known about specific genes that control functional traits in native plants, emerging techniques currently on the horizon may provide the capabilities for future association of functional traits with specific gene loci. There is strong evidence for the highly conservative nature of chloroplast deoxyribonucleic acid (DNA) and structure, pointing to the possibility of a less dynamic role in metabolic adaptation than structural changes driven by developmental genes (Clegg et al. 1995). It has also been suggested that the development of Kranz-type anatomy may have more potential for engineering traits of native species into crop species, rather than installation of C_4 metabolic capabilities in single cells (Sage 2004a/b).

Measurement Techniques

Essential to answering the fundamental questions described above is the improved capability for measuring photosynthetic light and CO_2 capture across a broad range of structural and spatial scales, for example, the chloroplast and cell (biochemistry) level, inside the leaf (fiber optic microprobes, laser beams, and photoacoustics), the whole-leaf, crown, and whole-plant (gas exchange chambers, fluorescence, and stable isotope techniques), and the community and landscape level (tower and aircraft eddy covariance, stable isotopes, and remote sensing) (e.g., Smith et al. 2003). The above "footprints" of spatial resolution for measuring photosynthetic properties range from nanometers to kilometers. However, these measurement capabilities are now producing the comparative data necessary for evaluating comprehensively the adaptive significance of emerging properties in plant form at higher structural/spatial scales that may strongly influence the efficiency of photosynthetic carbon assimilation. Coupled with rapidly improving genetic techniques for coupling molecular mechanisms with functional traits (Remington and Purugganan 2003), these capabilities for scaling photosynthesis will be a powerful tool for unraveling the adaptive importance of specific traits. This measurement capability will enable direct comparisons of the efficiency of photosynthesis across a broad scale of organizational and structural complexity found in the plant kingdom.

Global Change Effects

The above information will be fundamental to understanding global change impacts on carbon sources and sinks, carbon sequestration, and, ultimately, future changes in plant distribution and diversity (e.g. Sage 2001, 2004b). The same information will also be important for making future decisions about genomic manipulation of agricultural species to improve production capabilities, or to improve current efforts in agroecology and land-use management as tools for global management. Importantly, future efforts in genetic engineering will probably involve the transfer of genetic material derived from native plants into common agricultural species, for example, genes for drought resistance from desert plants, or genes for cold tolerance from alpine species. A more functional understanding of photosynthetic adaptation at all levels of botanical organization could be fundamental to the development of robust, predictive models of photosynthetic capabilities that have evolved so far, along with those of the future.

Previous Reviews

The scientific community is now acutely aware of the importance of understanding the biological systems from the "bottom-up and the top-down," especially with today's capability for altering and designing genomes, plus the realization that the global ecosystem has been seriously impacted by anthropogenic influences at a multitude of point sources across the globe. In contrast to the ecophysiological scaling approach provided in Ehleringer and Field (1993), as well as other books that focus on certain components of the photosynthetic process (e.g., Osmond et al. 1980, Schulze and Caldwell, 1995, Baker 1996, Hall and Rao 1999), the current volume focuses more specifically on the interpretation of photosynthetic adaptation across the broad spectrum of organizational scale. Just as the organism level is recognized as the fulcrum, or evolutionary unit, for understanding the ecological scaling process, the focus here is on the chloroplast as the photosynthetic unit. How chloroplasts are packaged to optimize sunlight and CO_2 assimilation at each level of the structural hierarchy is an underlying theme of this volume, although the individual, organism level is also recognized as the fundamental, evolutionary unit of adaptation within a population (see Fig. 1.1). We also recognize that any comprehensive understanding of the evolution of functional traits in plants must consider a host of variables beyond the photosynthetic performance of leaves, for example, numerous other life history, structural, and physiological traits that can have a major impact on a species' ecological fitness. It has been recognized for some time that any conclusive determination of the adaptive value of a given trait requires direct linkage to increased reproductive success and the production of viable offspring (see review by Ackerly and Monson 2003b). These data are not possible for extinct species, and, despite this recognition, data combining measurements of photosynthetic performance with changes in fitness are virtually nonexistent for extant species (Mooney 1991), with the exception of some studies that link physiological traits with certain reproductive efforts, such as flower production (e.g., Young and Smith 1982, Hill and Smith in press).

In addition to a general contrast, some interesting overlap with certain chapters of Ehleringer and Field (1993) occurs in this volume. For example, the chapter by Dawson and Chapin (Ehleringer and Field, Chapter 17) addresses the possible simplification of scaling between the leaf and landscape via leaf form-function relationships. These authors suggest that the "bottom-up" approach is essential, ultimately, for understanding the scaling process mechanistically, as well as from an evolutionary perspective. However, the "top down" approach also provides important information on feedback processes and also identifies critical processes and important, contributing factors at the ecosystem level. For either approach, the grouping of plants by form-function relationships will provide the most efficient approach to understanding the scaling process.

Another specific overlap with Ehleringer and Field (1993) is the discussion by Levine in the introductory chapter of that book concerning the integrity of the "basic unit of study." As already inferred above, this integrity degenerates as the level of organization and complexity increases, for example, chloroplasts must diverge in function when organized into cells, when cells are organized into cell layers, cell layers into leaves, leaves into crowns, and even when crowns are organized into canopies, and canopies into landscapes. This divergence in function at lower organizational levels is necessary because of new, emerging limitations to light and CO_2 processing at higher levels of organization. Finally, there are several other chapters in Ehleringer and Field (1993) that the present volume either complements or supplements by providing a more mechanistic evaluation of the relationship between plant form and photosynthetic function from the chloroplast to the landscape level. There are other edited books that also provide an excellent overview of photosynthesis research in specific areas (see Orr and Govindjee 2001 for a comprehensive listing), but without a systematic emphasis on identifying adaptations related to light and CO_2 capture and processing at different levels of the structural/spatial hierarchy (e.g., Schulze and Caldwell 1995, Baker 1996, Falkowski 1994, Leegood et al. 2000, Garab 1998).

Summary of Objectives

The chapters in this volume are designed to evaluate and compare comprehensively the specific adaptive mechanisms of terrestrial, vascular plants that have emerged at different organizational levels, and which may act to enhance photosynthetic efficiency in response to increasing botanical complexity in life form (see Fig. 1.1). Across the broad and diverse structural spectrum found in the plant kingdom, chloroplasts are progressively packaged into more and more complex levels of organization (i.e., inside cells, in leaves, along branches, within a crown, and in a canopy) as plants increase in size, and structural/spatial arrangements become more complex. Given this generalized scenario of selective pressure (evolution from wet, shaded, and disturbed habitats) some fundamental questions regarding evolutionary "strategies" become apparent (Smith et al. 1997, 1998). How do the metabolic capabilities of a single chloroplast respond to this increasing complexity, so that carbon gain per unit biomass does not diminish? Are photosynthetic adaptations at the metabolic and biochemical level of the chloroplast/cell potentially more important than changes in structure and form, or are they functionally integrated? In other words, what is the adaptive potential for changes in genes involved in photosynthetic metabolism, as opposed to changes in developmental genes regulating diversity in plant form? Important questions related to the past versus future adaptive venues are also apparent. Which plant genes should be targeted for engineering new agricultural (or native) varieties with increased capacity and efficiency in light and CO_2 assimilation/processing under current global change scenarios—those controlling the metabolic reactions of photosynthesis, or those controlling variation in plant form (e.g., structure, spatial arrangement, and orientation) that may strongly influence photosynthetic capability? The answer to these fundamental questions of adaptive capabilities will provide a basis for understanding past and future adaptations in the photosynthetic process.

The primary objective of the present volume is to initiate pursuit of a fundamental understanding of the basic evolutionary challenge of adaptively altering, organizing, packaging, and displaying chloroplasts for maximum sunlight/CO_2 capture and processing as structure, organization, and complexity increase in the plant kingdom. A secondary focus is to evaluate the potential for metabolic changes versus changes in plant form (developmental) for maintaining high photosynthesis per unit biomass invested. Simply expressed—Which genes have the most potential for influencing photosynthetic performance, developmental genes that control diversity in traits influencing plant form and life history, or metabolic genes that control the biochemistry of such important physiological processes as photosynthesis, or is integrative interaction among functional traits critical? Finally, if evolutionary breakthroughs have occurred, have they dictated adaptive strategies to lower levels of the botanical hierarchy, for example, metabolic changes in chloroplasts of sun/shade leaves, or in bundle sheath cells of species that first evolved Kranz anatomy (Sage 2004a).

References

Aber, J. D., Reich, P. B., and Goulden, M. L. 1996. Extrapolating leaf CO_2 exchange to the canopy: A generalized model of photosynthesis compared with measurements by eddy correlation. Oecologia 106:257–265.

Ackerly, D. D., and Monsoon, R. K. (eds.) 2003a. Evolution of functional traits in plants. Chicago: University of Chicago Press.

Ackerly, D. D., and Monson, R. K. 2003b. Waking the sleeping giant: The evolutionary foundation of plant function. Int. J. Plant Sci. 164(Suppl.):1–6.

Ackerly, D. D., Dudley, S. A., Sultan, S. E., Schmit J., Coleman, C. R., Linder, C. R., Sandquist, D. R., et al. 2000. The evolution of plant ecophysiological traits: Recent advances and future directions. BioScience 50:979–995.

Baker, N. R. (ed.) 1996. Photosynthesis and the Environment. Advances in Photosynthesis and Respiration Series. Dordrecht, The Netherlands: Kluwer Academic Publishers.

Baker, N. R., and Bowyer, J. R. 1994. Photoinhibition of Photosynthesis—From Molecular Mechanisms to the Field. Environmental Plant Biology Series. Oxford: Bios Scientific Publishers Limited.

Clegg, M. T., Learn, G. H., and Morton, B. R. 1995. Rates and patterns of chloroplast DNA evolution. In: Tempo and Mode in Evolution: Genetics and Paleontology 50 Years After Simpson, pp. 215–226. Washington D.C.: National Academy of Sciences.

Ehleringer, J. R., and Field, C. B. (eds.) 1993. Scaling Physiological Processes. New York: Academic Press.

Falkowski, P. G., 1994 The role of phytoplankton photosynthesis in global biogeochemical cycles. Photosyn. Res. 39:235–258.

Field, T. S., Arens, N. C., and Dawson, T. E. 2003. The ancestral ecology of angiosperms: Emerging perspectives from extant basal lineages. Int. J. Plant Sci. 164(Suppl.): S129–S142.

Garab, G. (ed.) 1998. Photosynthesis: Mechanisms and Effects, Vol. V. Dordrecht, The Netherlands: Kluwer Academic Publishers.

Geber, M. A., and Griffen, L. R. 2003. Inheritance and natural selection on functional traits. Int. J. Plant Sci. 164(Suppl.):1–6.

Gutschick, V. P., and BassiriRad, H. 2003. Extreme events as shaping physiology, ecology, and evolution of plants: Toward a unified definition and evaluation of their consequences. New Phytologist 160:21–42.

Hall, D. O., and Rao, K. K. (eds.) 1999. Photosynthesis, 6th ed. Cambridge: Cambridge University Press.

Hill, J. P., Willson, C. J., and Smith, W. K. 2004. Morphotype ecophysiology enhances reproductive effort in a shrub steppe. Plant Ecol. (in press).

Leegood, R. C., Sharkey, T. D., and von Cammerer, S. 2000. Photosynthesis: Physiology and Metabolism. Advances in Photosynthesis and Respiration Series. Dordrecht, The Netherlands: Kluwer Academic Publishers.

Monson, R. K. 2003. Gene duplication, neofunctionalization, and the evolution of C_4 photosynthesis. Int. J. Plant Sci. 164(Suppl.):43–54.

Mooney, H. A. 1991. Plant physiological ecology: Determinants of progress. Funct. Ecol. 5:127–135.

Orr, L., and Govindjee, R. 2001. Photosynthesis and the web: 2001. Photosyn. Res. 68:1–28.

Osmond, C. B., Björkman, O., and Anderson, D. J. (eds.) 1980. Physiological Processes in Plant Ecology. New York: Springer-Verlag.

Remington, D. L., and Purugganan, M. D. 2003. Candidate genes, quantitative trait loci, and functional trait evolution in plants. Int. J. Plant Sci. 164(Suppl.):7–20.

Sage, R. F. 2001. Environmental and evolutionary preconditions for the origin and diversification of the C_4 photosynthetic syndrome. Plant Biol. 3:202–213.

Sage, R. F. 2004a. The evolution of C_4 photosynthesis. New Phytologist 161:341–370.

Sage, R. F. 2004b. Atmospheric CO_2, environmental stress and the evolution of C_4 photosynthesis. In: A History of Atmospheric CO_2 and Its Effects on Plants, Animals, and Ecosystems. J. R. Ehleringer, T. E. Cerling, and D. Dearling, (eds.) (in press).

Schulze, E.-D., and Caldwell, M. M. 1995. Ecophysiology of Photosynthesis. New York: Springer-Verlag.

Smith, W. K., Vogelmann, T. C., Bell, D. T., DeLucia, E. H., and Shepherd, K. A. 1997. Leaf form and photosynthesis. BioScience 47:785–793.

Smith, W. K., Bell, D. T., and Shepherd, K. A. 1998. Associations between leaf orientation, structure and sunlight exposure in five western Australian communities. Am. J. Bot. 84:1698–170.

Smith, W. K., Kelly, R. D., Welker, J. M., Fahnestock, J. T., Reiners, W. A., and Hunt, E. R. 2003. Leaf-to-aircraft measurements of net CO_2 exchange in a sagebrush steppe ecosystem. J. Geophys. Res. 108(D3), 4122, doi: 10.1029/2002JD002512, 2003.

Young, D. R., and Smith, W. K. 1982. Simulation studies of the influence of understory location on transpiration and photosynthesis in *Arnica cordifolia* Hook. Ecology 63:1761–1771.

Part 2

Sunlight Capture

2
Chloroplast to Leaf

JOHN R. EVANS, THOMAS C. VOGELMANN, WILLIAM E. WILLIAMS,
AND HOLLY L. GORTON

Light in the Environment

Light that plants encounter in the environment originates mostly from the sun. The spectral properties of sunlight fundamentally reflect the surface temperature of the sun (~5800 K). However, the solar spectrum is modified by the earth's atmosphere due to absorption by water and CO_2, and by scattering. Light can either be expressed in terms of an energy flux or a quantum flux. The energy content of a quanta depends on its wavelength, decreasing as wavelength increases. Energy flux (E, W m^{-2}) can be calculated from quantum flux (I, mol m^{-2} s^{-1}) for a given wavelength (λ, m) with the following equation:

$$E = IN_0hc/\lambda$$

where N_0 is Avogadro's number (6.02×10^{23}), h is Planck's constant (6.63×10^{-34} J s photon^{-1}) and c is the speed of light (3×10^8 m s^{-1}). For example, 1 μmol quanta m^{-2} s^{-1} of 400 nm light is equivalent to 0.3 W m^{-2}, while for 700 nm light it is 0.17 W m^{-2}.

On a clear day, the amount of direct sunlight varies with time of day, date, and latitude. This is illustrated in Figure 2.1, where a sunmap is shown in panel A, calculated for a latitude of 35°S. The position of the sun is given by the azimuth (the compass direction to the sun) and the elevation (degrees above the horizon, concentric circles on the diagram). The three arcs show the daily traces at the summer and winter solstices and at the equinox, with symbols at hourly intervals. On the equinox, the sun rises due east and sets due west, with a 12-h daylength. At the winter solstice, the sun rises around 7 AM, closer to the northeast and only reaches an elevation of 45° at noon. For the summer solstice, the sun rises before 5 AM, toward the southeast and climbs to an elevation of about 80° at noon. Irradiance increases as solar elevation increases, so noon irradiance varies considerably through the year, especially at higher latitudes (see Fig. 2.1B). The combination of greater noon irradiance and longer daylength results in a threefold variation in potential daily irradiance through the year at this latitude. The mathematical description of solar position enables one to calculate the light environment in the absence of cloud, at a particular

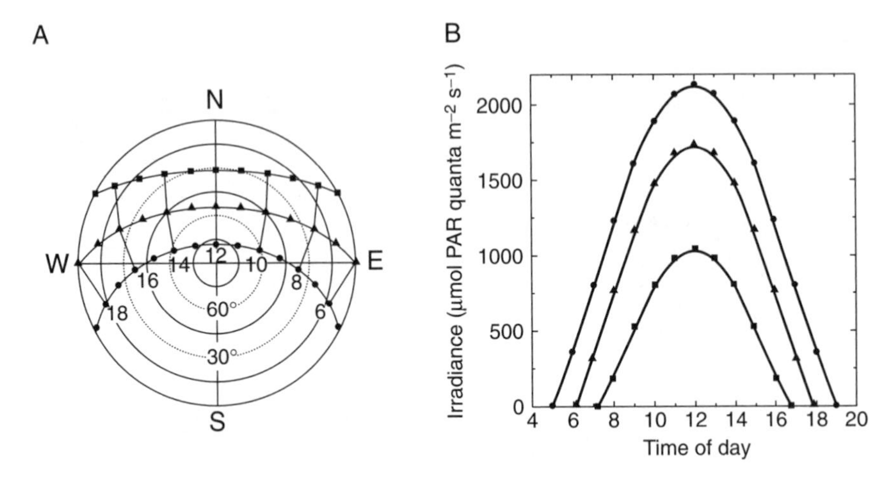

FIGURE 2.1. Solar map and diurnal variation in sunlight through the seasons. The figure is constructed for Canberra, Australia, 35°S for the summer solstice (●), equinox (▲) and winter solstice (■). Panel A shows a polar map of the arc of the sun, with symbols placed at hourly intervals. Concentric circles are placed at 15° intervals. Panel B shows that not only does the photoperiod vary from 9.5 to 14.5 h, but the maximum irradiance and daily irradiance vary twofold through the year.

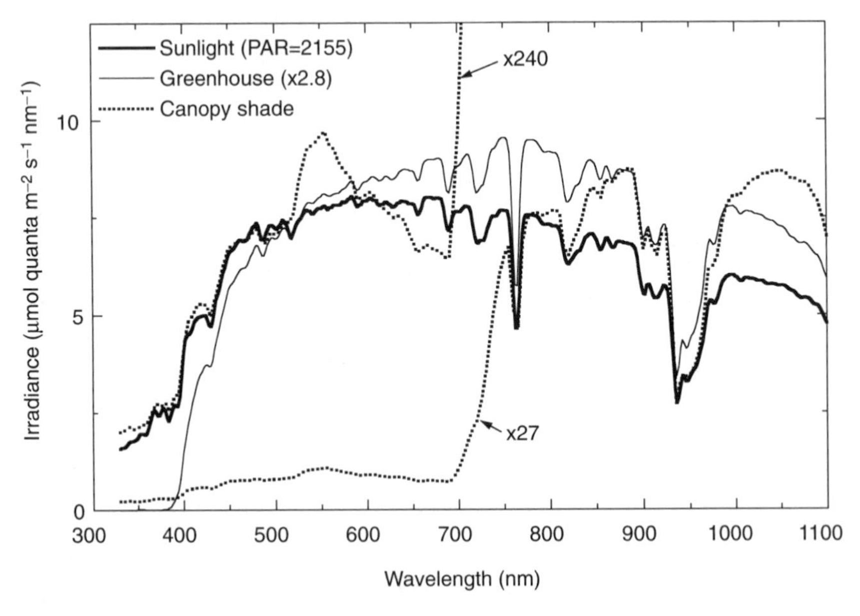

FIGURE 2.2. Spectra of three natural light environments. Sunlight (—), sunlight inside a greenhouse that screens out UV wavelengths below 400 nm (—), and shadelight at the forest floor where interception by foliage has largely removed light below 700 nm (···). The spectra have been multiplied by various factors to enable comparison in the PAR wavelength range. This emphasises that despite absorption of most of the light, shadelight has only slightly more green and less red than sunlight.

location from hemispheric photographs taken within plant canopies (Chazdon and Field 1987).

The spectrum of sunlight is relatively flat in the visible range. Three spectra are shown in Fig. 2.2 that have been normalized to have equal numbers of quanta between 400 and 700 nm. Sunlight is shown in bold. The thinner line shows a spectrum measured within a greenhouse, where the polycarbonate window absorbs all of the ultraviolet (UV) waveband. A spectrum of shade light was obtained at ground level within a rainforest (dotted line). Absorption of light by leaves reduces the relative amount of light at short wavelengths while increasing the relative amount of infrared.

When light encounters a leaf, parts are reflected, some is transmitted, and the remainder is absorbed. The relative balance between those three fates depends on the wavelength (Fig. 2.3). For wavelengths greater than 720 nm, about 50% of the light is reflected, 45% is transmitted, and only a few percent are absorbed. Between 400 and 700 nm, around 85% of the light is absorbed. The proportion of light absorbed depends strongly on the chlorophyll content per unit leaf area, while leaf reflectance varies with growth environment (Evans and Poorter 2001). Leaf absorptance is not directly influenced by leaf mass per unit area (Evans and Poorter 2001). Leaves are generally green because they reflect and transmit slightly more light around 550 nm than at other visible wavelengths. However,

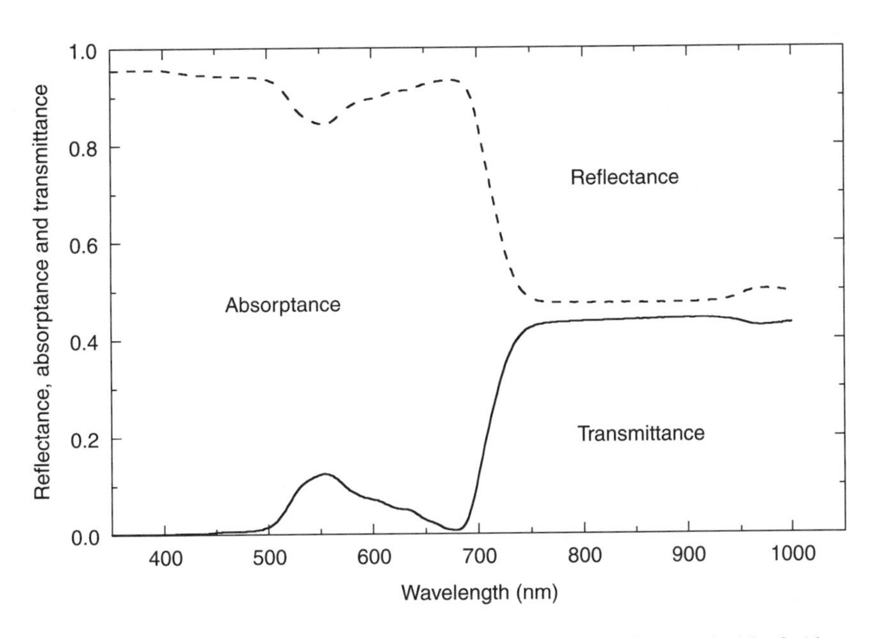

FIGURE 2.3. Reflectance, transmittance, and absorptance spectra for a typical leaf. About 85% of light between 400 and 700 nm is absorbed by the leaf, which appears green because those wavelengths are the main component of reflected and transmitted light. Beyond 700 nm, nearly 50% is reflected and 45% transmitted, so infrared light penetrates through and is scattered by leaf canopies. The steep shoulder around 700 nm provides the spectral signature used by satellite-based spectrometers to estimate vegetation cover.

they still absorb about 70% of green light. Due to the abrupt transition of leaves from efficient to poor absorbers around 700 nm, there is a rapid change in irradiance around these wavelengths between the sunlight and shade spectra shown in Figure 2.2. This is sensed by plants with their phytochrome system and is used in remote sensing to quantify leaf area index, such as the normalized difference vegetation index (NDVI). Despite the fact that leaves transmit and reflect more green light than red or blue light, the shade spectrum does not differ greatly from sunlight between 400 and 700 nm.

Light interception by leaves depends on their orientation relative to the sun, their position in a leaf canopy, the amounts of cloud and wind, and the cuticular properties of the leaf. For example, in environments where excess light is available, hairs and waxes on the leaf surface can increase the reflectance (Ehleringer and Björkman 1978). There have been many approaches to modeling light attenuation through leaf canopies. The most robust techniques follow both sunlit and shaded (receiving only diffuse light) leaf area (e.g., DePury and Farquhar 1997). As cloudiness or haze dramatically alters the balance between direct and diffuse light, it is advantageous to be able to account for leaf area in both classes. However, it is more difficult to deal with wind. The consequence of wind is that it causes leaves to flutter and plants to flex, so that the irradiance on a leaf can fluctuate rapidly. For example, within a poplar canopy, 80% of daily photosynthetically active radiation (PAR) came as sunflecks. The majority of these lasted less than 1 s (Roden and Pearcy 1993).

The efficiency with which lightflecks are converted to photosynthate depends on their frequency and duration as well as the plant species, photosynthetic induction state and stomatal conductance. Electron transport rate during brief lightflecks can exceed that in continuous light of the same irradiance because of the buffering influence of intermediate pools (Pearcy 1990, Pons and Pearcy 1992). The intervening low-light periods enable the "dark" reactions of photosynthesis to drain the pools, which is equivalent to increasing the capacity of carboxylation to electron transport.

Pigments and Thylakoid Membranes

In higher plants, light is captured for photosynthesis by pigment-protein complexes embedded in the thylakoid membranes within chloroplasts. There are four classes of complex, the reaction centers of photosystems I and II, and light-harvesting complexes I and II. The reaction center complexes contain a core antenna of chlorophyll a and carotenes around the electron donor pigment. Additional antenna molecules of chlorophyll a and b and xanthophylls are bound by the light-harvesting complexes. X-ray crystallography has revealed detailed structural information at the atomic level for the light-harvesting complex II (Kuhlbrandt et al. 1994) and the reaction center complexes (Jordan et al. 2001, Schubert et al. 1998, Zouni et al. 2001). An increasing number of the chlorophylls and carotenoids can be located to precise ligands with the orientations of the pigments specified.

Through the use of nondenaturing gels, it was possible to separate the pigment-protein complexes. This allowed the pigment and polypeptide composition of each complex to be determined. The absorption spectra for the major complexes are shown in Figure 2.4. The spectra for the reaction center complexes of photosystem II (CPa) and photosystem I (CP1) mainly reflect that of chlorophyll a and β carotene. CP1a represents the reaction center of photosystem I with additional light-harvesting complex I. Chlorophyll b in the light-harvesting complex (LHC) results in absorption peaks at 475 and 650 nm, while xanthophylls contribute to the shoulder at 500 nm. Just over half of the chlorophyll in a leaf is associated with light-harvesting complex II, while the reaction center complexes I and II and light-harvesting complex I each account for about 15% of the chlorophyll.

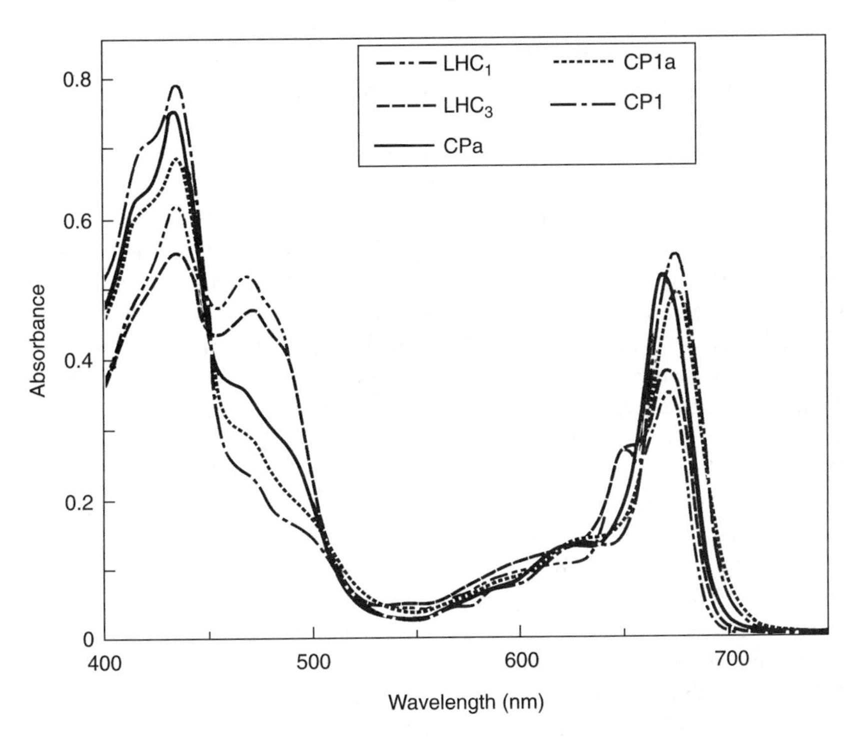

FIGURE 2.4. Absorption specta of five of the major pigment-protein complexes in leaves. About 40% of the chlorophyll is found in LHC_1, the light-harvesting chlorophyll a/b complex mainly associated with photosystem II. Another 15% of chlorophyll is associated with LHC_3. CP_a is the photosystem II reaction center complex, which contains about 11% of the chlorophyll. CP1 and CP1a are the complexes of the photosystem I reaction center, having 6 and 26% of the chlorophyll associated with each, respectively. CP1a has associated with it peripheral light-harvesting complexes that are absent from the CP1 complex. Spectra were obtained from spinach thylakoids that had their pigment-protein complexes separated by nondenaturing gel electrophoresis and are normalized to 10 μM solutions in a 1-cm pathlength (Evans and Anderson 1987).

In contrast to the atomic detail that has been revealed for the pigment-protein complexes, there is still uncertainty about both thylakoid structure and the distribution of complexes. Electron micrographs revealed that thylakoid membranes form structures like stacks of pancakes, called grana, linked together by single thylakoids, the stromal lamellae. In fact, grana can be seen through a microscope as bright spots of chlorophyll fluorescence within chloroplasts (Anderson 1999). Grana are more prominent in shade plants, having greater numbers of thylakoids per granum than those from sun leaves. The closely packed thylakoids in the grana exclude protein complexes that protrude into the stroma. The adenosine triphosphatase (ATPase) complex was found only on stroma exposed lamellae (Miller and Staehelin1977). Subsequently, physical separation of grana and stromal membrane fractions revealed lateral heterogeneity in the distribution of the photosystems (Andersson and Anderson 1980). Photosystem II was enriched in the grana while photosystem I was depleted. Complexity continues to be added because each photosystem consists of at least two types. Those found in the grana have a larger antenna than photosystems in the stroma lamellae, giving rise to α and β centers, respectively (Andreasson et al. 1988; Svensson et al. 1991).

The reaction center of photosystem I contains about 90 chlorophylls and is surrounded by a single layer of 8 light-harvesting complex I pigment-proteins (Lhca1 to 4), (Boekema et al. 1994), giving a total of 190 chlorophylls/PSI. By associating with an additional trimeric light-harvesting complex (Lhcb1 to 3) in the grana, PSIα has a 30% larger antenna than PSIβ (Andreasson and Albertsson 1993). The reaction center of photosystem II contains about 50 core chlorophyll a molecules with three inner chlorophyll a/b antenna proteins (Lhcb4 to 6) and several outer trimeric light harvesting complexes (Lhcb1 to 3). Together these dimerise to form a larger photosystem II complex that has been resolved by electron microscopy (Boekema et al. 1999, Boekema et al. 2000, Yakushevska et al. 2001). The outer trimeric complexes bind with varying stoichiometry depending on the packing structure analyzed.

The variable stoichiometry exists at multiple levels. Firstly, the thylakoids are fluid and dynamic, so that the complexes are forever jostling about and interacting with different neighbors. Secondly, lateral heterogeneity means that a given reaction center complex can have a different apparent antenna size depending on its location, due to the way it associates with nearby complexes. Thirdly, the stacking of the membranes can decrease due to phosphorylation of LHCII proteins (Stefansson et al. 1995) or in vitro when the concentration of cations is varied. This in turn alters the likelihood of interactions between complexes, with the antenna size of photosystem II decreasing as stacking decreases. Fourthly, the relative amounts of the complexes depend on growth conditions, principally the amount of light. As growth irradiance decreases, the proportion of LHCII increases. Under low irradiance, a shade-type phenotype is observed. This is characterized by a low chlorophyll a/b ratio due to the increased proportion of chlorophyll in LHCII, which results in greater numbers of thylakoids per granum. It can also be observed within a leaf, where the gradient in light results in sun- and shade-type chloroplasts being present in palisade and spongy mesophyll cells, respectively (Terashima and

Inoue 1985). Transgenic *Arabidopsis* plants have been constructed that lack the major LHCII genes Lhcb1 and Lhcb2. Surprisingly, this did not affect the formation of grana stacks (Andersson et al. 2003). This does not rule out the involvement of LHCII in thylakoid membrane stacking, because the plants compensated for the loss of Lhcb1 and Lhcb2 by increased expression of Lhcb5, which formed a novel alternative trimeric LHCII complex (Ruban et al. 2003).

Diagrams of thylakoids tend to misrepresent them because they generally portray an idealized electron transport chain with one of each complex. By combining quantitative measurements of each component of chloroplast thylakoids with estimates of their size, it is possible to calculate the average proportion of membrane surface area covered. An example is given for spinach in Table 2.1. Two thirds of the membrane surface is covered by pigment-protein complexes on average. The existence of appressed and stroma exposed membranes means that the proportions of surface area given in Table 2.1 are not exactly equivalent to the situation in the chloroplast. Freeze fracture analysis under the electron microscope has allowed visualization of some of the complexes in vivo. Photosystem II complexes in grana membranes occur with a density of 1500 to 2000 μm^{-2} (Staehelin and van der Staay 1996). Using the complex size, this equates to photosystem II occupying 16% of the surface, which is similar to the value presented in Table 2.1. The ATPase complex can occur with a density of 700 μm^{-2}, which suggests that the ATPase occupies about 2% of the stroma exposed membrane. However, given that stroma exposed membranes account for about 20% of thylakoid membrane surface, one cannot reconcile the values when expressed per unit thylakoid membrane.

The formation of grana creates regions in the thylakoid membrane that have different compositions of the protein complexes. Physical separation of photosystem II and photosystem I between appressed and non-appressed regions altered the way captured light could be transferred (Andersson and Anderson 1980). More recently, Albertsson (1995, 2001) has proposed that cyclic electron transport occurs at the stroma exposed thylakoids. Stroma lamellae consistently account for 20% of the total membrane surface and chlorophyll (Albertsson 1995).

TABLE 2.1. Relative stoichiometries between the major components of spinach thylakoids and the proportion of the membrane surface they occupy (Kirchhoff et al. 2002).

	mmol (mol Chl)$^{-1}$	% of total Chl	% of surface area
PSII[a]	2.18 (1.9–2.7)[d]	20	19
LHCII[b]	9	38	19
PSI[c]	2.25 (1.7)[d]	42	26
Cyt b/f	1.3 (1.6–2.6)[d]		2
ATPase	0.95		2
lipid			32

[a]Each complex was assumed to be dimeric with each PSII present with three inner LHC and a tightly bound LHC trimer.
[b]Assuming that these are all present as trimeric complexes.
[c]Each PSI was assumed to have a surrounding ring of 8 LHCI complexes.
[d]Range observed between shade and sun leaves of spinach (Terashima and Evans, 1988).

It is still not possible to measure what proportion of light is absorbed by the grana versus the stroma thylakoids, nor the distribution between the two photosystems, although it must approximately equate to the distribution of chlorophyll. However, the maximum quantum yield of photosynthesis varies little across C_3 species (Björkman and Demmig 1987) or with respect to sun/shade acclimation (Evans 1987). Quantum yield does however vary with wavelength (McCree 1971). The wavelength dependence has been interpreted to reflect the relative absorption spectra of the two photosystems (Evans 1987), but the precise distribution between all the complexes awaits the development of new methods.

Leaf Tissue Optics

Most leaves are sufficiently pigmented such that they absorb most of the light within the photosynthetically active region of the spectrum (400 to 700 nm). From the standpoint of maximizing photosynthetic rate at the level of the whole leaf, it would be advantageous to match the internal distribution of absorbed light energy with CO_2 supply and photosynthetic capacity (Evans 1999, Evans and Vogelmann 2003). Presumably, these quantities depend upon leaf anatomy, orientation, environment, and growth conditions (Smith et al. 1998). Detailed knowledge about how internal gradients of absorbed light and photosynthetic capacity influence whole-leaf photosynthetic performance depends upon the ability to measure these gradients, and recent advances have made this possible (Vogelmann et al. 1996). With respect to light, leaf anatomy and pigmentation play major roles in determining light propagation and absorption and there are subtleties that can exert a profound influence on the allocation of absorbed light energy to the photosynthetic tissues of a leaf.

Leaf tissue optics is strongly influenced by two optical phenomena: the sieve or package effect and light scattering. The sieve effect arises from the fact that light-absorbing pigments are not distributed homogeneously throughout the leaf, but rather they are packaged within chloroplasts. When light travels though a leaf, it may intercept a chloroplast and be absorbed; alternatively, it may miss the chloroplasts and pass through the leaf, hence the reference to a sieve. The net result of the sieve effect is that it decreases that amount of light absorption per unit of chlorophyll and per unit of leaf area (Fig. 2.5). If the photosynthetic pigments were released from the chloroplasts and distributed throughout the leaf, then leaves would absorb more light, even though the total amount of pigment remains constant. It follows that, in determining how much light the individual cell layers absorb within a leaf, it is not a simple matter of measuring the total amount of pigments but rather determining the location and density of chloroplasts. Moreover, chloroplasts move in response to how much light they receive and this movement can profoundly influence the internal distribution of absorbed quanta (see later section on chloroplast movements). Thus, the sieve effect is not just an optical curiosity but rather it is something that can be used by a leaf to dynamically control the internal distribution of absorbed light energy.

Light scattering is a second optical phenomenon that influences the global ab-

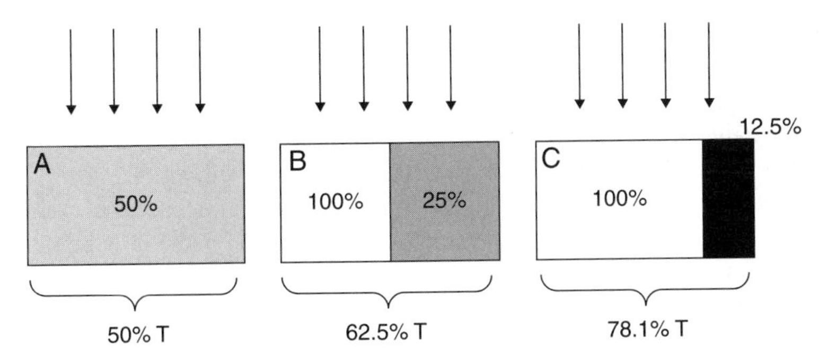

FIGURE 2.5. A graphical portrayal of the sieve effect. The same amount of pigment is present in cells A to C. The transmittance (T) of cell A was set to 50%. Moving the pigment to one half of the cell in B creates a region in which transmittance is 100% (no pigment) and 25% (pigment). The overall transmittance of the cell is 62.5%, an increase from cell A where the pigment is uniformly distributed. Concentrating the pigment further (C) leads to progressively higher transmittance.

sorption properties of a leaf. Light scattering in leaves is largely determined by the intercellular air spaces, which commonly comprise 20 to 50% of the leaf volume. Scattering arises from the jump in refractive index (n) between air (n = 1) and cells (n = 1.33 to 1.45), which creates mirror-like reflections. The path of light through a leaf can be thought of as a series of deflections between cells and air spaces (Fig. 2.6). Light scattering in this system is independent of wavelength and light is scattered predominantly in the forward direction. These deflections

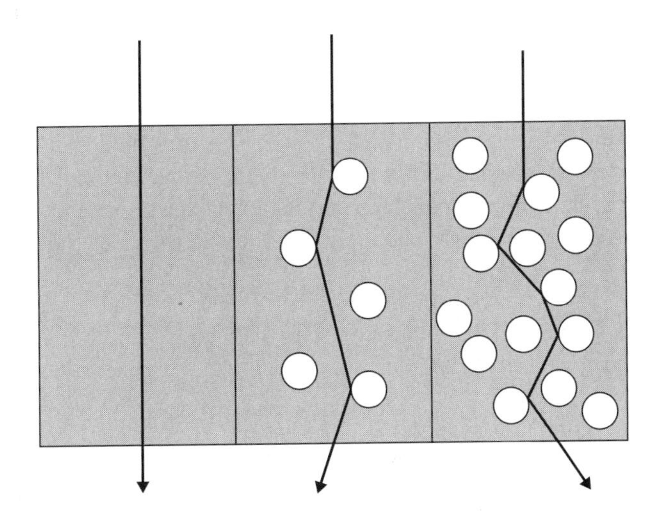

FIGURE 2.6. Light scattering and increase in pathlength of light. Light scattering in leaves is caused primarily by reflection between the intercellular air spaces and cells. This mirror-like reflection deflects light at relatively small angles so that light is scattered mostly in the forward direction. Also, this type of scattering is relatively independent of wavelength.

increase pathlength, which increases the probability for light capture and absorption. Light scattering counteracts the sieve effect; it increases the amount of light absorption per unit chlorophyll and per unit leaf area.

The extent to which light is scattered in a leaf is determined by the number of times that it encounters an air-cell interface. Total intercellular air space volume is important but so is the three-dimensional geometry of that air space. Given an equal amount of air space, many small pockets of air will potentially scatter light more than a few large air channels. Unlike chloroplast movement, which can alter internal light absorption within minutes, that spatial distribution of air space is fixed during leaf development and its three-dimensional geometry is probably irrevocably linked to the type of mesophyll tissue (palisade versus spongy) and the number of cell layers for the life of the leaf. Estimates of pathlength enhancement by scattering in leaves usually range from 2 to 4 (Vogelmann 1993). However, quantifying the relationship between leaf anatomy and pathlength has been confounded by difficulties in directly measuring pathlength enhancement and how it is related to the three-dimensional organization of the intercellular air spaces. Because of these methodological limitations, more quantitative research is necessary to evaluate how leaf anatomy may control light absorption through scattering.

Optical Properties of Individual Leaf Tissues

The epidermis is the first cell layer that light encounters as it strikes a leaf. The epidermis is usually transparent to visible light and it frequently contains screening pigments that filter ultraviolet-B (UV-B) radiation (280 to 315 nm). Even though the epidermis is transparent to PAR, it can affect the irradiance within the underlying photosynthetic tissue at the microscopic level because each epidermal cell can act as a lens that focuses light. Epidermal focusing is quite common in plants and focal intensifications can reach 20 times incident light levels, although in most leaves, light focusing intensifies the light only severalfold. Whether epidermal focusing serves any adaptive purpose is unknown. It has been suggested that the extreme convex epidermal cell shape of some tropical understory plants may enhance light capture by minimizing specular reflection of diffuse light.

The underlying photosynthetic tissue of many leaves is frequently comprised of tubular palisade cells and then a layer of more randomly arranged irregular-shaped spongy mesophyll. The tubular shape may facilitate the penetration of light because these cells have a central vacuole, which serves as a transparent channel that appresses the chloroplasts against the cell periphery. The elongate shape could also serve other purposes such as enhancement of short-distance transport of sugars and other metabolites. The more random tissue arrangement of the underlying spongy mesophyll may enhance light scattering such that unabsorbed light that passes through the palisade is redirected back into the interior of the leaf. The contrasting light scattering properties of the palisade and spongy mesophyll, along with various epicuticular structures (e.g. waxes and trichomes), create the commonly observed phenomenon of leaf bicoloration where the lower (abaxial) surface of leaves often appears less green than the upper surface.

Absorption Profiles Within Leaves Determine the Energy Input to Photosynthesis

In order to ascertain how much each mesophyll layer contributes to whole-leaf photosynthesis, a first step is to know how much light is absorbed within each layer. In combination with photosynthetic capacity, this sets an upper boundary on the amount of carbon that can be fixed within each tissue layer (Evans 1995, Evans and Vogelmann 2003) and ultimately the whole leaf. Owing to optical complications introduced by light scattering and the sieve effect, it has been difficult to estimate light absorption profiles. Complex mathematical models have been constructed using radiation transport (Richter and Fukshansky 1996a,b, 1998) and ray tracing (Ustin et al. 2001), and absorption profiles have been calculated for different leaves subjected to varying irradiation regimes. Until recently, experimental verification of these models has not been possible owing to a lack of an experimental approach to measure absorption profiles. A new method measures chlorophyll fluorescence profiles from a cross sectional view of a leaf irradiated on its adaxial or abaxial surface (Koizumi et al. 1998, Takahashi et al. 1994, Vogelmann and Han 2000, Vogelmann and Evans 2002). Assuming constant quantum efficiency for chlorophyll fluorescence throughout the leaf, absorption profiles can be determined from the chlorophyll fluorescence profile.

In spinach, chlorophyll fluorescence profiles are determined by wavelength and leaf orientation. When leaves were irradiated on their upper (adaxial surface), the amount of fluorescence typically increased to a maximum beneath the epidermis and then it declined (Fig. 2.7). The rate of decline was related to wavelength, being greatest for the 450-nm blue light and less for 650 nm (red) and 550 nm (green) (see Fig. 2.7A). For inverted leaves, the maximum fluorescence occurred closer to the leaf surface and it declined more rapidly with depth (see Fig 2.7B). Chlorophyll fluorescence profiles are exponential in shape and they can be approximated by the Beer-Lambert/Bouguer laws (Vogelmann and Evans 2002).

Infiltrating spinach leaves with water greatly reduces light scattering between the air and cell wall interfaces (see Fig. 2.7C,D). Under these conditions, light absorption profiles decline more gradually with depth compared with native leaves and the effect is larger at wavelengths where there is weaker absorption by chlorophyll (e.g., green) than where there is stronger absorption (blue and red). By comparing the profiles of control and water-infiltrated leaves, it is possible to calculate pathlength within the tissues of a leaf (Fig. 2.8). For a spinach leaf irradiated on the adaxial surface, initially pathlength was 1.5 for red and blue, and 1.8 for green light. Going deeper into the leaf, the value for red light was constant across the leaf whereas it increased to around 2.1 for green light. For abaxial illumination, initially the values for red and blue light were similar to the adaxial surface, but they declined with depth suggesting that for these wavelengths, spongy and

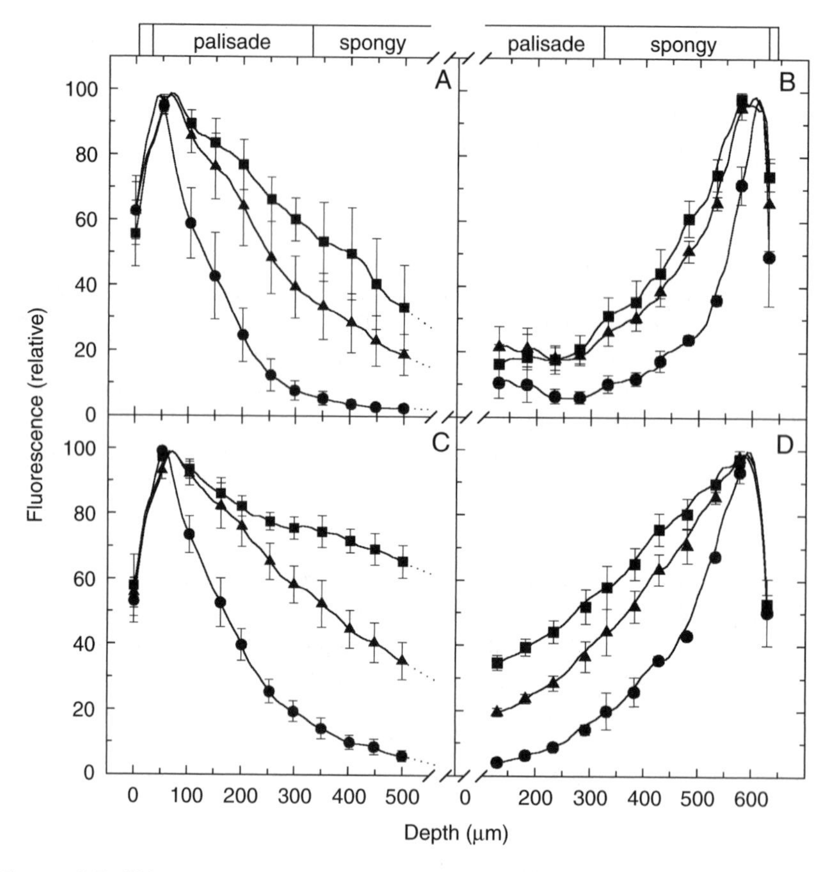

FIGURE 2.7. Chlorophyll fluorescence in spinach leaves irradiated with monochromatic light. Both native (A,B) and water-infiltrated leaves (C,D) were measured. Leaves were irradiated on their adaxial (A,C) or abaxial surface (B,D) with broad band blue (450 nm) (●), green (550 nm) (■) or red (650 nm) (▲) light. Symbols identify the lines; actual data points are spaced 6.5 μm apart. Error bars show SE, $n = 4$. (From Vogelmann and Evans 2002.)

palisade tissue scattered light to a similar extent. Green light had a pathlengthening value around 2 in the spongy tissue, which declined steadily toward the adaxial surface, suggesting greater scattering in spongy than palisade tissue. In view of the contrasting optical properties of the palisade and spongy tissues, it is curious that pathlength was relatively uniform in these tissues.

Despite the fundamental relationships between leaf optics, light absorption profiles in leaves, and photosynthetic performance, we have a way to go in elucidating the important relationships. Some of the details have been worked out for spinach but it remains to be demonstrated whether leaves from other species have similar profiles of light absorption and photosynthetic capacity. The same can be said for sun and shade leaves of the same plant. Current data present an average

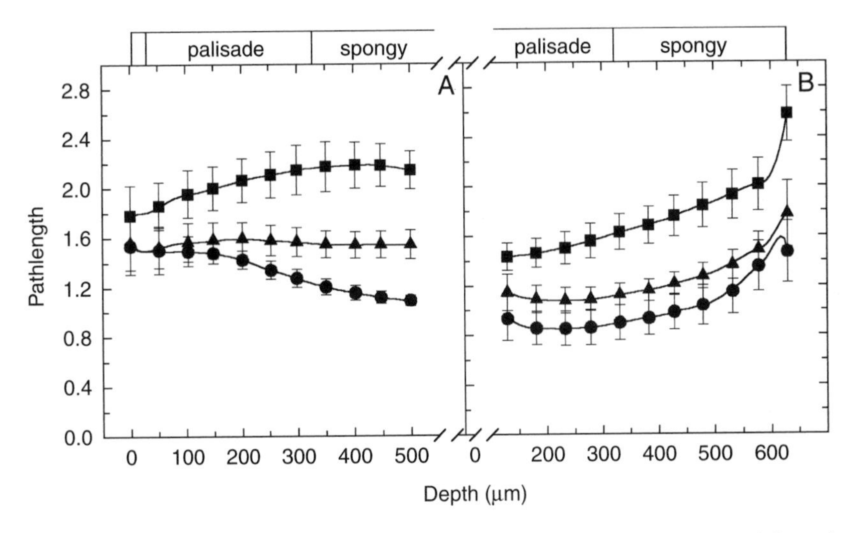

FIGURE 2.8. Optical pathlengthening by intercellular air spaces was calculated from the data shown in Fig. 2.7 as a function of depth for each wavelength with adaxial (A) or abaxial (B) leaf orientation. Symbols as in Fig. 2.7. (From Vogelmann and Evans 2002.)

view of a leaf but it must be kept in mind that the anatomical heterogeneity of a typical leaf adds another layer of complexity. Vascular tissues of most C_3 leaves are curiously devoid of photosynthetic pigmentation, and the transparent nature of bundle sheath extensions has been proposed to help light travel deeper in the leaf than is possible through mesophyll tissues (Karabourniotis et al. 2000). On the other hand, matching light absorption between chloroplasts within the bundle sheath and mesophyll tissues of C_4 plants is critical for efficient photosynthetic performance. Finally, little is known about how light direction interacts with leaf anatomy to set up internal absorption profiles. Current experimental data have been collected when light strikes the leaf perpendicularly, yet this is rarely the case in nature. Indeed, it is well known that photosynthesis within canopies is enhanced when the light is diffuse (Farquhar and Roderick 2003, Roderick et al. 2001) yet nothing is known regarding how diffuse light interacts with leaf tissues and how absorption profiles under these conditions compare with those in leaves irradiated with collimated light. Given recent advances in experimental techniques, essential information about light and photosynthesis within leaves no longer comprises a missing link in scaling photosynthetic processes from the chloroplast to the canopy level.

Chloroplast Movements

Chloroplasts are usually closely appressed to the cell membranes of mesophyll cells, facilitating gas exchange. However, their position within cells is dynamic and they may shift along the cell membrane to assume positions adjacent to periclinal (parallel to the leaf surface) or anticlinal (perpendicular to the leaf surface) walls under different circumstances. These movements change the optical properties of the leaf and have the potential to alter photosynthetic performance and perhaps whole-plant productivity.

Light is the dominant signal controlling chloroplast movements in higher plants and light-driven chloroplast movements have attracted attention for more than a century, largely because of the optical changes they engender. There are several reviews that describe the phenomenon (Britz 1979, Haupt, 1999, Haupt and Scheuerlein 1990, Wada and Kagawa 2001, Wada et al. 1993a, Yatsuhashi 1996). Patterns of chloroplast movement differ among various algae, mosses, ferns, and higher plants. We focus here on higher plants. If light is dim, chloroplasts are generally found along the periclinal walls, perpendicular to the incident light, where they intercept and absorb light maximally. In contrast, when light is bright, chloroplasts move to anticlinal walls, forming an optically dense cylinder in the peripheral cytoplasm but leaving a chloroplast-free path through which light can penetrate to deeper cell layers. This movement causes whole-leaf transmittance and reflectance to increase, so leaves in high light appear paler than in low light, a change that is fully reversible as light fades and chloroplasts return to periclinal walls (Fig. 2.9).

Chloroplast movements have generally been studied either directly, using microscopy, or indirectly, using optical techniques exploiting changes in transmittance that accompany chloroplast movement. Using these techniques, light-driven chloroplast movements have been observed in most groups of photosynthetic organisms, including algae, bryophytes, ferns, and higher plants (Table 2.2). They occur in hydrophytes (e.g., *Elodea*), mesophytes (e.g., *Alocasia*), xerophytes (e.g., *Sedum*) and evergreen sclerophytes (e.g., *Ilex*). The extent of light-mediated chloroplast movement varies with species; it may be hampered by chloroplast crowding (McCain 1998) as well as by large chloroplast size (Jeong et al. 2002). There are reports of plants that do not show chloroplast movements, but some (Lechowski 1974) are based on transmittance changes alone, which may not always detect chloroplast movement. Thick leaves with multiple cell layers would show negligible change in overall leaf transmittance. If chloroplasts move to anticlinal walls, light might penetrate one cell layer only to be absorbed by the next. For example, chloroplast movements are evident in micrographs of pea leaves, but corresponding changes in leaf transmittance are small (Park et al. 1996). A few plants, such as the moss *Fontinalis antipyretica*, do not show significant chloroplast movement detectable using either technique (Zurzycki 1961), but such reports are rare. Light-driven chloroplast movements, while perhaps not ubiquitous, are extremely widespread in photosynthetic organisms.

Mechanism

Studies on the mechanism of chloroplast movements have focused on two areas, light perception and force generation. In higher plants, chloroplast movement is

FIGURE 2.9. Striking changes in leaf appearance accompany chloroplast movements. The sunburst pattern visible here was shaded while this *Alocasia* leaf was irradiated, so chloroplasts in shaded cells remained in their low-light position along periclinal walls; absorbance was therefore relatively high, so the shaded area appears dark. In the irradiated area, chloroplasts moved to the periclinal walls, absorbance was lower, and the tissue appears a lighter shade of green. Diagrams represent chloroplast position in cells in the two areas of the leaf.

mediated by two members of the phototropin family of plasma membrane-associated blue-light photoreceptors. Movement to anticlinal walls in dim light is mediated by both phototropin 1 (phot1) and phototropin 2 (phot2), while movement to periclinal walls in high light appears to be mediated by phot2 alone (Briggs et al. 2001, Jarillo et al. 2001, Kagawa et al. 2001, Sakai et al. 2001). In some algae, mosses, and ferns, red/far-red reversible control mediated by phytochrome is involved as well (Sato et al. 2001, Wada et al. 1993b). In most plants, chloroplast movement is microfilament based and can be blocked by actin antagonists such as cytochalasin or m-maleimidobenzoic acid *N*-hydroxysuccinimide ester (Gorton et al. 1999, Izutani et al. 1990, Malec et al. 1996, Tlalka and Gabrys 1993, Wagner et al. 1972). In some cases, however, microtubules may be involved, as in red-light mediated chloroplast movement in the moss *Physcomitrella* (Sato et al. 2001). Little is known about signal transduction for chloroplast movement; clearly ATP is necessary (Slesak and Gabrys 1996), and Ca^{2+}, probably from internal stores, appears to be involved (Tlalka and Gabrys 1993, Tlalka and Fricker 1999).

TABLE 2.2. Widespread occurrence of light-driven chloroplast movement in algae and the pant kingdom. Chloroplast movement was demonstrated either directly by microscopy (M), indirectly by monitoring changes in the optical properties of the tissues (O), or, in one case, by nuclear magnetic resonance spectroscopy (NMR). Some observations of two of the authors (HLG and WEW) are included.

	Family	Species	Technique	Reference
Green algae	Chlorophyceae	*Eremosphaera viridis*	M	Weidinger and Ruppel 1985
	Mesotaeniaceae	*Mesotaenium caldariorum*	M	Haupt and Thiele 1961
	Ulotrichaceae	*Hormidium flaccidum*	M	Scholz 1976
	Zygnemataceae	*Mougeotia scalaris*	M	Haupt 1959
Chromophytes	Dictyotaceae	*Dictyota dichotoma*	M, O	Rüffer et al. 1981
	Vaucheriaceae	*Vaucheria sessilis*	M	Zurzycki and Lelatko 1969
Mosses	Brachytheciaceae	*Cirriphyllum piliferum*	M	Zurzycki and Lelatko 1969
	Funariaceae	*Funaria hygrometrica*	M, O	Zurzycki 1961
	Funariaceae	*Physcomitrella patens*	M	Sato et al., 2001
	Mniaceae	*Mnium cuspidatum*	M	Zurzycki and Lelatko 1969
	Mniaceae	*Mnium medium*	M	Zurzycki and Lelatko 1969
Club mosses	Selaginellaceae	*Selaginella martensii*	M	Mayer 1964
Ferns	Dryopteridaceae	*Cyrtomium flacatum*	O	Brugnoli and Björkman 1992
	Pteridaceae	*Adiantum candatums*	M, O	Augustynowicz and Gabrys 1999
	Pteridaceae	*Adiantum capillus-vereris*	M,O	Augustynowicz and Gabrys 1999; Kadota et al. 1989
	Pteridaceae	*Adiantum diaphanum*	M, O	Augustynowicz and Gabrys 1999
	Pteridaceae	*Pteris cretica*	M, O	Augustynowicz and Gabrys 1999
Aquatic monocots	Hydrocharitaceae	*Vallisneria spiralis*	M	Seitz 1967
	Hydrocharitaceae	*Elodea canadensis*	M, O	Zurzycki 1961
	Lemnaceae	*Lemna trisulca*	M, O	Zurzycki 1961
	Lemnaceae	*Spirodela polyrrhiza*	M, O	Inoue and Shibata 1973
	Potamogetonaceae	*Potamogeton crispus*	M	Zurzycki and Lelatko 1969
Terrestrial monocots	Araceae	*Alocasia macrorrhiza*	M, O	Gorton et al. 1999
	Commelinaceae	*Commelina communis*	O	Inoue and Shibata 1974
	Commelinaceae	*Tradescantia albiflora*	M, O	Gabrys and Walczak 1980
	Poaceae	*Atra caryophyllea*	O	Inoue and Shibata 1974
	Poaceae	*Coix lacryma*	O	Inoue and Shibata 1974
	Poaceae	*Digitaria sanguinalis*	O	Inoue and Shibata 1974
	Poaceae	*Eleusine indica*	O	Inoue and Shibata 1974
	Poaceae	*Oplismenus undulatifolius*	O	Inoue and Shibata 1973
	Poaceae	*Oryza sativa*	O	Inoue and Shibata 1974
	Poaceae	*Setaria viridis*	O	Inoue and Shibata 1973
	Poaceae	*Triticum aestivum*	M, O	Inoue and Shibata 1973
	Poaceae	*Zea mays*	O	Inoue and Shibata 1973
	Liliaceae	*Asparagus sprengeri*	M	Zurzycki and Lelatko 1969

	Liliaceae	*Chlorophytum elatum*	M, O	Gabrys and Walczak 1980
	Liliaceae	*Polygonatum odoratum*	M	Zurzycki and Lelatko 1969
	Liliaceae	*Smilacena stallata*	O	Personal observation, HLG and WEW
	Musaceae	*Musa sapientum*	O	Personal observation, HLG and WEW
Aquatic dicots	Lentibulariaceae	*Utricularia vulgaris*	M	Zurzycki and Lelatko 1969
	Primulaceae	*Hottonia palustris*	M	Zurzycki and Lelatko 1969
Terrestrial dicots	Aceraceae	*Acer platanoides*	NMR	McCain 1998
	Araliaceae	*Hedera canariensis*	O	Brugnoli and Björkman 1992
	Asteraceae	*Erigeron annus*	O	Inoue and Shibata 1974
	Asteraceae	*Erigeron sumatorensis*	O	Inoue and Shibata 1974
	Asteraceae	*Helianthus annuus*	O	Brugnoli and Björkman 1992
	Asteraceae	*Lactuca laciniata*	O	Inoue and Shibata 1974
	Asteraceae	*Taraxacum officinale*	O	Inoue and Shibata 1974
	Balsaminaceae	*Impatiens capensis*	O	Personal observation, HLG and WEW
	Begoniaceae	*Begonia semperflorens*	M, O	Inoue and Shibata 1973
	Boraginaceae	*Symphytum officinale*	M, O	Lelatko 1970; Zurzycki and Lelatko 1969
	Brassicaceae	*Arabidopsis thaliana*	M, O	Trojan and Gabrys 1996
	Caprifoliaceae	*Sambucus nigra*	M, O	Lelatko 1970; Zurzycki and Lelatko 1969
	Caryophyllaceae	*Melandrium album*	O	Lelatko 1970
	Chenopodiaceae	*Spinacia oleracea*	O	Inoue and Shibata 1974
	Crassulaceae	*Sedum maximum*	M, O	Lelatko 1970; Zurzycki and Lelatko 1969
	Cucurbitaceae	*Marah fabaceus*	O	Brugnoli and Björkman 1992
	Fabaceae	*Medicago sativa*	O	Personal observation, HLG and WEW
	Fabaceae	*Pisum satuvum*	M, O	Park et al. 1996
	Fabaceae	*Trifolium* sp.	O	Personal observation, HLG and WEW
	Fagaceae	*Fagus silvatica*	O	Lechowski 1972
	Geraniaceae	*Pelargonium* sp.	O	Personal observation, HLG and WEW
	Lamiaceae	*Ajuga reptans*	M, O	Gabrys and Walczak 1980
	Lamiaceae	*Galeopsis tetrahit*	O	Lelatko 1970
	Malvaceae	*Gossypium hirsutum*	O	Brugnoli and Björkman 1992
	Oxalidaceae	*Oxalis oregana*	O	Brugnoli and Björkman 1992
	Oxalidaceae	*Oxalis acetosella*	O	Hoel and Solhaug 1998
	Paeoniaceae	*Paeonia officinalis*	O	Personal observation, HLG and WEW
	Phytolaccaceae	*Phytolacca americana*	O	Inoue and Shibata 1974
	Polygonaceae	*Polygonum hydropiperoides*	O	Inoue and Shibata 1974
	Polygonaceae	*Rumex crispus*	O	Inoue and Shibata 1974
	Rannunculaceae	*Aquilegia flavescens*	O	Personal observation, HLG and WEW
	Solanaceae	*Nicotiana tabacum*	M, O	Jeong et al., 2002
	Vitaceae	*Vitis vinifera*	O	Lelatko 1970

Chloroplast Movements Under Natural Conditions

It is simple to mask leaves in the field when they are exposed to the sun and later remove the masks to see if chloroplast movements have caused a change in leaf appearance as in Figure 2.9. However, available instrumentation has restricted kinetic measurements of chloroplast movements to the laboratory. It has been difficult to assess, for example, how laboratory-derived light-response curves might relate to actual chloroplast movements with changing solar irradiance throughout the day. Now, with a field-portable device that monitors leaf transmittance using a pulsed measuring beam and lock-in detection, it is possible to follow chloroplast movements in the field (Fig. 2.10)(Williams et al. 2003). Data are shown for *Aquilegia flavescens* (columbine), a plant that grows at high altitude, in a high-PAR, high-UV environment. The transmittance signal obtained in the field is not directly comparable with that obtained in the laboratory because of changes in parameters such as angle and collimation of irradiation and because of differences in instrument design. However, observed transmittance changes are consistent with light-mediated chloroplast movement. In this example, leaf transmittance tracks PAR, decreasing during the late afternoon as PAR drops and chloroplasts move to periclinal walls, remaining low overnight, and increasing again the next day as chloroplasts move to the anticlinal walls.

Functionality

Chloroplast movements are widespread yet require energy; they occur in the field as well as under controlled laboratory conditions. These observations lead one to consider the possible adaptive advantage they may have for plants. Hypotheses about the possible benefits of chloroplast movement generally have been based on the changes they cause in optical properties of the leaves. The magnitude of changes in whole-leaf transmittance, reflectance, and absorptance vary with species and with leaf thickness; they are generally small, but leaf absorptance may change 15 to 20% in some thin leaves such as *Lemna trisulca* (Zurzycki 1961) or *Oxalis oregana* (Brugnoli and Björkman 1992). Chloroplast movements under dim light and under bright light are distinct responses, cause leaf absorptance changes in different directions, and likely serve different adaptive roles.

One adaptive advantage of chloroplast movement to periclinal walls under dim light seems clear: increased leaf absorptance correlates with increased photosynthesis under light-limiting conditions (Lechowski 1974, Zurzycki 1955). However, changes in the rate of photosynthesis may be greater than expected based on changes in whole-leaf absorptance alone (Zurzycki 1961). Several hypotheses, perhaps complementary, have been suggested to explain chloroplast movement to anticlinal walls in bright light.

Hypothesis 1: Chloroplast movement in high light protects chloroplasts from photodamage. This hypothesis is based on the idea that fewer chloroplasts would be exposed to potentially damaging rays; mutual shading would protect most of

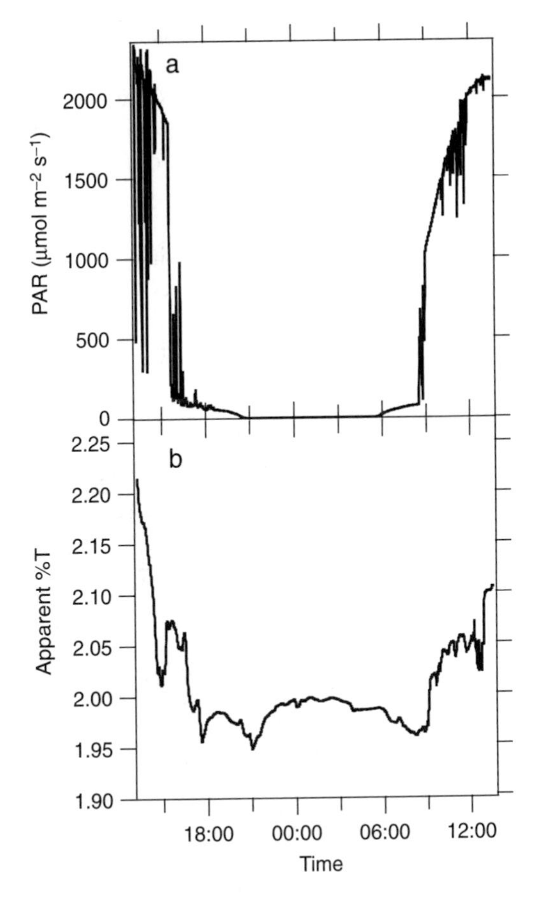

FIGURE 2.10. Incident solar irradiance (a) and apparent transmittance changes (b, measuring beam penetrating the leaf at a 40° angle) in an *Aquilegia flavescens* leaf over a 24-h period, midday July 16 to midday July 17, 2001. Measurements were made in the field in the Snowy Mountains of southwestern Wyoming at about 3700-m elevation.

the chloroplasts in the palisade. It is an early hypothesis (Zurzycki 1957), and there is recent evidence to support it. The greater tolerance of light stress in *Tradescantia* than in *Pisum* has been attributed to light-induced chloroplast movement in *Tradescantia* (Park et al. 1996). However, although absorptance changes were greater in the thinner-leafed *Tradescantia*, both species showed marked chloroplast movement, so there may be other factors involved. Another study employed transgenic tobacco plants deficient in plastid division with only a few large chloroplasts per cell rather than many small ones (Jeong et al. 2002). Although the large chloroplasts did move in response to light, transmittance changes were smaller than for wild-type leaves, and the transgenic plants were more susceptible to photodamage. Both of these studies depend on correlation between transmittance changes and susceptibility to photodamage, and for both, factors other than chloroplast movement could cause the observed differences in photosensitivity. Experiments using other ways to block chloroplast movement, such as with the *phot* mutants that do not show chloroplast movement, will be important to confirm these conclusions. In addition, little is known about how differences in sensitivity to photodamage found in the laboratory translate to differences for plants growing in the field, where irradiation regimes are more variable and where small differences accumulate over time.

Hypothesis 2: Chloroplast movement to anticlinal walls under high light opens channels of light penetration to deeper cell layers within the leaf and relieves light limitation of photosynthesis in those deeper layers (Brugnoli and Björkman 1992, Terashima and Hikosaka 1995). One prediction from this hypothesis is supported: light penetration does increase as chloroplasts move to anticlinal walls, and more light is absorbed by deeper cell layers (Fig. 2. 11). However, there is no evidence that the observed redistribution of light absorption leads to any increase in photosynthesis, at least in the short term (Lechowski 1974, Zurzycki 1955).

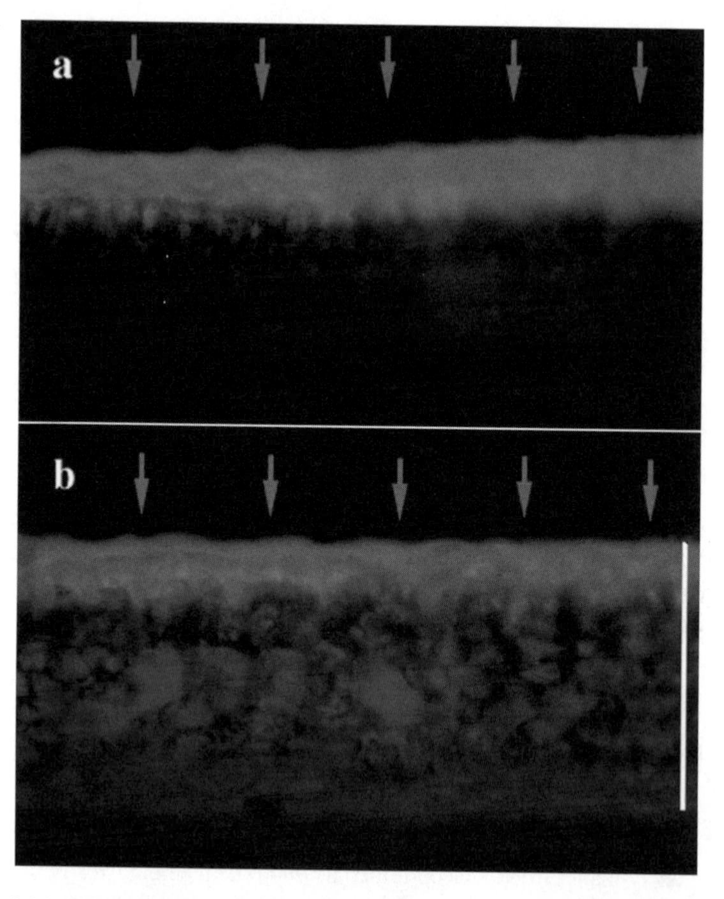

FIGURE 2.11. Chlorophyll fluorescence indicates where light is absorbed within *Alocasia macrorrhiza* leaves with chloroplasts along periclinal walls in the dark (a) and along anticlinal walls in bright light (b). Fluorescence was induced by blue light normal to the adaxial surface (arrows).

Hypothesis 3: Chloroplast movement in bright light reduces CO_2-diffusion distance and enhances CO_2 uptake. Under this hypothesis, chloroplasts would move to their low-light position to maximize light absorption when light was limiting, and to their high-light position to maximize the rate of CO_2-diffusion when CO_2 was limiting. Both distance in the liquid phase and surface area of chloroplasts adjacent to intracellular air spaces affect the rate of CO_2 diffusion from the intercellular air spaces to the chloroplasts (Evans 1999, Evans and Vellen 1996). Thus, chloroplast movement in high light could increase CO_2 diffusion by moving the chloroplasts more tightly against the plasma membrane, or if it flattened them so that more chloroplast surface was against the membrane. This hypothesis is consistent with the light requirements for chloroplast movement to anticlinal walls; saturation of the response generally requires at least as much light as saturation of photosynthesis and can require much more (Gorton et al. 1999). However, photoacoustic experiments designed to assess the diffusion rate of CO_2 into chloroplasts do not show that chloroplast movement under bright light enhances gas diffusion (Gorton et al. 2002).

Although much of its specific function remains to be elucidated by future experiments, it seems clear that chloroplast movement is an important part of the armamentarium that includes leaf shape, display angle, active movement, and wilting, that plants have evolved for keeping them precisely adapted to a varying light environment.

Summary and Conclusions

Despite tremendous diversity in leaf anatomy and light environments, there are many features that allow generalization. Leaves exist to increase the efficiency with which a plant can capture light and gain CO_2. The diversity of solutions reflect the different environmental pressures, and diversity appears to increase as one scales up from the photosystem reaction center to the chloroplast, cell, leaf, and plant.

There is little difference between species in the pigment-protein complexes of chloroplast thylakoids. For shade chloroplasts (either those deep within a leaf, or from a leaf growing under low irradiance), there is an increased proportion of chlorophyll associated with light-harvesting complexes compared to sun or high-light chloroplasts. This is associated with an increased number of thylakoids per granum, but curiously, does not seem to alter the ratio of stroma exposed to appressed thylakoid membrane surface. We still lack experimental techniques to probe the fate of light as it is absorbed in the thylakoids to definitely know how much is absorbed by each photosystem. The complexity in the distribution and function of the various types of each photosystem, whilst known, is not properly understood. A key question that awaits definitive explanation is the wavelength dependence of quantum yield. Why do plants in white light only achieve 85% of the quantum yield that they achieve in red light?

Perhaps, rather than waiting for the invention of new techniques, progress can be made by using mutant and transgenic plants. While multiple interpretations of such experiments can be made, having independent ways of asking the same question improves our chances of correct understanding. Unfortunately, what is often revealed is the flexibility a plant has in bypassing the problem. For example, removing the two major LHCII proteins was compensated by increased expression of another gene. Lhcb5 is usually a minor component restricted to a place close to the reaction center complex of photosystem II. However, in transgenic plants lacking Lhcb1 and Lhcb2, the expression of Lhcb5 increased sufficiently for it to be able to assemble into trimeric complexes, a role previously thought to be specific for Lhcb1 and Lhcb2 (Ruban et al. 2003).

Chloroplast movement is a widespread phenomenon and yet it has been difficult to demonstrate any substantial benefit to performance in terms of protection from damage in bright light, enabling light to penetrate more deeply into a leaf, or reducing the resistance to CO_2 diffusion into the chloroplast. Here too, mutants and transgenic plants may assist in revealing the function of chloroplast movement.

By far the greatest diversity for light capture by the leaf exists among plants in their anatomy. It is now possible to infer the profile of light absorption through the leaf by measuring chlorophyll fluorescence (Vogelmann and Evans 2002). This revealed the dominant role that chlorophyll distribution played, as well as the impact of light scattering by the intercellular air spaces. Considerable work remains to be done in examining a broad range of foliage, from thick needles to thin leaves. The problem of dealing with diffuse light and light striking the leaf at different angles also requires effort. It is known that the photosynthetic capacity of chloroplasts can vary and adapt. Usually it is thought that chloroplast photosynthetic capacity will relate to the amount of light absorbed by the chloroplast. However, are there limits to the flexibility of a given chloroplast? What are the lower and upper bounds in photosynthetic capacity per unit chlorophyll? It is not known whether different chloroplasts within a given cell have to share the same capacity per unit chlorophyll. It is also only possible to match one profile in light absorption, which means that it is not perfectly matched to other profiles that may occur due to changing angle of incidence of the light, altered proportion of diffuse light, or light falling onto the other surface. The mismatch between light absorption and photosynthetic capacity reduces the efficiency of the leaf because less photosynthesis is achieved for a given investment in resources. Within the leaf, this is the area where the greatest potential exists to improve photosynthesis. The huge diversity in leaf anatomy shows that there are many solutions to the problems faced by a plant. This also provides us with the ability to tease apart the different contributions that tissue, cell, and chloroplast morphology make toward light capture by a leaf.

References

Albertsson, P. A. 1995. The structure and function of the chloroplast photosynthetic membrane—A model for the domain organization. Photosyn. Res. 46:141–149.

Albertsson, P. A. 2001. A quantitative model of the domain structure of the photosynthetic membrane. Trends Plant Sci. 6:349–354.

Anderson, J. M. 1999. Insights into the consequences of grana stacking of thylakoid membranes in vascular plants: A personal perspective. Aust. J. Plant Physiol. 26:625–639.

Andersson, B., and Anderson, J. M. 1980. Lateral heterogeneity in the distribution of chlorophyll-protein complexes of the thylakoid membranes of spinach chloroplasts. Biochim. Biophys. Acta 593:427–440.

Andersson, J., Wentworth, M., Walters, R. G., Howard, C. A., Ruban, A. V., Horton, P., and Jansson, S. 2003. Absence of the Lhcb1 and Lhcb2 proteins of the light-harvesting complex of photosystem II—Effects on photosynthesis, grana stacking and fitness. Plant J. 35:350–361.

Andreasson, E., and, Albertsson, P. A. 1993. Heterogeneity in photosystem-I—The larger antenna of photosystem-I-alpha is due to functional connection to a special pool of LHCII. Biochim. Biophys. Acta 1141:175–182.

Andreasson, E., Svensson, P., Weibull, C., and, Albertsson, P. A. 1988. Separation and characterization of stroma and grana membranes—Evidence for heterogeneity in antenna size of both photosystem-I and photosystem-II. Biochim. Biophys. Acta 936:339–350.

Augustynowicz, J., and Gabrys, H. 1999. Chloroplast movements in fern leaves: Correlation of movement dynamics and environmental flexibility of the species. Plant Cell Environ. 22:1239–1248.

Björkman, O., and, Demmig, B. 1987. Photon yield of O_2 evolution and chlorophyll fluorescence characteristics at 77K among vascular plants of diverse origins. Planta 170:489–504.

Boekema, E. J., Boonstra, A. F., Dekker, J. P., and Rogner, M. 1994. Electron microscopic structural analysis of photosystem-I, photosystem-II, and the cytochrome-b6/f complex from green plants and cyanobacteria. J. Bioenerg. Biomemb. 26:17–29.

Boekema, E. J., van Roon, H., van Breemen, J. F. L., and Dekker, J. P. 1999. Supramolecular organization of photosystem II and its light-harvesting antenna in partially solubilized photosystem II membranes. Europ. J. Biochem. 266:444–452.

Boekema, E. J., van Breemen, J. F. L., van Roon, H., and Dekker, J. P. 2000. Arrangement of photosystem II supercomplexes in crystalline macrodomains within the thylakoid membrane of green plant chloroplasts. J. Mol. Biol. 301:1123–1133.

Briggs, W. R., Beck, C. F., Cashmore, A. R., Christie, J. M., Hughes, J., Jarillo, J. A., Kagawa, T., Kanegae, H., Liscum, E., Nagatani, A., Okada, K., Salomon, M., Rüdiger, W., Sakai, T., Takano, M., Wada, M., and Watson, J. C. 2001. The phototropin family of photoreceptors. Plant Cell 13:993–997.

Britz, S. J. 1979. Chloroplast and nuclear migration. In: Encyclopedia of Plant Physiology, New Series, Vol. 7, M. Feinleib (ed.), pp. 170–205. New York: Springer-Verlag.

Brugnoli, E., and Björkman, O. 1992. Chloroplast movements in leaves: Influence on chlorophyll fluorescence and measurements of light-induced absorbance changes related to DpH and zeaxanthin formation. Photosynth. Res. 32:23–35.

Chazdon, R. L., and Field, C. B. 1987. Photographic estimation of photosynthetically active radiation: Evaluation of a computerized technique. Oecologia 73:525–532.

DePury, D. G. G., and Farquhar, G. D. 1997. Simple scaling of photosynthesis from leaves to canopies without the errors of big-leaf models. Plant Cell Environ. 20:537–557.

Ehleringer, J. R., and Björkman, O. 1978. A comparison of photosynthetic characteristics of *Encelia* species possessing glabrous and pubescent leaves. Plant Physiol. 62: 185–190.

Evans, J. R. 1987. The dependence of quantum yield on wavelength and growth irradiance. Aust. J. Plant Physiol. 14:69–79.

Evans, J. R. 1995. Carbon fixation profiles do reflect light absorption profiles in leaves. Aust. J. Plant Physiol. 22:865–873.
Evans, J. R. 1999. Leaf anatomy enables more equal access to light and CO_2 between chloroplasts. New Phytologist 143:93–104.
Evans, J. R., and Anderson, J. M. 1987. Absolute absorption spectra for the five major chlorophyll-protein complexes and their 77K fluorescence excitation spectra. Biochim. Biophys. Acta 892:75–82.
Evans, J. R., and Poorter, H. 2001. Photosynthetic acclimation of plants to growth irradiance: the relative importance of specific leaf area and nitrogen partitioning in maximizing carbon gain. Plant Cell Environ. 24:755–767.
Evans, J. R., and Vellen, L. 1996. Wheat cultivars differ in transpiration efficiency and CO_2 diffusion inside their leaves. In: Crop Research in Asia: Achievements and Perspective, R. Ishii and T. Horie (eds.), pp. 326–329. Tokyo: Asian Crop Science Association.
Evans, J. R., and Vogelmann, T. C. 2003. Profiles of C-14 fixation through spinach leaves in relation to light absorption and photosynthetic capacity. Plant Cell Environ. 26:547–560.
Farquhar, G. D., and Roderick, M. L. 2003. Atmospheric science: Pinatubo, diffuse light, and the carbon cycle. Science 299:1997–1998.
Gabrys, H., and Walczak, T. 1980. Photometric study of chloroplast phototranslocations in leaves of land plants. Acta Physiol. Plantarum 2:281–290.
Gorton, H. L., Williams, W. E., and Vogelmann, T. C. 1999. Chloroplast movement in *Alocasia macrorrhiza*. Physiol. Plantarum 106:421–428.
Gorton, H. L., Herbert, S. K., and Jogelmann, T. C. 2003. Photoacoustic analysis indicates that chloroplast movement does not alter liquid-phase CO_2 diffusion in leaevs of *Alocasia brisbanensis*. Plant Physiol. 132:1–11.
Haupt, W. 1959. Die Chloroplastendrehung bei *Mougeotia*. 1. Über den quantitativen Lichtbedarf der Schwachlichtbewegung. Planta 53:484–501.
Haupt, W. 1999. Chloroplast movement: From phenomenology to molecular biology. Prog. Botany 60:1–36.
Haupt, W., and Scheuerlein, R. 1990. Chloroplast movement. Plant Cell Environ. 13:595–614.
Haupt, W., and Thiele, R. 1961. Chloroplastenbewegung bei *Mesotaenium*. Planta 56:388–401.
Hoel, B. O., and Solhaug, K. A. 1998. Effect of irradiance on chlorophyll estimation with the Minolta SPAD-502 leaf chlorophyll meter. Ann. Bot. 82:389–392.
Inoue, Y., and Shibata, K. 1973. Light-induced chloroplast rearrangements and their action spectra as measured by absorption spectrophotometry. Planta 114:341–358.
Inoue, Y., and Shibata, K. 1974. Comparative examination of terrestrial plant leaves in terms of light-induced absorption changes due to chloroplast rearrangements. Plant Cell Physiol. 15:717–721.
Izutani, Y., Takagi, S., and Nagai, R. 1990. Orientation movements of chloroplasts in *Vallisneria* epidermal cells: Different effects of light at low- and high-fluence rate. Photochem. Photobiol. 51:105–111.
Jarillo, J. A., Gabrys, H., Capel, J., Alaonso, J. M., Ecker, J. R., and Cashmore, A. R. 2001. Phototropin-related NPL1 controls chloroplast relocation induced by blue light. Nature 410:952–954.
Jeong, W. J., Park, Y.-I., Suh, K., Raven, J. A., Yoo, O. J., and Liu, J. R. 2002. A large population of small chloroplasts in tobacco leaf cells allows more effective chloroplast movement than a few enlarged chloroplasts. Plant Physiol. 129:112–121.

Jordan, P., Fromme, P., Witt, H. T., Klukas, O., Saenger, W., and Krauss, N. 2001. Three-dimensional structure of cyanobacterial photosystem I at 2.5 angstrom resolution. Nature 411:909–917.

Kadota, A., Kohyama, I., and Wada, M. 1989. Polarotropism and photomovement of chloroplasts in the protonema of the ferns *Pteris* and *Adiantum*: Evidence for the possible lack of dichroic phytochrome in *Pteris*. Plant Cell Physiol. 30:523–531.

Kagawa, T., Sakai, T., Suetsugu, N., Oikawa, K., Ishiguro, S., Kato, T., Tabata, S., Okada, K., and Wada, M. 2001. *Arabidopsis* NPL1: A phototropin homolog controlling the chloroplast high-light avoidance response. Science 291:2138–2141.

Karabourniotis, G., Bornman, J. F., and Nikolopoulos, D. 2000. A possible optical role of the bundle sheath extensions of the heterobaric leaves of *Vitis vinifera* and *Quercus coccifera*. Plant Cell Environ. 23:423–430.

Kirchhoff, H., Mukherjee, U., and Galla, H. J. 2002. Molecular architecture of the thylakoid membrane: Lipid diffusion space for plastoquinone. Biochemistry 41:4872–4882.

Koizumi, M., Takahashi, K., Mineuchi, K., Nakamura, T., and Kano, H. 1998. Light gradients and the transverse distribution of chlorophyll fluorescence in mangrove and camellia leaves. Ann. Bot. 81:527–533.

Kuhlbrandt, W., Wang, D. N., and Fujiyoshi, Y. 1994. Atomic model of plant light-harvesting complex by electron crystallography. Nature 367:614–621.

Lechowski, Z. 1972. Action spectrum of chloroplast displacements in leaves of land plants. Acta Protozool. 11:202–209.

Lechowski, Z. 1974. Chloroplast arrangement as a factor of photosynthesis in multilayered leaves. Acta Soc. Bot. Pol. 43:531–540.

Lelatko, Z. 1970. Some aspects of chloroplast movement in leaves of terrestrial plants. Acta Soc. Bot. Pol. 39:453–468.

Malec, P., Rinaldi, R. A., and Gabrys, H. 1996. Light-induced chloroplast movements in *Lemna trisulca*. Identification of the motile system. Plant Sci. 120:127–137.

Mayer, F. 1964. Lichtorientierte Chloroplasten-Verlagerungen bei *Selaginella martensii*. Z. Bot. 52:346–381.

McCain, D. 1998. Chloroplast movement can be impeded by crowding. Plant Sci. 135:219–225.

McCree, K. J. 1971. The action spectrum, absorptance and quantum yield of photosynthesis in crop plants. Agric. For. Meteorol. 9:191–216.

Miller, K. R., Staehelin, L. A. 1977. Analysis of the thylakoid outer surface: Coupling factor is limited to unstacked membrane regions. Biochim. Biophys. Acta 68:30–47.

Park, Y. I., Chow, W. S., and Anderson, J. M. 1996. Chloroplast movement in the shade plant *Tradescantia albiflora* helps protect photosystem II against light stress. Plant Physiol. 111:867–875.

Pearcy, R. W. 1990. Sunflecks and photosynthesis in plant canopies. Annu. Rev. Plant Physiol. Plant Mol. Biol. 41:421–453.

Pons, T. L., and Pearcy, R. W. 1992. Photosynthesis in flashing light in soybean leaves grown in different conditions. 2. Lightfleck utilization efficiency. Plant Cell Environ. 15:577–584.

Richter, T., and Fukshansky, L. 1996a. Optics of a bifacial leaf. 1. A novel combined procedure for deriving the optical parameters. Photochem. Photobiol. 63:507–516.

Richter, T., and Fukshansky, L. 1996b. Optics of a bifacial leaf. 2. Light regime as affected by the leaf structure and the light source. Photochem. Photobiol. 63:517–527.

Richter, T., and Fukshansky, L. 1998. Optics of a bifacial leaf. 3. Implications for photosynthetic performance. Photochem. Photobiol. 68:337–352.

Roden, J. S., and Pearcy, R. W. 1993. Effect of leaf flutter on the light environment of poplars. Oecologia 93:201–207.

Roderick, M. L., Farquhar, G. D., Berry, S. L., and Noble, I. R. 2001. On the direct effect of clouds and atmospheric particles on the productivity and structure of vegetation. Oecologia 129:21–30.

Ruban, A. V., Wentworth, M., Yakushevska, A. E., Andersson, J., Lee, P. J., Keegstra, W., Dekker, J. P., Boekema, E. J., Jansson, S., and Horton, P. 2003. Plants lacking the main light-harvesting complex retain photosystem II macro-organization. Nature 421:648–652.

Rüffer, U., Pfau, J., and Nultsch, W. 1981. Movements and arrangements of *Dictyota* phaeoplasts in response to light and darkness. Z. Pflanzenphysiol. 101:283–293.

Sakai, T., Kagawa, T., Kasahara, M., Swartz, T. E., Christie, J. M., Briggs, W. R., Wada, M., and Okada, K. 2001. *Arabidopsis* nph1 and npl1: Blue light receptors that mediate both phototropism and chloropalst relocation. Proc. Natl. Acad. Sci. 98:6969–6974.

Sato, Y., Wada, M., and Kadota, A. 2001. Choice of tracks, microtubules and/or actin filaments for chloroplast photo-movement is differentially controlled by phytochrome and a blue light receptor. J. Cell Sci. 114:269–279.

Scholz, A. 1976. Lichtorientierte Chloroplastenbewegung bei *Hormidium flaccidum*: Perception der Lichtrichtung mittels Sammellinseneffekt. Z. Pflanzenphysiol. 77:406–421.

Schubert, W. D., Klukas, O., Saenger, W., Witt, H. T., Fromme, P., and Krauss, N. 1998. A common ancestor for oxygenic and anoxygenic photosynthetic systems—A comparison based on the structural model of photosystem I. J. Mol. Biol. 280:297–314.

Seitz, K. 1967. Wirkungsspektren für die Starklichtbewegung der Chloroplasten, die Photodinese und die lichtabhängige Viskositätsänderung bei *Vallisneria spiralis* ssp. *torta*. Z. Pflanzenphysiol. 56:246–261.

Slesak, I., and Gabrys, H. 1996. Role of photosynthesis in the control of blue light-induced chloroplast movements. Inhibitor study. Acta Physiol. Plantarum 18:135–145.

Smith, W. K., Vogelmann, T. C., Bell, D. T., DeLucia, B. H., and Shepherd, K. A. 1997. Leaf form and photosynthesis. BioScience 47:785–793.

Staehelin, L.A., and van der Staay, G. W. M. 1996. Structure, composition, functional organization and dynamic properties of thylakoid membranes. In: Oxygenic Photosynthesis: The Light Reactions. D. R. Ort and C. F. Yocum eds. pp. 11–30. Dordrecht, The Netherlands: Kluwer Academic Publishers.

Stefansson, H., Wollenberger, L., Yu, S. G., and Albertsson, P. A. 1995. Phosphorylation of thylakoids and isolated subthylakoid vesicles derived from different structural domains of the thylakoid membrane from spinach chloroplast. Biochim. Biophys. Acta Bioenerget. 1231:323–334.

Svensson, P., Andreasson, E., and Albertsson, P. A. 1991. Heterogeneity among Photosystem-I. Biochim. Biophys. Acta 1060:45–50.

Takahashi, K., Mineuchi, K., Nakamura, T., Koizumi, M., and Kano, H. 1994. A system for imaging transverse distribution of scattered light and chlorophyll fluorescence in intact rice leaves. Plant Cell Environ. 17:105–110.

Terashima, I., and Evans, J. R. 1988. Effects of light and nitrogen nutrition on the organization of the photosynthetic apparatus in spinach. Plant Cell Physiol. 29:143–155.

Terashima, I., and Hikosaka, K. 1995. Comparative ecophysiology of leaf and canopy photosynthesis. Plant Cell Environ. 18:1111–1128.

Terashima, I., and Inoue, Y. 1985. Vertical gradient in photosynthetic properties of spinach chloroplasts dependent on intra-leaf light environment. Plant Cell Physiol. 26:781–785.

Tlalka, M., and Fricker, M. 1999. The role of calcium in blue-light-dependent chloroplast movement in *Lemna trisulca* L. Plant J. 20:461–473.

Tlalka, M., and Gabrys, H. 1993. Influence of calcium on blue-light-induced chloroplast movement in *Lemna trisulca* L. Planta 189:491–498.

Trojan, A., and Gabrys, H. 1996. Chloroplast distribution in *Arabidopsis thaliana* (L.) depends on light conditions during growth. Plant Physiol. 111:419–425.

Ustin, S. L., Jacquemoud, S., and Govaerts, Y. 2001. Simulation of photon transport in a three-dimensional leaf: Implications for photosynthesis. Plant Cell Environ. 24:1095–1103.

Vogelmann, T. C. 1993. Plant tissue optics. Annu. Rev. Plant Physiol. Plant Mol. Biol. 44:231–251.

Vogelmann, T. C., Nishio, J. N., and Smith, W. K. 1996. Leaves and light capture: Light propagation and gradients of carbon fixation within leaves. Trends in Pl. Sci. 1:65–71.

Vogelmann, T. C., and Evans, J. R. 2002. Profiles of light absorption and chlorophyll within spinach leaves from chlorophyll fluorescence. Plant Cell Environ. 25:1313–1323.

Vogelmann, T. C., and Han, T. 2000. Measurement of gradients of absorbed light in spinach leaves from chlorophyll fluorescence profiles. Plant Cell Environ. 23:1303–1311.

Wada, M., and Kagawa, T. 2001. Light-controlled chloroplast movement. In Photomovement, Vol. 1. M. Lebert (ed.), pp. 897–924. New York: Elsevier.

Wada, M., Grolig, F., and Haupt, W. 1993a. Light-oriented chloroplast positioning. Contribution to progress in photobiology. J. Photochem. Photobiol. 17:3–25.

Wada, M., Grolig, F., and Haupt, W. 1993b. Light-oriented chloroplast positioning. Contribution to progress in photobiology. J Photochem Photobiol. 17:3–25.

Wagner, G., Haupt, W., and Laux, A. 1972. Reversible inhibition of chloroplast movement by cytochalasin B in the green alga *Mougeotia*. Science 176:808–809.

Weidinger, M., and Ruppel, H. 1985. Ca^{2+}-requirement for a blue-light-induced chloroplast translocation in *Eremosphaera viridis*. Protoplasma 124:184–187.

Williams, W. E., Gorton, H. L., and Witiak, S. M. 2003. Chloroplast movements in the field. Plant Cell Environ. 26:2005–2014.

Yakushevska, A. E., Jensen, P. E., Keegstra, W., van Roon, H., Scheller, H. V., Boekema, E. J., and Dekker, J. P. 2001. Supermolecular organization of photosystem II and its associated light-harvesting antenna in *Arabidopsis thaliana*. Eur. J. Biochem. 268: 6020–6028.

Yatsuhashi, H. 1996. Photoregulation systems for light-oriented chloroplast movement. J. Plant Res. 109:139–146.

Zouni, A., Witt, H. T., Kern, J., Fromme, P., Krauss, N., Saenger, W., and Orth, P. 2001. Crystal structure of photosystem II from *Synechococcus elongatus* at 3.8 angstrom resolution. Nature 409:739–743.

Zurzycki, J. 1955. Chloroplasts arrangement as a factor in photosynthesis. Acta Soc. Bot. Pol. 24:27–63.

Zurzycki, J. 1957. The destructive effect of intense light on the photosynthetic apparatus. Acta Soc. Bot. Pol. 26:157–175.

Zurzycki, J. 1961. The influence of chloroplast displacements on the optical properties of leaves. Acta Soc. Bot. Pol. 30:503–527.

Zurzycki, J., and Lelatko, Z. 1969. Action dichroism in the chloroplasts rearrangements in various plant species. Acta Soc. Bot. Pol. 38:493–506.

3
Leaf to Landscape

Alessandro Cescatti and Ülo Niinemets

Introduction

Temporal dynamics and structural complexity of plant canopies strongly affect light harvesting, generating variable spatio-temporal distributions of the irradiance on leaf area (Baldocchi and Collineau 1994). Leaf light interception scales linearly with incident irradiance, but plant photosynthesis and photomorphogenesis typically exhibit a saturating response to light. Because of the inherent nonlinearity in light responses, estimates of the photosynthetic rate at canopy scale cannot be obtained from mean irradiance values, but require a full description of the radiative field. This means that scaling of light harvesting from leaf to landscape is a central issue for the prediction and understanding of plant canopy processes (Asner and Wessman 1997). Because of the strong linkage between photosynthesis and plant water use, canopy radiative field is not only relevant for primary plant productivity, but also for the partitioning of ecosystem energy fluxes between sensible and latent heat. Thus, architecture of plant stands and resulting light environment exert a major control over the meteorology of plant communities (Baldocchi and Harley 1995, Lai et al. 2000).

Detailed upscaling methodologies are required both in ecological modeling and remote sensing to deal with the spatio-temporal distribution of the irradiance on leaf surface (Myneni and Ross 1991). The increasing awareness of the role of terrestrial ecosystem in global bio-geochemical cycles enhances the interest in the application of soil-vegetation-atmosphere transfer (SVAT) models. This class of process-oriented models has typically a hierarchical structure, scaling physical and physiological processes from a single leaf to higher hierarchical orders such as single crown, vegetation canopies, and landscape mosaics (e.g., Baldocchi and Harley 1995). The predictions of SVAT models are strongly affected by the scaling strategies adopted for the integration of the leaf-level photosynthetic response with the light climate (Law et al. 2001).

There is a plethora of architectural strategies for light harvesting, and the efficiency of these strategies is evident in numerous physiological characteristics, for example, compensation and saturation points in the light response curve. This wealth of biological details implies that modes of light environment have to tackle with extremely complex plant geometries. With development of computer technology, very detailed models can be developed to described light environment of single leaves and shoots (Valladares and Pearcy 1999), or even clonal plant stands (Casella and Sinoquet 2003) with very high degree of predictability. However, extensive parameterization requirements limit application of fine-detailed schemes to larger scales. Thus, architectural drivers of light harvesting must be simplified to scale up plant physiological processes in larger scale models. Such a model simplification requires a full understanding of the relative impact of different architectural levels on light capture as well as the interactions between the structural and temporal scales of light harvesting and plant functioning. For instance, most SVAT models rely on the assumption of homogeneous canopy to simplify the scaling of light interception, that is, leaves are considered to be randomly dispersed in a series of horizontal layers (Baldocchi and Wilson 2001). However, only a few sensitivity analyses have been conducted to test for the appropriateness of such important assumptions.

In parallel, scaling canopy processes at larger scales, from canopy to the landscape, region, and biome levels, is often based on remote sensing techniques to retrieve structural and physiological information of the canopies from the spectra of reflected radiation. To predict intensity and angular distribution of reflected radiation, these procedures are based on direct or inverse application of detailed radiative transfer models that link leaf-level optics with canopy-scale architecture (Pinty et al. 2001). In the last decades, the increasing attention on these issues and the development of instrumentation, theories, and models have significantly improved the capability to upscale light harvesting through architectural, spatial, and temporal scales. However, recent evidence of major differences in direct and diffuse irradiance use-efficiency at the ecosystem level (Roderick et al. 2001; Gu et al. 2002, 2003) demonstrates that the current models may represent an oversimplification of vegetation vs. light interactions.

In this chapter, we first review new developments in the description of plant architecture and radiative field that allow accounting for the effect of canopy heterogeneity on leaf irradiance (Gastellu-Etchegorry et al. 1996, North 1996, Cescatti 1997a, Nilson and Ross 1997, Law, Cescatti and Baldocchi 2001, Kuusk, Nilson and Paas 2002). The other task of this chapter is to outline a series of challenging objectives, such as simulation of penumbra and clustering of light-harvesting elements at various hierarchical scales, that should be included in the future generation of global-scale simulation models.

Methods in Scaling the Light Climate

The major sources of strong temporal and spatial dynamics of light in the vegetation are (1) changes in the radiative field above the canopy, for example, due to modifications in solar elevation, cloudiness, or angular distribution of diffuse radiation; (2) complexity of canopy structure, for example, angular and spatial distribution of the leaves in shoots and crowns, and (3) the dynamic nature of the vegetation, for example, seasonal trends in plant phenology and leaf fluttering (Baldocchi and Collineau 1994). The variability of irradiance on leaf surface, resulting from the interplay of radiative field and canopy optical and geometrical assets, largely affects the scaling of physiological and biophysical processes from the leaf to the landscape (photosynthesis, transpiration, energy balance). The variation in canopy light climate can be directly investigated with experimental measurements of incoming photosynthetic photon flux density (PPFD), or can be predicted by radiative transfer models starting from the quantitative description of canopy architecture and optics.

Direct Measurements of the Radiative Regime

The experimental description of the light climate within plant canopies is particularly complex because larger variations in the geometry of the radiation field above the canopy are further amplified by canopy structure and optics, implying an inherently large variability in the distribution of irradiance on leaf surface. The structure of the radiative field can experimentally be investigated in time and space, and separating between angular and spectral variation of irradiance (Norman and Jarvis 1974, Campbell and Norman 1989, Baldocchi and Collineau 1994, Ross et al. 1998).

Because of the complexity of the radiative field within canopies, direct measurements of light climate or canopy reflectance generally aim to indirectly estimate canopy architecture parameters by the inversion of canopy radiative transfer models (Welles 1990, Weiss et al. 2004). In addition, direct light measurements are used to verify the predictive capacity of radiative transfer models (RTMs) or to investigate the modification of the radiative field by plant characteristics such as leaf fluttering and movement due to wind as well as penumbra, which are not included in most of the existing radiative transfer models (Roden and Pearcy 1993, Palmroth et al. 1999). The high spatio-temporal variability of radiation in plant canopies severely limits the possibilities to scale light climate by direct measurements, and it is essential to clearly define which features of the radiative field should be measured and for which aims. Overall, three general goals can be pursued: description of radiative field for scaling photosynthesis, estimation of canopy transmittance or reflectance to indirectly parameterize RTMs, and generation of data sets for the validation of RTMs.

When the radiative field has to be directly investigated by measurements, first the number of sensor/sampling points required for a target precision should be estimated. The number of independent sensors primarily depends on the variability of the light regime in the investigated canopy type and the time scale of interest. For instance, uniform broadleaf canopies show a lower spatial variability compared with conifers or sparse canopies (Baldocchi and Collineau 1994), while the spatial variability largely decreases with increasing averaging time. When the ultimate goal is upscaling photosynthesis from leaf to canopy, it is essential to consider that sensors should be distributed in space and oriented in angles to mimic leaf distribution (Palva et al. 2001). Very often cosine-corrected flat sensors are erroneously assumed to correctly represent the irradiance distribution on leaf surface.

Spatial and temporal sampling can be combined using mobile sensors. This methodology is also applied to estimate gap size distribution (Norman et al. 1971, Chen 1996). At a smaller scale, the direct description of radiative regime generated by conifer shoots has been performed by Palmroth et al. (1999) using a novel multipoint measuring system. In addition, Palva et al. (2001) applied an intensive sampling scheme at the tree scale for experimental description of the spatial and temporal variability of radiation along the vertical profile of a *Pinus sylvestris* canopy (Fig. 3.1). These detailed light measurements have been ultimately applied to upscale photosynthetic cuvette measurements from shoot to canopy (Vesala et al. 2000).

A separate description of the long-term radiative fields for direct and diffuse components can be achieved using hemispherical images of the canopy. For now,

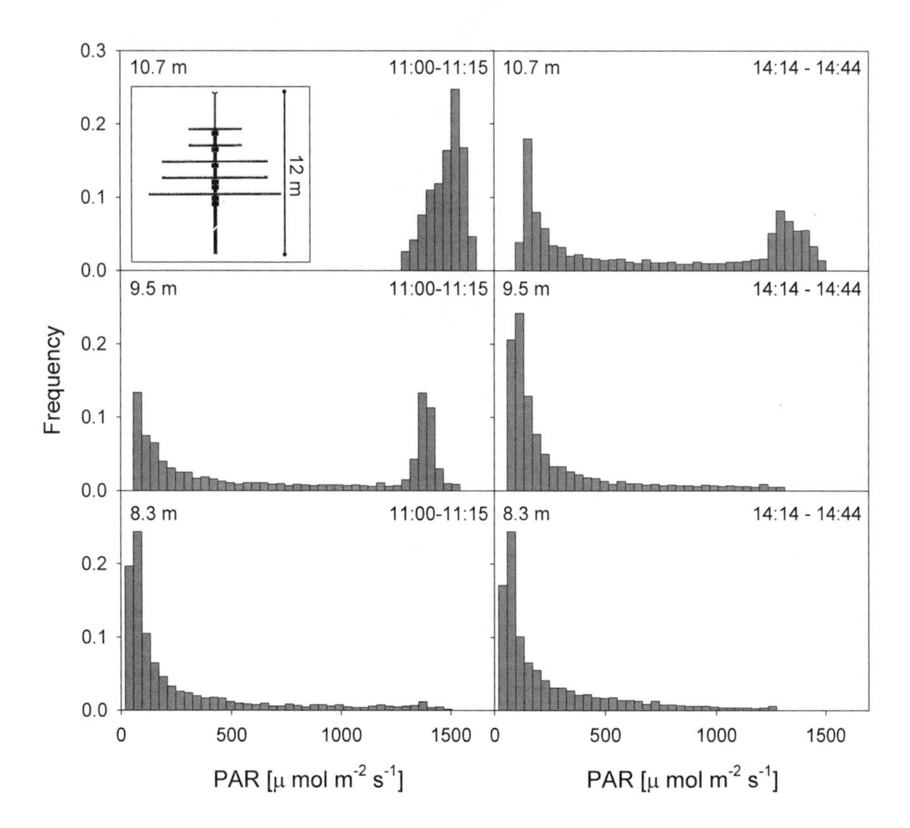

FIGURE 3.1. Within-canopy variation in the frequency distribution of photosynthetic quantum flux density (PPFD) at two different periods during a day and three heights in the canopy. Light measurements were taken in a *Pinus sylvestris* canopy with a unique measuring system (modified from Vesala et al. 2000). The inset illustrates the tree-scale light measurement system that consists of 168 quantum sensors (modified from Palva et al. 2001).

this methodology has been applied for more than 40 years (Evans and Coombe 1959, Anderson 1964) and is under constant development of both the improvement of hardware from film to digital cameras (Hale and Edwards 2002) and analysis techniques (van Gardingen et al. 1999). The largest uncertainty of this methodology, especially in closed canopies, is the conversion of gray tones images into corresponding binary black and white images (Macfarlane et al. 2000). Wagner (1998, 2001) proposed a detailed methodology to improve the conversion process and reduce the subjective effect of choosing a certain gray threshold. Recently, the possibility to use CCD-radiometers has greatly increased the possibility to obtained angular high-resolution measurements in different wavebands both for canopy reflectance (Nandy, Thome and Biggar 2001) and transmittance (Fig. 3.2) (Kuusk et al.).

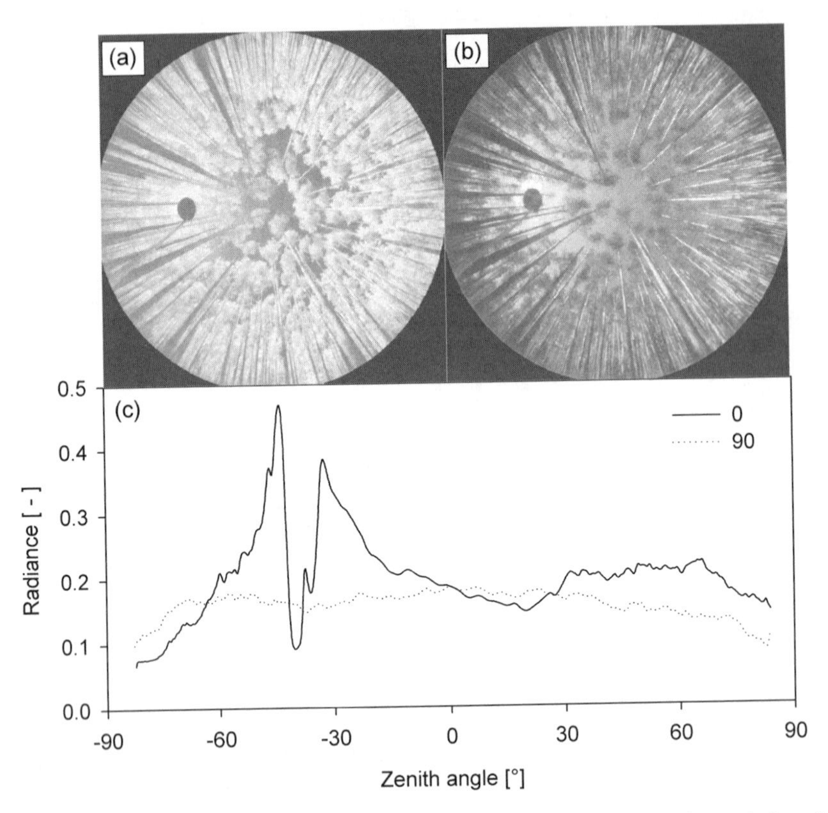

FIGURE 3.2. Hemispherical radiometric measurements in the red (a) and near-infrared (b) wavebands obtained under clear sky conditions with a novel CCD-based radiometer and a fish-eye lens in a *P. sylvestris* canopy. Hemispheric CCD-radiometer measurements provide extremely detailed information on radiance distribution as illustrated in (c) for red radiance in the principal (0° relative to the sun azimuth) and perpendicular (90° relative to the sun azimuth) planes (modified from Kuusk et al. 2002).

Radiative Transfer Models

Since the pioneering work of Monsi and Saeki (1953), the complex interplay between radiative field and canopy structure has been widely investigated with the development and application of RTMs (Myneni and Ross 1991). The earliest RTMs postulated an exponential decay of transmittance in a random turbid medium (Monsi and Saeki 1953). In the last decades, different models with increasing levels of complexity both in representing canopy architecture and in describing the radiative field have been developed (Myneni and Ross 1991).

As far as the scaling of light harvesting across architectural levels is concerned, RTMs can be classified according to the following characteristics (Table 3.1):

- spatial dimensions (0 to 3)
- representation of spatial pattern in leaf distribution
- description of irradiance distribution on the leaf area
- representation of the leaves as turbid medium (statistical models) or geometrical objects (ray tracing)
- simulation of multiple scattering (black leaf approaches versus models accounting for leaf optical characteristics)

Spatial Dimensions

Radiative transfer models rely on a statistical or geometrical representation of the canopy structure to account for the spatial and angular distribution of the phytoelements (Ross 1981). In the case of a statistical representation of the canopy, zero to three spatial dimensions may be used to describe the spatial heterogeneity of the structural canopy parameters (leaf area, leaf angle distribution, spatial pattern) (Campbell and Norman 1989). The simplest, zero-dimensional models, known also as "big-leaf models," represent the canopy as a single layer of leaves that is horizontally homogeneous and without a vertical dimension. Big-leaf models are widely used in global circulation models because of the simplified parameterization and the computational efficiency (Wang and Leuning 1998).

Vertical canopy light profiles can be simulated with one-dimensional (1-D) models that describe the canopy space as a vertical pile of horizontally homogeneous layers (Fig. 3.3a). This class of models has been widely applied for simulation of radiative and mass transfer processes in agricultural and forest canopies and in remote sensing (Norman and Jarvis 1975, Lai et al. 2000, Baldocchi and Wilson 2001).

Two-dimensional (2-D) models describe the variability both in the vertical and in one horizontal dimension and have been developed specifically for hedgerow systems or row crops such as apple orchards (Fig. 3.3b). In such horticultural systems, the canopy in one of the two horizontal dimensions can be considered homogeneous (Gijzen and Goudriaan 1989, Palmer 1989, Friday and Fownes 2001).

Three-dimensional (3-D) models finally describe the variability of structural parameters in all three spatial dimensions and can be used to simulate the vast het-

TABLE 3.1. Main features of the different classes of light models for plant canopies (Y = yes, N = no). Models based on the turbid medium analogy are ranked according to the number of dimensions: 0 for big-leaf or two-leaf models, 1 for multilayer models for homogeneous canopies, 2 for row systems, and 3 for discontinuous plant canopies. In ray-tracing models, plant elements are represented with geometrical shapes in 3-D and the radiative field is simulated by Monte Carlo methods.

Model type	Reference	Spatial dimensions	Spatial pattern	Irradiance classes	Penumbra	Scattered fluxes	Canopy reflectance
Big leaf	Wang and Leuning 1998	0	Y, N	2	N	Y, N	Y, N
Multi-layer	Norman and Jarvis 1975	1	Y, N	2-n	Y, N	Y, N	Y, N
Stochastic turbid medium	Shabanov et al. 2000	1	Y	2-n	N	Y	Y, N
2-D row model	Friday and Fownes 2001	2	Y	2-n	N	Y, N	Y, N
3-D crown geometry	Cescatti 1997a	3	Y	2-n	Y, N	Y, N	Y, N
3-D voxel based	Gastellu-Etchegorry et al. 1996	3	Y	2-n	Y, N	Y	Y
Ray tracing	North 1996	3	Y	N	Y, N	Y	Y

erogeneity of leaf area distribution due to branches or crowns, or due to the planting scheme (Fig. 3.3c). Three-dimensional models represent canopy elements both using geometrical objects (cones, spheroids, cylinders) filled with a turbid medium (Norman and Welles 1983, Wang and Jarvis 1990, Cescatti 1997a) or with a regular 3-D grid (Kimes and Kirchner 1982, Myneni 1991, Gastellu-Etchegorry et al. 1996). More recently, the effect of the canopy heterogeneity has been described with stochastic models of the radiative field (Shabanov et al. 2000). In the case of ray-tracing models, phytoelements are described as solid objects representing single leaves or branch segments (Chen, S. G., Ceulemans and Impens 1994, North 1996, Pearcy and Yang 1996) and the radiative field is reconstructed with a direct numerical simulation based on Monte Carlo procedures (Fig. 3.3d).

Spatial Dispersion of Phytoelements

The spatial dispersion of phytoelements refers to the level of regularity or clumping of leaves in space (Fig. 3.4). The default pattern assumed in most RTMs is the random distribution, for which the position of a phytoelement is independent of the position of other phytoelements. In statistical 1-D models, the probability of light transmittance (T) in a random canopy can be estimated by the Poisson distribution:

$$T(\varphi, z) = \exp[-K(\varphi)LAI(z)], \qquad \text{(Eq. 3.1)}$$

where K is the extinction coefficient, φ is the zenith angle, and z is the height in the canopy.

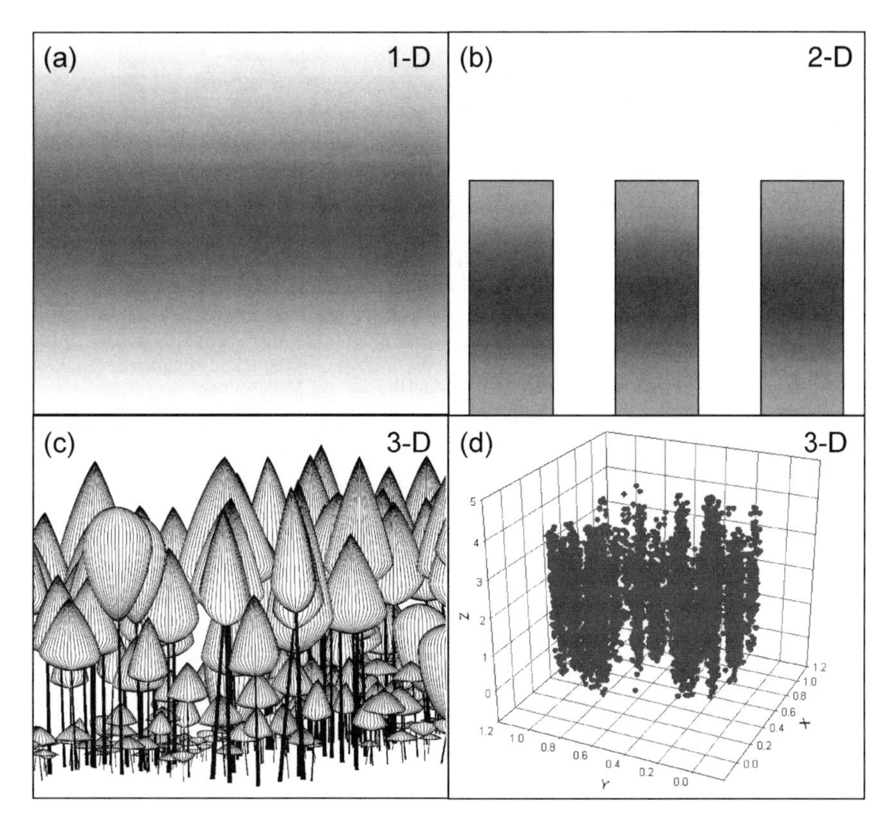

FIGURE 3.3. Spatial dimensions in statistical turbid medium model and in geometrical ray-tracing RTMs. One-dimensional (1-D) statistical models (a) assume that the canopy is horizontally homogeneous, while two-dimensional (2-D) models (b) distinguish between canopy and intercanopy space and assume regular distribution of these elements (row models). Three-dimensional (3-D) models (c and d) are not limited by any of these constraints. Depending on the approach, 3-D models can be divided between models based on the turbid medium approximation (c) and on geometrical objects and ray-tracing (d).

In many cases the spatial dispersion of phytoelements in the canopy is not random but either regular or clumped (Nilson 1971). At a common leaf area index (LAI), the canopy with regular dispersion of foliage intercepts more light than a random canopy, while the canopy with clumped foliage dispersion intercepts less radiation. The canopy gap fraction for a regular dispersion can be modeled using the positive binomial model:

$$T(\varphi,z) = \exp\left[\frac{LAI(z)}{\Delta L(z)}\ \ln(1 - K(\varphi)\Delta L(z))\right], \qquad \text{(Eq. 3.2)}$$

where ΔL is the thickness of a single canopy layer in LAI units ($m^2\ m^{-2}$), in which only one contact between a solar ray and leaves may occur.

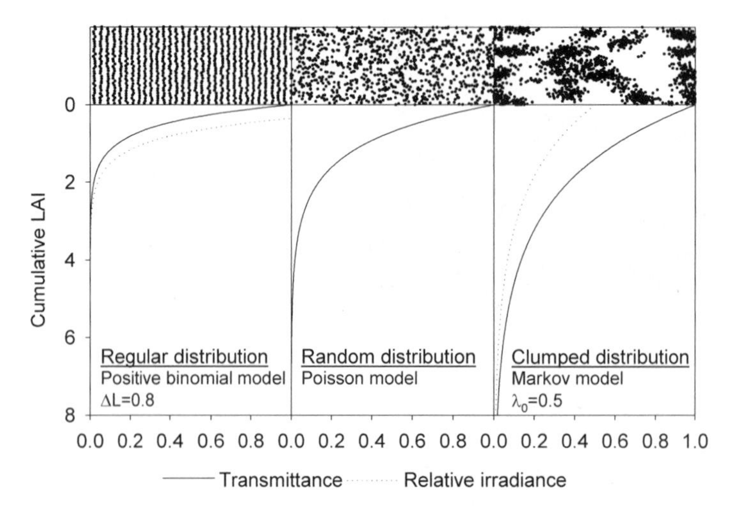

FIGURE 3.4. Representation of the spatial distribution of phytoelements in statistical 1-D models. The probability of light transmittance and the mean relative irradiance on leaf area have been estimated using the positive binomial model for the regular foliage dispersion (see Eq. 3.1), the Poisson model for random dispersion (see Eq. 3.2) and Markov model for clumped dispersion (see Eq. 3.3). The underlying assumptions of these models are: ΔL in the positive binomial model is a measure of foliage regularity, while λ_0 measures the degree of foliage aggregation. ΔL increases with increasing foliage regularity, while λ_0 decreases with increasing foliage clumping. For $\Delta L = 0$ and $\lambda_0 = 1$, the two models match the Poisson model for random distribution.

Very often plant canopies show a clustered structure due to clumping of phytoelements in crowns, branches, or shoots (Baldocchi and Hutchison 1986, Chen 1996, Cescatti 1998). The probability of interception in clumped canopies can be modeled with the negative binomial model or with conditional probability models such as Markov chain models, in which the position of a phytoelement in a layer is dependent on the probability of photon interception in the previous layer (Nilson 1971). Compared with the Poisson interception model, Markov models use an additional parameter, λ_0, which expresses the degree of dependence of element position in neighboring layers. Thus, λ_0 ranges from 1 in the case of a random dispersion and approaches 0 for a completely clumped dispersion:

$$T(\varphi,z) = \exp[-K(\varphi)\lambda_0\Delta LAI(z)] \qquad \text{(Eq. 3.3)}$$

Figure 3.4 illustrates the vertical profiles of canopy transmittance and mean relative leaf irradiance for regular, random and clumped dispersions, represented with positive binomial, Poisson, and Markov models, respectively. For a random canopy, the mean relative irradiance on the leaf area is equal to the canopy transmittance along the cumulative LAI profile, while for regular dispersions, the relative irradiance is higher than the transmittance. On the contrary, a clumped leaf distribution generates a mean leaf irradiance largely lower than canopy transmittance because of the higher self-shading between leaves. As a consequence of these different spatial patterns, light is more linearly distributed along the vertical profile of a clustered canopy, while regular canopies show higher efficiency in light interception per unit leaf area in the upper canopy layers.

Irradiance Distribution and Penumbra

In scaling photosynthetic responses from single leaf to a canopy, particular attention should be paid to the frequency distribution of the irradiance on the leaf area (Gutschick 1991). Because of the nonlinear leaf photosynthetic response to light, canopy level responses determined using mean leaf irradiances will be systematically overestimated (Fig. 3.5) (Spitters 1986, Roderick et al. 2001).

The bias induced by the use of average instead of actual irradiance distribution in photosynthetic response functions can be simply shown in mathematical terms. Leaf photosynthesis depends on light according to an increasing hyperbolic response function, $f(x)$. Thus, the second derivative of $f(x)$, $f''(x) < 0$ over the range $f''(x)$ is defined. It follows that the average response for every irradiance distribution is smaller that the response to the average irradiance $f(\bar{x})$ (see Fig. 3.5):

$$f''(\bar{x}) = \lim_{\Delta x \to 0} \frac{f(x_2) - 2f(\bar{x}) + f(x_1)}{\Delta x^2} < 0 \qquad \text{(Eq. 3.4)}$$

$$\frac{f(x_2) + f(x_1)}{2} < f(\bar{x})$$

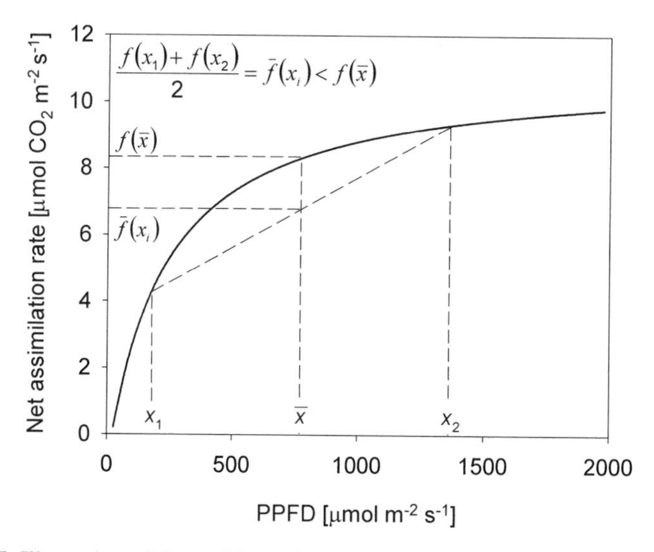

FIGURE 3.5. Illustration of the problem of linearity versus nonlinearity in integrating leaf physiological responses to light. Leaf light interception scales linearly with incident PPFD, while photosynthesis starts to saturate at relatively low PPFD values. If average PPFD ($\bar{x}$) is used to upscale photosynthesis from the leaf to the canopy, the nonlinearity in the leaf photosynthetic light response generates an overestimation of the canopy photosynthetic response [$f(\bar{x})$]. For unbiased estimation of canopy photosynthesis [$\bar{f}(x_i)$], actual PPFD frequency distribution on leaf area (x_i) should be used.

The main sources of variation in leaf irradiance are related to mutual shading, variable leaf orientation, simultaneous occurrence of diffuse and direct radiation fluxes with different angular distribution, and the occurrence of penumbra (Fig. 3.6). The most common approximation to describe the irradiance distributions is to separate the canopy foliage between two leaf classes: shaded leaves, receiving only diffuse radiation, and sunlit leaves, receiving both direct and diffuse radiation (Wang and Leuning 1998). Because of the unidirectional distribution of direct radiation, the direct fluxes should be cosine corrected to account for the variable leaf orientations. This approach is widely applied in 1-D turbid medium models and leads to a vertical profile of diffuse irradiance (neglecting the within-layer variability in diffuse radiation fluxes) and to different frequency distributions of irradiances on the sunlit leaf fraction in each layer.

The ranking of leaves in sunlit and shaded classes is based on the assumption that the sun is a true point source of light. In reality, the solar disk as seen from the earth has a radius of about 0.27 degrees. For "purely" geometrical reasons, the partial obstruction of the solar disk by phytoelements generates a series of intermediate irradiances between the fully sunlit and the fully shaded leaf areas, defined as penumbral areas (see Fig. 3.6). Considering that an object can completely obscure the sun up to a distance of about 100 times its minimum dimension, penumbra has an increasing importance in tall canopies with small leaves, such as conifers, or on canopies with tall trees and numerous gaps (Smith et al. 1989).

Direct measurements of the penumbral effects are difficult and pose specific requirements on the dimension of the sensors and the time response of the measuring system (Figure 3.1) (Ross et al. 1998, Palmroth et al. 1999, Palva et al. 2001). An

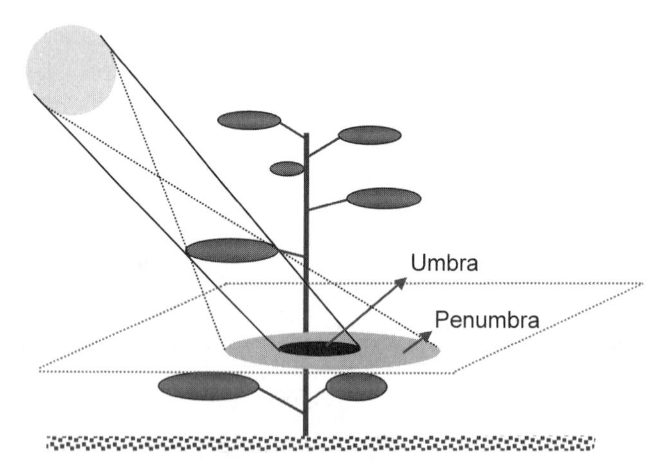

FIGURE 3.6. Leaf area may either be sunlit or shaded (umbra). In addition, a certain fraction of leaf area is neither fully sunlit nor fully shaded, a situation called penumbra. Penumbral areas in plant canopies result from the circumstance that the sun is not a point source, but has a finite size as seen from the earth. The occurrence of penumbra due to the partial shading of the sun by foliage elements increases with the beam pathlength in the canopy and decreases with the leaf size.

analytical description of the penumbra is possible only for simplified canopy structure (e.g., horizontal circular leaves), for which the probability of bidirectional gap fraction can be used to estimate the standard deviation of the visible fraction of the solar disk (Stenberg 1995). Radiative transfer models based on the turbid medium approximation allow simplified simulation of penumbra (Wang and Baldocchi 1989, Stenberg 1995). On the contrary, direct numerical simulation of the penumbra effects can be achieved for any canopy architecture adopting a Monte Carlo ray-tracing approach (Oker-Blom 1984, Myneni and Impens 1985, Stenberg 1995).

Scattered Fluxes

Because of marked absorption peaks in different spectral bands of photosynthetic pigments, water, cellulose, and other leaf constituents, leaves show typical absorption spectra with a maximum in the blue and red, and a minimum in the infrared region. Average leaf absorptance (1 − reflectance − transmittance) is on the order of 88% in the photosynthetically active radiation (PAR) (400 to 700 nm) and 17% in the near-infrared (NIR) (700 to 1100 nm) (Walter-Shea and Norman 1991). For the sake of simplicity and computational efficiency, several RTMs applied in ecological studies often assume black body behavior for leaves (absorptance = 1), leading to overestimation of light harvesting of plant canopies.

The overall effect of scattering is, in fact, a reduction of light absorption in the upper canopy layers and an increase of leaf irradiance in the lower layers. The simulation of scattered fluxes is achieved both in 1-D and 3-D models with different levels of accuracy. An empirical approach in 1-D has been proposed by Cowan (1968), with the modification of the extinction coefficient. More robust approximations of the radiative transfer equation based on the two-stream approximation were proposed by Norman and Jarvis (1974) and lately by Kuusk (1992). Different iterative solutions of the radiative transfer equation can be adopted in 1-D models to simulate the absorption profiles and the canopy reflectance (Knyazikhin and Marshak 1991).

Light scattering has been investigated in 3-D turbid medium models by Norman and Welles (1983), Wang and Jarvis (1990) and Cescatti (1997a,b) using geometrical objects (ellipsoids, cylinders, etc.) to represent tree crowns and the "equivalent canopy" concept. This methodology is based on the assumption that scattered fluxes in a 3-D heterogeneous canopy may be approximated by fluxes estimated in an "equivalent" 1-D canopy with the same gap fraction of the 3-D canopy (Norman and Welles 1983). In this way, the radiative field is estimated in 3-D concerning light interception and in 1-D for light scattering.

An alternative approach to the calculation of light scattering in 3-D heterogeneous turbid media is based on the separation of the scene in cubic cells (voxel) to simplify the numerical solution of the radiative transfer equation (Kimes and Kirchner 1982, Myneni 1991, Gastellu-Etchegorry et al. 1996). More recently these complex 3-D models have been numerically inverted in order to recover canopy architectural characteristics from canopy reflectance measurements obtained by remote sensing (Kimes, Gastellu-Etchegorry and Estéve 2002). The application of

Monte Carlo ray-tracing models has led to detailed description of the scattered fluxes (Disney, Lewis and North 2000). Ray-tracing models avoid most of the assumptions required in statistical models, such as assumptions concerning foliage spatial dispersion, angular distribution of leaves, approximation of the numerical solution, and bi-Lambertian behavior. Detailed measurements of leaf shape and displacement may be applied in this class of models to reconstruct the radiative field of single plants (Pearcy and Yang 1996, 1998; Valladares and Pearcy 1998).

Multiple scattering has to be carefully considered for the description of the radiative field in the NIR, where the large scattering coefficient strongly affects the leaf-level energy balance. In addition, detailed modeling of the angular distribution of scattered fluxes is required in remote sensing studies when multi-angular measurements of canopy reflectance data are available (Pinty et al. 2001).

Morphological and Spatial Scales of Light Interception

Light harvesting in plant canopies is determined by a hierarchy of morphological and spatial scales, from leaf to landscape (Ross 1981). At every hierarchical level, the architectural, geometrical, and optical properties of the canopy may affect the overall distribution of light on the photosynthetic surfaces, thereby influencing the light response of the whole canopy (Campbell and Norman 1989, Baldocchi and Collineau 1994). In this section, we investigate main processes occurring at different spatial levels, combining experimental data and model simulations to (1) identify the limiting processes at each spatial scale of organization, and (2) estimate the light distribution on leaf surfaces. Determination of the sensitivity of light interception to various morphological constraints allows identification of major limiting process at lower hierarchical levels in larger-scale RTMs, and thus, opens an important avenue for process-based scaling of photosynthesis from leaf to landscape level.

Leaf Scale

Leaf characteristics may affect light harvesting because of geometrical reasons (cross-sectional geometry, leaf size, and leaf inclination) and due to differences in leaf optical properties. Although simulation studies to analyze the efficiency of specific leaf shapes and optics are generally conducted for hypothetical situations without the influence of shoot- or canopy-level architectural constraints, it is important that canopy attributes at higher hierarchical levels may significantly influence the effects of leaf properties on the radiative field. Thus, importance of leaf-level attributes on light-harvesting efficiency should be assessed in the canopy context.

Leaf Cross-Sectional Geometry

There is a broad variation of leaf shapes among various life forms. Broadleaf flat foliar elements have very high efficiency for light interception and gaseous exchange with the atmosphere. For such leaves, the foliar light-harvesting characteristics are essentially driven by their area and inclination, and are insensitive to moderate changes in the thickness. In contrast, there is a wide array of needle-leaved species with the thickness of foliar elements approximating their width (Fig. 3.7). In these species, leaf cross-sectional geometry may significantly modify the light-harvesting efficiency of the foliage, because it determines the fraction of leaf surface relative to the total surface area exposed for various incidence angles of incoming radiation.

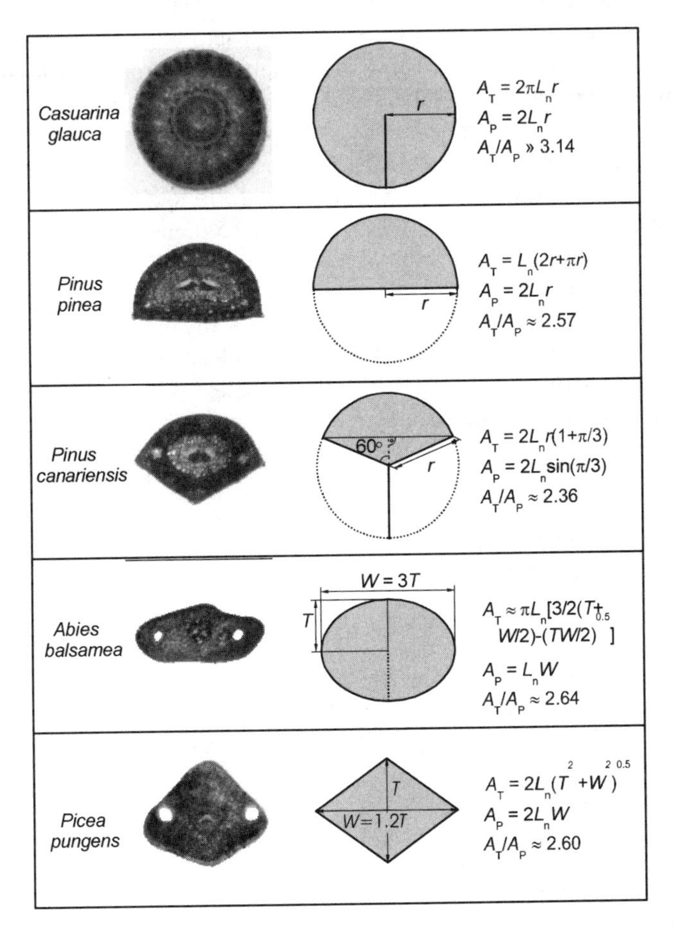

FIGURE 3.7. Typical cross sections of photosynthetic organs in needle-leaved species (unpublished data of Niinemets), along with geometrical shape approximations and formulas for the calculation of total (A_T) and projected (A_P) leaf area from total leaf length (L_n), radius of leaf cross section (r) or thickness (T) and width (W). Free-hand leaf cross sections were colored by fluoroglucinol that specifically stains lignified tissues red. The photosynthetic organs of *Casuarina glauca* are stems, while in all other species they are the leaves.

In plant species with circular leaf cross sections, like species of *Casuarina, Genista, Retama*, and conifer *Pinus monophylla*, the ratio of exposed to total leaf surface area is the same independently of the direction of illumination (see Fig. 3.7) (Jordan and Smith 1993). For leaves with noncircular cross sections (see Fig. 3.7), the efficiency of light interception depends on the beam angle of incident radiation. Geometrical considerations show that needle-leaved species exhibit a higher efficiency in the interception and photosynthetic utilization of sunlight at low incidence angles compared with a laminar leaf (Jordan and Smith 1993). However, these simulations are only valid for a fixed leaf position in space. In plant canopies, the angular distribution of leaf surfaces is often spherical, that is, with equal probability of leaf exposure to various angles of incidence. For example, needle twisting in conifers and morphological constraints on leaf packing in shoots contribute to the spherical distribution of leaf surface. According to Cauchy's theorems, the appropriate leaf area to model radiative transfer is half of the total area if the foliage angular distribution is spherical. This is because the spherically averaged projected to total area ratio for convex objects, such as most conifer needles, is equal to 0.5 (Lang 1991, Chen and Black 1992). As an important implication of the concept of "half of the total" leaf area, leaves of various cross-sectional geometries will have a similar efficiency of light interception per unit surface area if they are spherically distributed in the canopy.

Cauchy's theorems do not hold for concave objects such as the needles of some *Abies* species (see Fig. 3.7). Concave needles would intercept less light than convex needles of equal total area, due to an increased self-shading. The

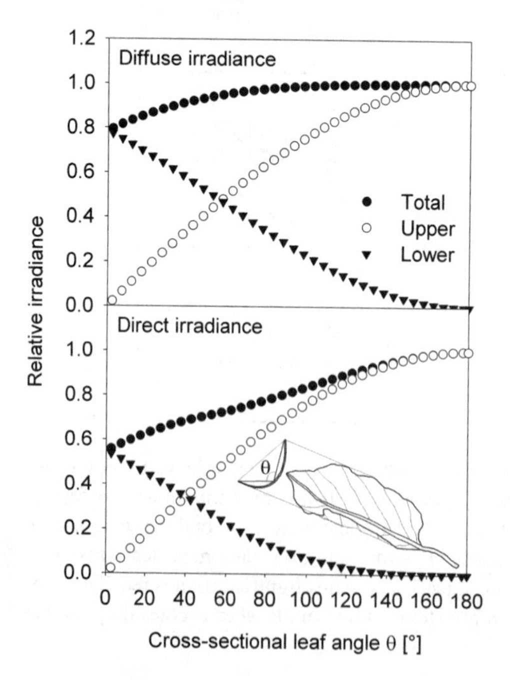

FIGURE 3.8. Simulation of the dependence of leaf irradiance, relative to a flat leaf, on the degree of leaf rolling for the upper and lower sides, and for direct and diffuse radiation field in temperate deciduous broad-leaved species *Fagus sylvatica*. As shown in the inset, leaf rolling was characterized by the cross-sectional leaf angle (θ) (modified from Fleck et al. 2003).

situation is similar for rolled or twisted leaf laminas of broad-leaved species. Leaf laminas of many grasses (Alcocer-Ruthling et al. 1989, Fernandez and Castrillo 1999) and broad-leaved species are often rolled (Fig. 3.8) (Farque, Sinoquet and Colin 2001, Fleck et al. 2003). Depending on the degree of leaf rolling, this response may severely reduce leaf light interception (Fig. 3.9). However, leaf rolling mostly affects the interception efficiency of direct irradiance, and only moderately influences the interception of diffuse irradiance, which generates a lower mean irradiance and is more efficiently used in photosynthesis (see Fig. 3.9). Leaf rolling becomes especially significant in response to water limitations and high irradiance (Alcocer-Ruthling et al. 1989), and overall represents an important mechanism to avoid interception of excess irradiance of stressed leaves.

Most plant species have a large plasticity in leaf cross-sectional geometry in response to variation in environmental factors such as light and water availabilities (Roderick et al. 2000). In conifers, the ratio of total to projected leaf area (A_T/A_P) generally decreases with decreasing light availability in the canopy, im-

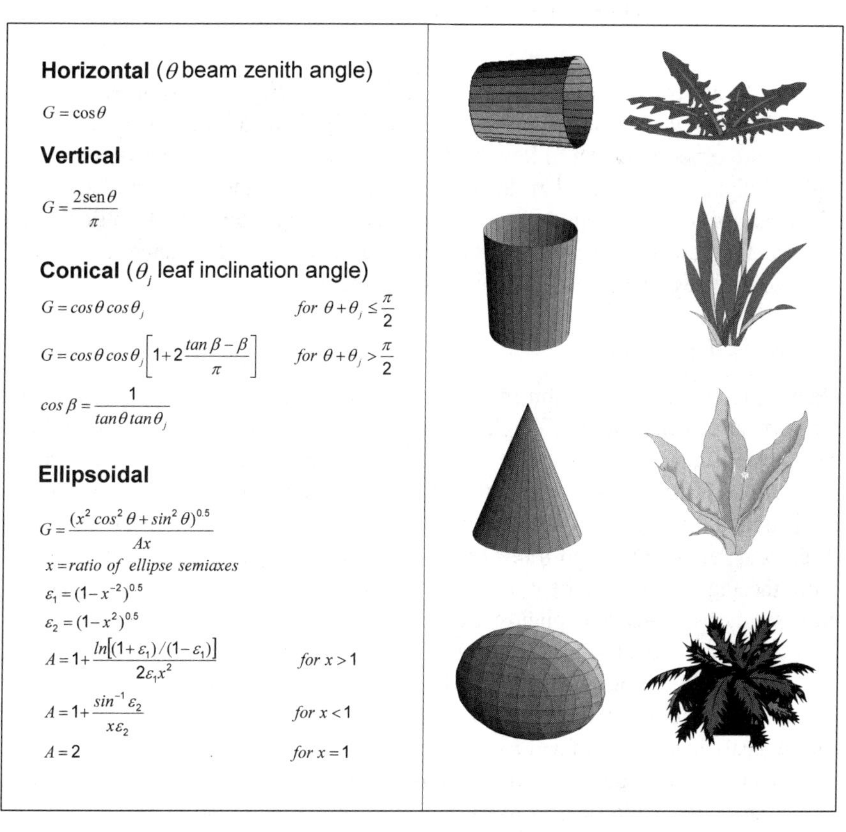

FIGURE 3.9. Common models of LAD assume that leaves are oriented according to the surface of geometrical objects. The analytical expression of the extinction coefficient is presented for the different distributions according to Campbell and Norman (1989).

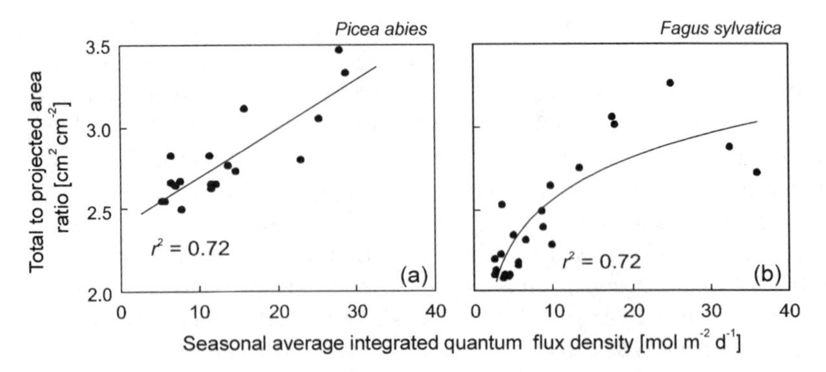

FIGURE 3.10. Variation of total to projected leaf area ratio (A_T/A_P) in response to the light gradient within (a) a conifer *Picea abies* canopy (modified from Niinemets and Kull 1995b), and within (b) a broad-leaved *F. sylvatica* tree (modified from Fleck et al. 2003). A_T/A_P varies with growth irradiance due to changes in the thickness of foliar elements (see Fig. 3.1) or due to modifications of cross-sectional leaf angle (see Fig. 3.8) and leaf inclination with respect to an horizontal surface.

plying a larger fraction of exposed leaf surface area (Fig. 3.10) (Niinemets et al. 2001, 2002b, Cescatti and Zorer 2003). Similarly, leaf laminas of broad-leaved species are essentially flat in low irradiance, while they may be heavily rolled in high irradiance (see Fig. 3.10). Such plastic changes enhance the light-harvesting efficiency of lower canopy leaves. In the upper canopy, high A_T/A_P ratios result in greater investment of photosynthetic resources in high irradiance in conifers, and avoid excess irradiance in broad-leaved species.

Leaf Angle Distribution

Plants exhibit a large variation in leaf orientation both between species and along the vertical crown profile of single specimens. Leaf angle distribution (LAD) has been experimentally investigated for many canopies (Ross 1981). These labor-intensive architectural measurements have been largely enhanced by the recent introduction of 3-D digitizers that have been employed for herbaceous species canopies (Rakocevic et al. 2000) and forest tree canopies (Utsugi 1999, Farque et al. 2001). Since the angular displacement of leaf surface strongly affects light interception and photosynthesis, LAD is explicitly included in RTMs as the extinction coefficient (K, Eq. 3.1) or as the G-function ($G = K/\cos(\varphi)$; where φ is the solar zenith angle). The G-function, defined by Ross (1981), is given as the ratio between the leaf area and the leaf silhouette area projected on a surface orthogonal to the solar beam, and can be analytically determined for a series of LAD models (see Fig. 3.9).

The angular distribution is modeled assuming that leaves are oriented parallel to the surface of solid objects like cone, cylinder, sphere, and ellipsoid (see Fig. 3.9) (Campbell and Norman 1989). Among different geometrical models, the ellipsoidal LAD model has been largely used because of its versatility and simple parameterization (Campbell 1986). More recently, a rotated ellipsoidal distribution model has been proposed (Thomas and Winner 2000) to improve the estimation of the G-function in plant canopies with horizontal leaves.

The importance of LAD on light interception is high at the leaf scale but may vanish when the effects are integrated over the entire leaf population at the crown or canopy level, especially at high LAI values. Sensitivity analysis performed with the 3-D canopy model MAESTRO (Wang and Jarvis 1990) suggests that the effect of LAD on light interception in a spruce forest is moderate compared with the effect of leaf spatial dispersion, while the effect on daily photosynthesis is highly relevant due to the effects of LAD on the vertical distribution of solar radiation. For instance, a vertical leaf orientation reduces radiation interception of a leaf at the canopy top, but increases the light penetration into deeper canopy layers, increasing the overall photosynthesis of dense canopies (Oker-Blom and Kellomäki 1982). Furthermore, the functional variability in LAD within the canopy may involve a balance between the regulation of light interception, photosynthesis, and photoprotection (Werner et al. 2001). There is growing body of information that in arid environments, vertical leaf orientations represent a specific strategy to reduce drought damage and photoinhibition (Valladares and Pearcy 1998, Valladares and Pugnaire 1999, Werner et al. 2001).

Because LAD affects the mutual shading between individuals, LAD importantly alters the light competition between co-existing plants. The application of a game theoretic model on competing plants has demonstrated that the evolutionarily stable light extinction coefficient is generally larger than the value that would maximize canopy photosynthesis (Hikosaka and Hirose 1997).

Influence of Leaf Size on Canopy Radiative Regime

Due to the finite angular size of the solar disk (radius 0.27 deg), the distribution of the direct irradiance in plant canopies is characterized by sunflecks (fully visible sun disk), shaded areas (no visible sun) and intermediate penumbral areas where the solar disk is only partly visible (see Fig. 3.6) (Miller and Norman 1971). The frequency of penumbra decreases with increasing leaf size and increases with the distance of the leaf from the target point, that is, with increasing canopy height. Therefore, penumbra is predominant in forest canopies that are characterized by deep crowns and narrow needles (Stenberg 1995).

The relevance of penumbral effects on radiative regime is mostly determined by the beam path length between two consecutive contacts in the canopy that varies with canopy height, sun angle, and spatial pattern of the phytoelements, and by the dimensions of the leaves relative to canopy height (Stenberg 1995). Penumbra may also be enhanced by leaf lobing and dissection. Such modifications in lamina architecture do not confer any specific benefit in light interception (Niklas 1989) when one does not account for the enhanced penumbra.

Due to inherent limitations of turbid medium models in characterizing penumbra, we investigated the implication of penumbra on the photosynthetic response of a canopy using a ray-tracing model. Leaves were represented by circular disks with a spherical leaf LAD. For simplicity, the spatial dispersion was assumed to be random, even if the ray-tracing framework makes it possible to simulate different angular and spatial distributions. We compared three canopies with the same LAI (3.6), but with varying ratio of canopy height to leaf width (30-3000). The irradiance on each leaf was numerically determined by tracing 200 rays from the leaf center to the sun disk, and finally the frequency distribution of the cosine-corrected irradiance on the leaves of the three canopy types was computed (Fig. 3.11). This simulation shows that in the case of a spherical LAD, the frequency distribution of the direct irradiance on the leaves is never bimodal (as predicted by the sunlit/shaded approach), and is clearly unimodal in tall canopies with small leaves. This distribution of the light is determined both by the angular distribution of the leaves and by penumbral effect.

Given a non-rectangular hyperbolic response of photosynthesis to light for single leaves, it is possible to calculate the canopy-scale light response from the irradiance distribution (Fig. 3.12). Assuming that the fraction of diffuse irradiance

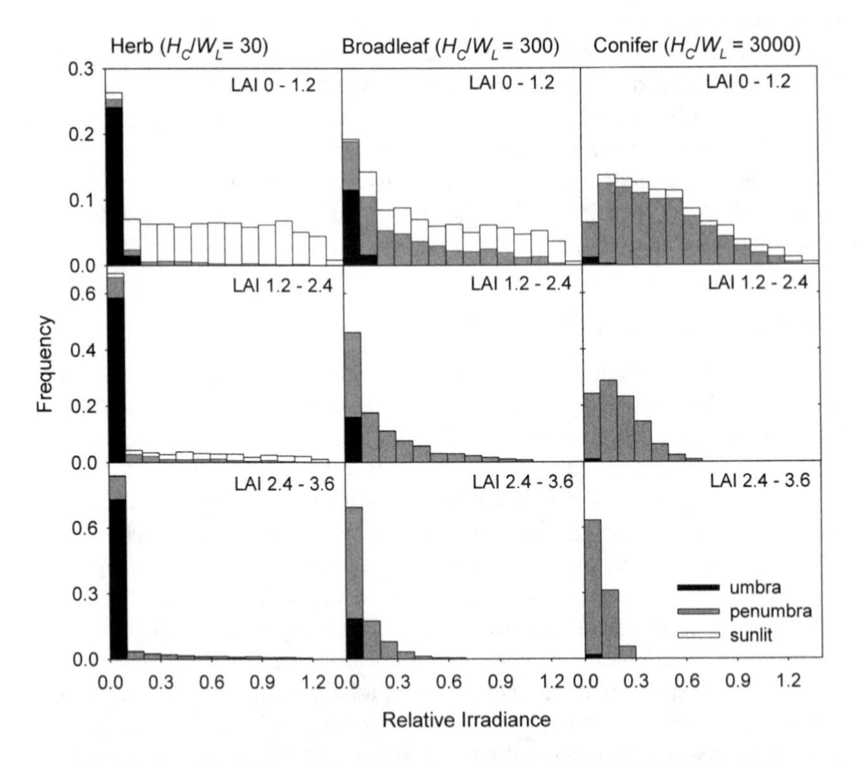

FIGURE 3.11. Simulated frequency distribution of the relative irradiance on leaf area in three typical canopies of contrasting canopy height (H_C) to leaf width (W_L) ratio. The fraction of shaded leaf area is given in black, the fully sunlit leaves in white, while gray refers to partially shaded (penumbra) leaves. The leaf irradiance relative to the irradiance on a flat surface above canopy was simulated with a ray-tracing radiative transfer model with 10^6 flat circular leaves randomly dispersed in the canopy space. Solar elevation angle was fixed at 45° and the fraction of diffuse radiation in the incoming flux was set at 0.15.

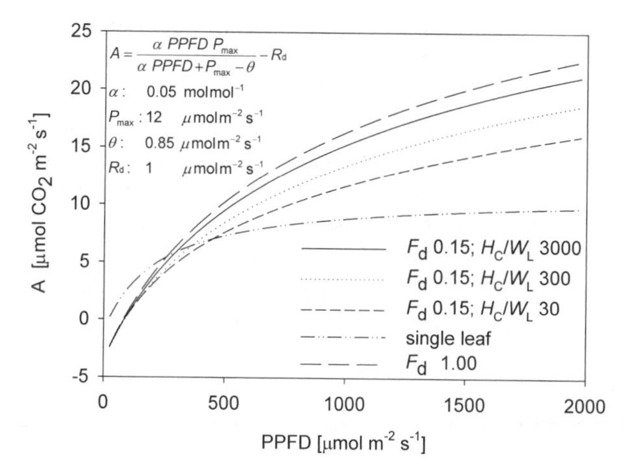

FIGURE 3.12. Dependence of canopy photosynthetic response to light upon the occurrence of penumbra. Leaf and canopy light response curves are compared for three canopies of differing canopy height to leaf size ratio (H_C/W_L). Leaf net assimilation rates were simulated by a nonrectangular hyperbola that was parameterized using typical values for temperate forest trees at 25 °C. The fraction of diffuse PPFD in total irradiance (F_d) was fixed at either 0.15 or 1.0. Computations were performed with the frequency distribution of irradiances reported in Figure 3.11 for a herb ($H_C/W_L = 30$), broad-leaf ($H_C/W_L = 300$), and conifer ($H_C/W_L = 3000$) canopy, and a common LAI of 3.6. All canopies behave in the same manner for $F_d = 1.0$ since the ratio H_C/W_L does not affect the distribution of diffuse radiation.

in incoming radiation is 0.15 (clear sky conditions), and taking as the reference case a canopy with a canopy height to leaf width ratio (H_C/W_L) of 30, the improvement of the canopy photosynthetic response due to penumbra is about 15% for a canopy with $H_C/W_L = 300$, and 30% for a canopy with $H_C/W_L = 3000$. For fully overcast sky conditions, the effect of leaf size on penumbra is negligible because the irradiance is incoming from the entire sky hemisphere, and the canopy response is on the order of 40% higher than that observed for the reference canopy in direct light. These results are comparable with the experimental estimates obtained by Gu et al. (2002,2003) from eddy covariance measurements of CO_2 net ecosystem exchange in different forest canopies.

Leaf Optics and Light Harvesting

Large variations in leaf absorptance are due to leaf pubescence, waxy leaf surfaces as well as modifications in foliar pigment content and leaf anatomical characteristics (Walter-Shea and Norman 1991). Thus, leaves are not perfect black bodies, as assumed in simple canopy light-harvesting models based on Lambert-Beer. Such inherent differences in light harvesting of real leaves versus optically black leaves are considered in canopy light interception models accounting for multiple scattering (see Table 3.1).

A survey of leaf optical characteristics across different plant functional types demonstrates that average leaf absorptances (θ) in the spectral band of the PAR (400 to 700 nm) vary between 0.8 and 0.9, with the succulents having the lowest θ and conifers the highest θ (Fig. 3.13). The range of variation is the same for various biomes, with desert plants having the lowest θ and tropical rainforest species the highest θ (see Fig. 3.13). Overall, the range in average absorptance across different plants and biomes is relatively small, demonstrating convergent evolution in foliar optical characteristics.

Although leaf absorptance is a critical parameter that determines the availability of light for photosynthesis at common values of incident light, the importance of scattering on the radiative field at canopy and landscape levels largely depends on canopy architecture as well as on total LAI. In the case of a homogeneous canopy with a spherical LAD, the dependence of the radiative field on leaf absorptance rate and LAI has been simulated with a ray-tracing model (Fig. 3.14). The analysis shows that the largest variation induced by the leaf-level scat-

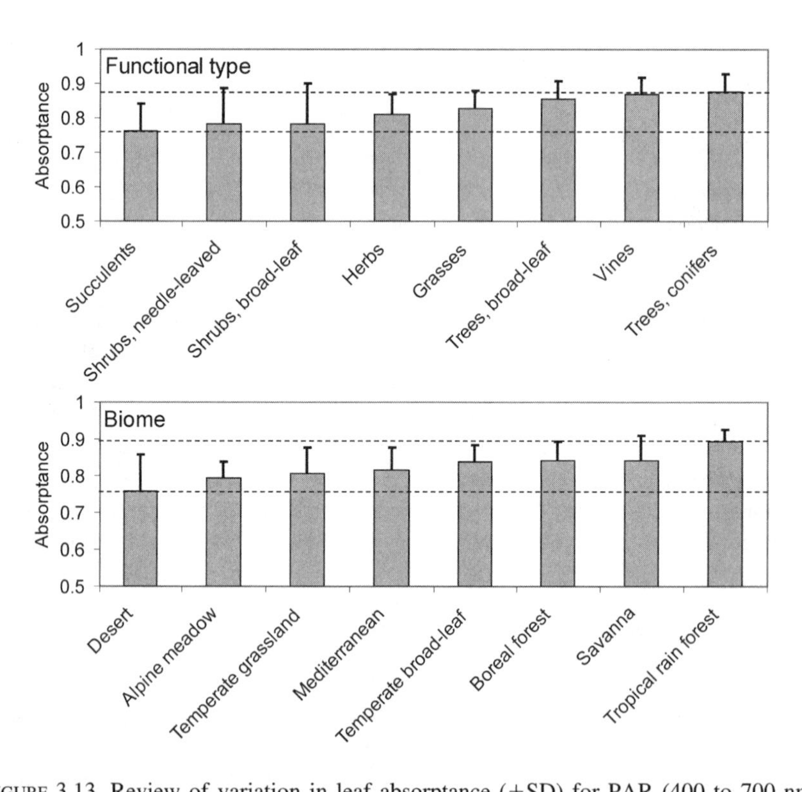

FIGURE 3.13. Review of variation in leaf absorptance (+SD) for PAR (400 to 700 nm) among different life forms and biomes based on 41 published studies reporting spectral data for 427 species (695 observations). The bars are ranked according to increasing average leaf absorptance values. The database along with literature references is available from the authors upon request.

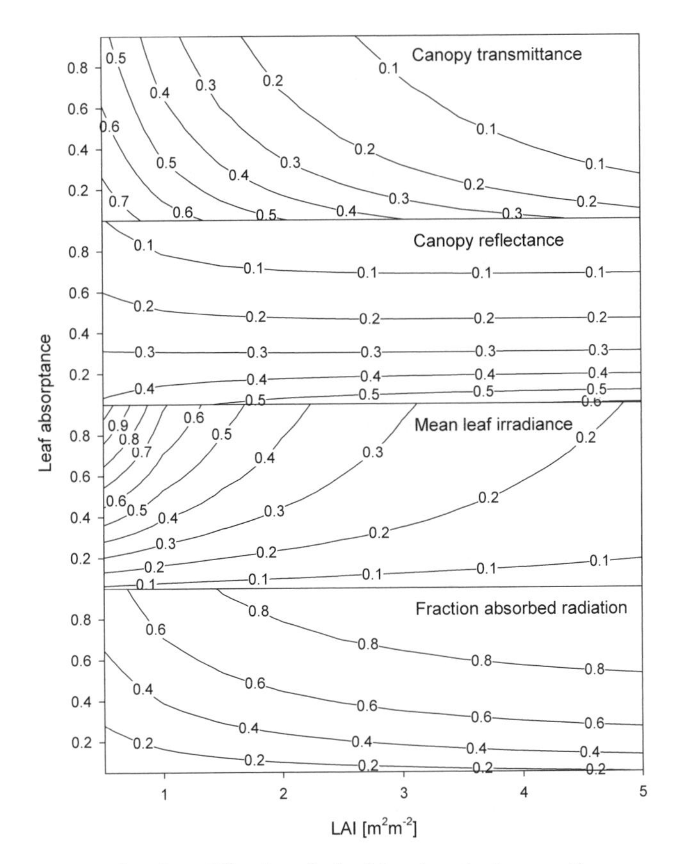

FIGURE 3.14. The role of modifications in leaf-level optical properties on canopy-scale light-harvesting efficiency in canopies with varying LAI (0.5-5 $m^2\ m^{-2}$). The simulation was performed with a ray-tracing approach assuming a random dispersion of 10^6 circular leaves. Solar elevation angle was fixed at 45°, the fraction of background scattering at 0.3, and imposing a fraction of diffuse light of 0.15.

tering on the canopy radiative field occurs at the lower LAI values. Despite the similarity in average θ, this simulation indicates that there may be important between-biome differences in light use efficiency induced by differences in LAI, which may significantly modify the whole ecosystem carbon gain.

Interestingly, among the different plant characteristics affecting the radiative field, canopy reflectance is the least dependent on LAI. In particular, at values of LAI larger than 3, canopy reflectance does not show any significant response to LAI, due to a saturation of the canopy space. This phenomenon is particularly relevant in remote sensing applications, when LAI is tentatively retrieved from canopy reflectance at different wavelengths (Turner et al. 1999). Furthermore, it is relevant that within each community and for each plant functional type, there is a large variation in leaf absorptance (see Fig. 3.14). This large variation may result from species-to-species differences in leaf surface properties (Walter-Shea and Norman 1991, Holmes and Keiller 2002), as well as temporal and site-to-site trends in leaf optics. For instance, leaf optical properties vary during the season in herb and broad-leaved deciduous woody species (Tanner and Eller 1986, Masoni et al. 1994), and with leaf age in evergreen species (Kursar and Coley 1992). Thus, seasonality as well as fractional leaf area composition of leaves of different age classes may significantly modify the scattered fluxes. In addition, variation in θ is often associated with site-to-site variability in soil nutrient availability that modifies the availability of nitrogen for the formation of chlorophyll-protein complexes (Ercoli et al. 1993). Depending on the acclimation of foliar chlorophyll contents to local leaf irradiance, there may or may not be a trend in θ with long-term leaf irradiance (St-Jacques et al. 1991, Niinemets 1997).

In special cases, large effects of scattering on canopy light climate may be observed. In Figure 3.15, the effects of leaf surface characteristics on canopy light climate were modeled for a South African succulent shrub *Cotyledon orbiculata*. The leaves of this species present a waxy surface with $\theta = 0.63$ (see inset in Figure 3.15), while θ rises to 0.84 after the artificial removal of waxes. As this simulation demonstrates, the presence of a wax layer on the leaves reduces the light interception of the upper canopy, and allows more light to penetrate into lower

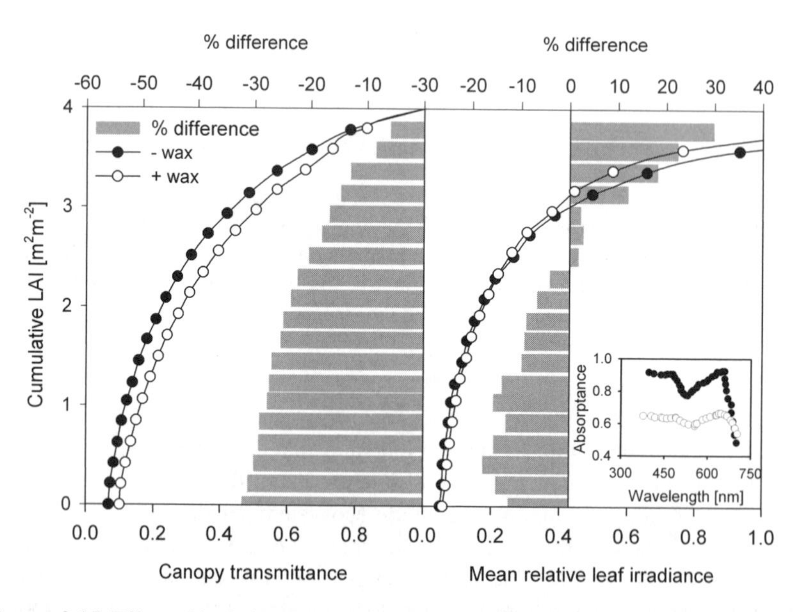

FIGURE 3.15. Effect of leaf absorptance on the vertical profile of canopy transmittance and on the mean leaf irradiance in a succulent plant *Cotyledon orbiculata* with intact waxy leaves, and for a hypothetical situation without the leaf wax layer. Simulations were conducted for a random leaf dispersion and setting the sun elevation angle at 45° and the fraction of diffuse radiation at 0.15. The inset demonstrates leaf spectral data for intact leaves and for leaves without wax (data from Sinclair and Thomas 1970).

canopy layers. Although the waxy leaves have low absorptance, a larger canopy transmittance essentially compensates for that in the lower canopy, leading to an opposite effect of leaf waxiness on the absorbed irradiance in the upper and lower canopy. Hence, reducing θ with waxy surfaces is an effective strategy to avoid excess irradiance absorption by the upper canopy leaves, without reducing light absorption in the shaded canopy layers. Similar effects are also expected for hairy leaf surfaces. For instance, removal of leaf hairs increases θ from 0.43 to 0.86 in the desert species *Encelia farinosa* (Ehleringer and Björkman 1978), further highlighting the importance of specific adaptations of leaf surface properties.

Shoot Scale

Due to architectural constraints, foliar elements in plant canopies are aggregated around branches (Fig. 3.16). For long leaves or leaves with long petioles or for shoots with long internodal distances, light-harvesting characteristics may apparently differ relatively little between the neighboring leaves on a specific shoot. More often, there is a strong interaction between the adjacent leaves within a shoot (Carter and Smith 1985, Smith and Carter 1988). Therefore, the shoot rather than individual leaves may be considered as the functional unit for light interception in coniferous canopies, and shoot architecture plays a fundamental role in the generation of the canopy radiative field (Leverenz and Hinckley 1990, Cescatti 1998).

Modeling Shoot Architecture and Radiative Regime

There is a wide range of morphological strategies in shoot architecture across different plant species as outlined in Figure 3.16. Depending on leaf size, leaf angle with respect to shoot axis, leaf number per unit of shoot length, and angular distribution of leaf surface, light interception of shoots varies widely. To summarize the influence of various shoot morphological features on light interception efficiency, a radiative transfer model based on the theory of light penetration in a turbid medium (Nilson 1971) may be applied to the shoots as described in Cescatti and Zorer (2003), and Niinemets et al. (2002a). According to the model, the probability of photon interception, $F(\omega,\gamma)$, in the shoot volume is a function of the extinction coefficient (G-function) (Ross 1981) of the Markov coefficient for spatial aggregation (λ_0) (Nilson 1971) of the leaf area density in the shoot volume (ρ), and of the beam path-length in the shoot for specific azimuth (ω) and zenith (γ) angles, $L(\omega,\gamma)$:

$$F(\omega,\gamma) = 1 - \exp[-G(\gamma,c)\lambda_0 \rho L(\omega,\gamma)]. \qquad \text{(Eq. 3.5)}$$

The Markov coefficient measures the within shoot spatial aggregation and decreases from 1 (random dispersion) toward 0 (aggregated dispersion). The G-function is defined as the ratio between the projected area on a surface orthogonal to the direction of projection and the total leaf surface (Ross 1981), and it depends on the angular distribution of leaf surface in the shoot volume. The ellipsoidal LAD model (Campbell and Norman 1989) which assumes that leaf an-

	0°, 0°	90°, 0°	0°, 90°	
Abies alba	S_S = 0.346	S_S = 0.117	S_S = 0.066	A_T/A_P = 2.80 $\overline{S_S}$ = 0.14 c = 4.67 λ_0 = 0.60
Picea abies	S_S = 0.198	S_S = 0.132	S_S = 0.097	A_T/A_P = 2.73 $\overline{S_S}$ = 0.14 c = 2.88 λ_0 = 0.61
Pinus sylvestris	S_S = 0.240	S_S = 0.142	S_S = 0.074	A_T/A_P = 2.57 $\overline{S_S}$ = 0.14 c = 1.86 λ_0 = 0.89
Nothofagus fusca	S_S = 0.344	S_S = 0.150	S_S = 0.150	A_T/A_P = 2.00 $\overline{S_S}$ = 0.24 c = 1.56 λ_0 = 1.01
Fagus sylvatica	S_S = 0.870	S_S =0.348	S_S =0.293	A_T/A_P = 2.00 $\overline{S_S}$ = 0.19 c = 1.87 λ_0 = 1.00

gles are distributed as the surface of an oblate or prolate spheroid, has previously been used to describe the needle area distribution in shoots (Niinemets et al. 2002a, Cescatti and Zorer 2003). This distribution function is very flexible, allowing to parameterization of both erectophile and planophile LAD, using only one parameter, c (the ratio of the horizontal to vertical ellipse semiaxis).

The shoot model (see Eq. 3.5) may be parameterized using shoot silhouette areas measured for various view directions and allows to indirectly derive val-

FIGURE 3.16. Examples of shoot projections for two different rotation angles ([cph] = 0° vs. [cph] = 90°) and an inclination angle (ϕ) of 0°, and for ϕ = 90°, and [cph] = 0° in three conifer and two broadleaved species. A Markov model of light interception was fitted to the shoots as described in Cescatti and Zorer (2003). According to the shoot model, shoot light interception depends on leaf area density in the shoot volume, spatial clumping of leaf area (λ_0), and angular distribution of leaf surface (parameter c of elliptical leaf angle distribution (Campbell and Norman 1989). Leaf angular distribution becomes more horizontal with increasing c; $\lambda_0 < 1$ for aggregated foliage and $\lambda_0 = 1$ for random foliage dispersion (see Fig. 3.4 and Eq. 3.3). Leaf total to projected area ratio (A_T/A_S), shoot silhouette to total leaf area ratios for each specific shoot angle (S_S), and spatial average of S_S (S_S) are also depicted. [Data sources: *A. alba*, Cescatti and Zorer (2003); *P. abies* and *P. sylvestris*, Cescatti and Zorer (unpublished); *N. fusca*, Niinemets et al. (2004); *F. sylvatica*, Fleck, Cescatti and Zorer (unpublished)].

ues of λ_0 and c. Once the shoot model is parameterized it is possible to calculate the frequency distribution of the irradiance on the needle area, given a specific radiative field (proportion of direct radiation, sun angle). Assuming the shoot to be lying horizontally, and the sun to be at the zenith, the distribution of direct irradiance is presented in Figure 3.17 for two *Abies* shoots typical of sunlit and shaded conditions (Cescatti and Zorer 2003).

Species Differences and Acclimation of Shoot Light-Harvesting Efficiency

Leaf silhouette to total area ratio [$S_S(\varphi,\phi)$, where φ is the shoot rotation angle and ϕ the inclination angle], is often used as a morphological index of the shoot efficiency in light interception (Oker-Blom and Smolander 1988, Niinemets and Kull 1995a, Leverenz et al. 2000, Cescatti and Zorer 2003). The light interception efficiency for isotropic sky conditions scales with the spherical average of S_S (Stenberg et al. 1998). Species differences in mean S_S are primarily affected by the level of leaf clumping and, therefore, this coefficient may be similar in species with flat shoots (such as *Abies* and *Picea*) or axially symmetric shoots (such as *Pinus*) (see Fig. 3.16). On the contrary, different needle angle distributions have widely different efficiencies in light interception for anisotropic distribution of solar radiation, for example, in heterogeneous plant canopies. In particular, as the shoots become flatter, they intercept more efficiently light from low zenith angles than the spherical shoots (Leverenz and Hinckley 1990). Hence, species with flat shoots (*Abies*, *Picea*) are more competitive in homogeneous forest understories, where most of the light comes from large vertical angles. In such conditions, spherical shoots (*Pinus*) are less competitive because of the higher self-shading in the prevalent direction of illumination. However, spherical needle arrangement, as well as large degree of clumping, allows higher concentration of foliar biomass per unit stem length, thereby maximizing the shoot photosynthetic potential at high irradiance (Smolander et al. 1987, Cescatti 1998).

Shoot morphology is not a fixed species characteristic, but strongly acclimates to the local shoot light environment in plant canopies (Niinemets and Kull 1995a).

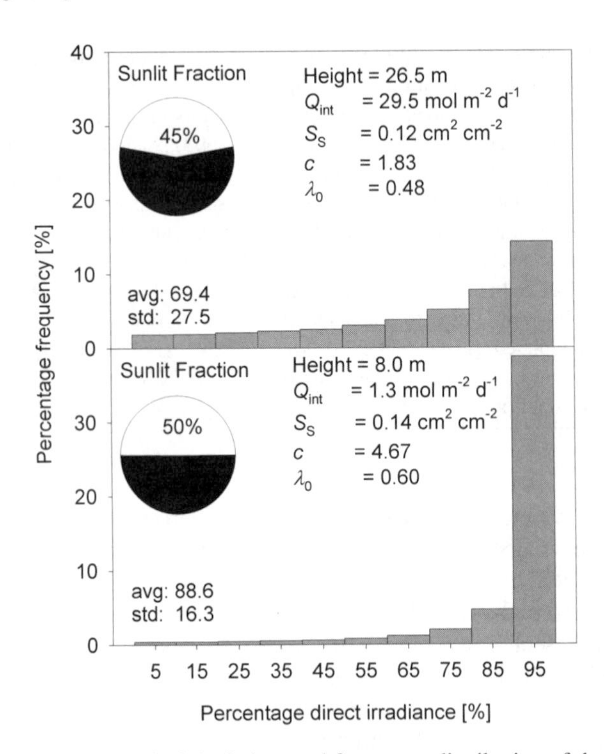

FIGURE 3.17. Average, standard deviation, and frequency distribution of the direct irradiance on the needle surface and sunlit leaf area fraction (white area in the pie charts) of two *Abies alba* shoots at different canopy heights and relative irradiances (Q_{int} – seasonal average daily integrated quantum flux density above the shoot) modeled with a turbid medium shoot model (modified from Cescatti and Zorer 2003). Definition of the characteristics as in Figure 3.16.

The maximum S_S (generally observed for the view direction $\varphi = 0°$, $\phi = 0°$) (see Fig. 3.16), decreases at increasing irradiance in most species (Fig. 3.18), implying a lower self-shading within the shoot. However, the plasticity for modification of exposed foliar area largely differs among species with the shade-tolerant *Abies* exhibiting larger plasticity than the shade-intolerant conifer *Pinus* (see Fig. 3.18). In *Abies*, changes in maximum S_S result both from a lower aggregation (larger λ_0), and from flatter angular distributions of leaf surface (larger *c*), such that shoots in the upper canopy are spherical and shoots in the lower canopy are flat in this species (Cescatti and Zorer 2003). Given that the distribution of solar radiation becomes increasingly vertical at increasing canopy depth, the flatter shoots (see Fig. 3.16) become progressively more advantageous in low irradiance. In *Pinus*, irradiance does not affect *c* (Niinemets et al. 2002a), leading to overall lower plasticity in shoot light interception efficiency (see Fig. 3.18). Such shoot-level differences in plasticity may significantly affect the whole-tree light interception capacity (Cescatti 1998). It has been hypothesized that species'

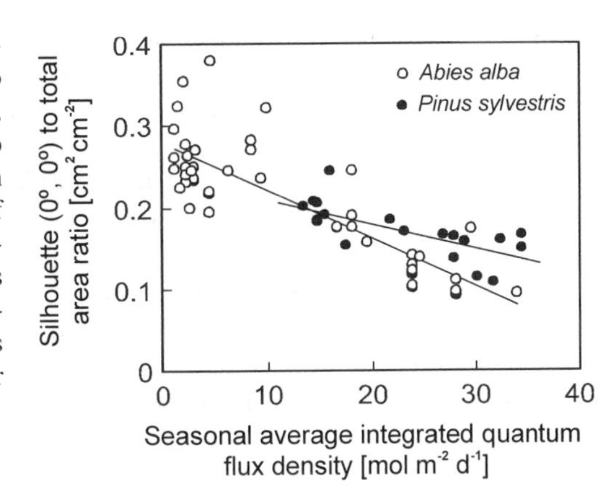

FIGURE 3.18. Modification of silhouette to total needle area ratio (projection [cph] = 0°, $\phi = 0°$, see Fig. 3.16) in response to canopy variation in irradiance in conifers *Abies alba* (data of Cescatti and Zorer 2003) and *Pinus sylvestris* (data of Niinemets et al. 2002a). The explained variances (r^2) of the linear regressions were 0.72 for *A. alba* and 0.54 for *P. sylvestris*.

ability to achieve a high S_S in shade is the primary determinant of stand LAI (Leverenz and Hinckley 1990, Leverenz et al. 2000).

Although shoot architecture has been primarily investigated in conifers (e.g., Niinemets and Kull 1995a, Stenberg et al. 1998), differences in the architecture may significantly modify radiative transfer also in canopies of deciduous trees (see Fig. 3.16) (Planchais and Sinoquet 1998, Valladares and Pearcy 1998, Farque et al. 2001). As in conifers, the light interception efficiency of broad-leaved shoots increases with decreasing irradiance, both as the result of decrease in leaf overlapping and in the leaf inclination angle with respect to horizontal (Planchais and Sinoquet 1998, Farque et al. 2001). Previous studies have primarily focused on irradiance effects on leaf inclination angle (e.g., Hikosaka and Hirose 1997), but recent work demonstrates that changes in foliage clumping may be at least as important as the modifications in leaf angular distribution (Planchais and Sinoquet 1998, Farque et al. 2001).

Constraints on Shoot Architecture

Since mechanical structures become progressively inefficient as they increase in size, there are often trade-offs between light harvesting and investment in leaf support. For instance, increases in lamina and petiole length can decrease self-shading within the shoot (Takenaka 1994, Pearcy and Yang 1998). However, for certain leaf inclination angles, the biomass investment in supporting tissues scales with the cube of leaf length (Niklas 1999), implying that requirements for support may often conflict with those for efficient light harvesting. In fact, temperate broad-leaved trees may invest more than 30% of leaf biomass in petioles and major veins (Niinemets and Fleck 2002). Often, needles of conifers (Niinemets et al. 2002b) and petioles and laminas of broad-leaved trees (Pearcy and Yang 1998, Niinemets and Fleck 2002) bend under their own weight, suggesting that leaf support investments may be limited in natural canopies. Such bending also means that various leaf parts are intercepting light with differing efficiency at different times of the day. Especially in understory conditions with strongly anisotropic irradiance distribution, leaves with a large curvature, for example, those in rosette plants such as tree ferns and palms, may have large leaf fractions that are always intercepting light with a low efficiency.

Leaf phyllotaxy, that is, leaf arrangement in genetically predetermined spirals on stem, may also limit the shoot light interception capacity. Depending on the angles of adjacent leaves on the stem (divergence angle) as well as relative leaf dimensions, fixed spiral arrangement may lead to an extensive leaf overlap (Niklas 1988, Pearcy and Yang 1998), in particular, for decussate leaves (Valladares and Brites 2004) and for vertical leaf inclination angles (Niklas 1988). Such phyllotactic constraints may also limit the plastic adjustment of shoot architecture to long-term irradiance, leading to a constant leaf angle distribution with irradiance as it has been observed in *Pinus* species (Niinemets et al. 2002a).

Crown and Canopy Scales

Crown Architecture and Light Interception

A single tree crown represents a fundamental level of spatial organization of phytoelements that corresponds to the smallest scale of ecosystem dynamic and management. Crown architecture is the outcome of conflicting plant requirements, that is, optimization of light harvesting by an efficient leaf display, reduction of the cost of the leaf support framework, and provision of an adequate water supply to the leaves. Because of multiple contrasting requirements, crowns cannot be simply optimized for one specific task.

For light harvesting, crowns represent a level of aggregation of phytoelements in space, which reduces the overall light interception by increasing the within c-tree self-shading compared with a horizontally homogeneous canopy (Nilson 1992, Chen et al. 1993, Stenberg et al. 1994, Niinemets et al. 2004). Analysis of the radiative field of a spruce forest with a 3-D RTM shows that canopy transmittance increases with the aggregation of leaves in crowns (Fig. 3.19) (Cescatti 1998). The observed dispersion of tree crowns resulting from the between-tree competition and forest management leads to a higher efficiency in intercepting light compared to a random tree dispersion. In addition, crown architecture reduces the relative irradiance on leaf area at the canopy top and increases it above a LAI value of 3.5, producing a more even distribution of the irradiance in dense canopies (see Fig. 3.19).

The overall effect of crowns on the canopy light climate is largely dependent on crown shape and on stand density (Kuuluvainen and Pukkala 1989, Nilson 1992). Assuming ellipsoidal crowns, the shape can be characterized as the ratio of the width to the height. By varying this ratio, the effect of crown shape on the canopy light climate is reported in Figure 3.20. Crown shape, similarly to LAD, affects the vertical distribution of the canopy gap fraction. Long and narrow crowns with a low value of the width/length ratio generate a higher gap fraction at the steepest vertical angles, and a lower gap fraction at low angles, increasing the fraction of intercepted radiation at the low sun angles typical of the northern

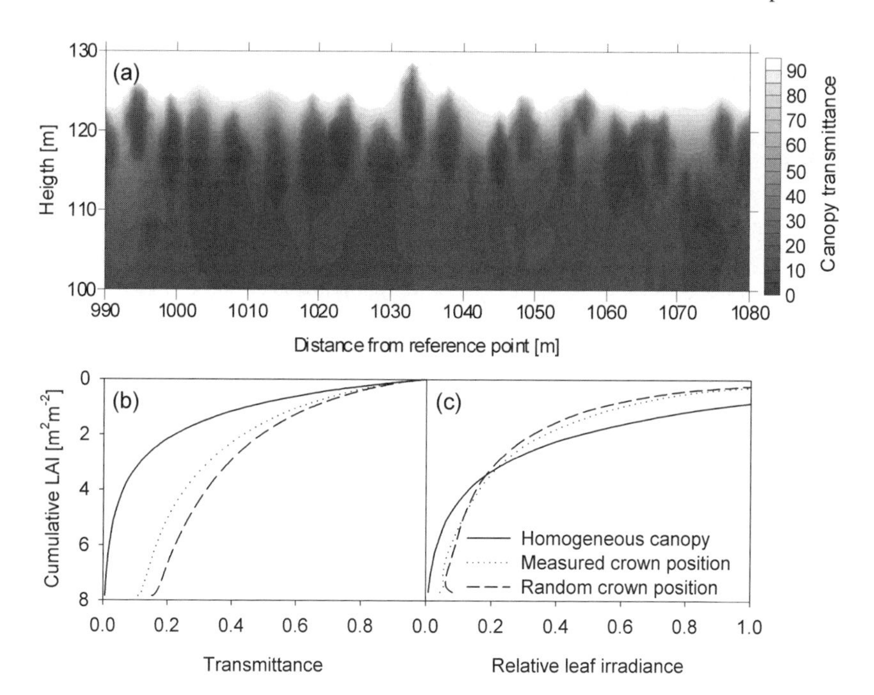

FIGURE 3.19. Effects of leaf clustering in crowns and crown distribution in the canopy space on the canopy light climate in a *Picea abies* forest (modified from Cescatti 1998). Upper panel shows the vertical profile of canopy transmittance to diffuse radiation, and the lower panel the canopy transmittance and leaf relative irradiance assuming a homogeneous leaf area distribution (no crown clumping) (—), the observed spatial position of tree crowns in the sample area (- - -) and a random dispersion of tree crowns in the stands (···).

latitudes. In contrast, flat crowns optimize light interception at steep angles, a situation that predominantly occurs at low latitudes (see Fig. 3.20).

Canopy Light Climate

All different morphological features of plant architecture at various scales ultimately have an integrated effect on the light climate of the canopy. Most broadleaf species with sympodial branching are characterized by a flexible tree architecture such that the spatial arrangement of tree crowns, and the acclimation gradient in leaf area densities and orientation, generally produce dense homogeneous canopies with a low ratio of within-tree to between-tree shading (e.g. oak, maple, beech forest). In contrast, plant architecture is less flexible in conifers with monopodial branching, resulting in open canopies that are characterized by a high ratio of within-tree/between-tree shading, as in pine and spruce (Kuuluvainen and Pukkala 1989). Mixed forests have complex canopy structures with more than one layer and different architectures in different layers. Therefore, mixed forests are characterized

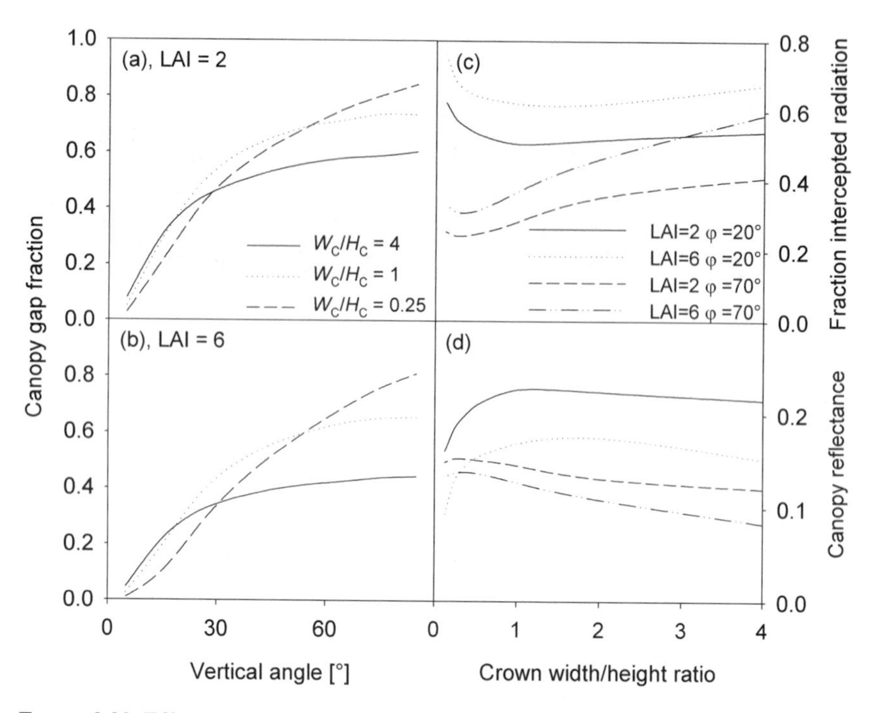

FIGURE 3.20. Effect of crown elongation (crown width to height ratio, W_C/H_C) on canopy light harvesting efficiency. Left panels show the angular dependence of the canopy gap fraction for two canopies with a LAI = 2 $m^2 m^{-2}$ (a) and LAI = 6 $m^2 m^{-2}$ (b) and for three different W_C/H_C ratios. Right panels illustrate the fraction of intercepted radiation (c) and canopy reflectance (d) at varying W_C/H_C ratios and LAI values, and for sun elevation angles of 20° and 70°. The simulations were performed by a ray-tracing model assuming ellipsoidal crowns with circular leaves and a constant crown volume of 3.6 m^3m^{-2}. The fraction of diffuse radiation was set at 0.15.

by higher space saturation and larger values of light interception (e.g., tropical rainforests, mixed temperate deciduous forests) than monotypic forests.

A fundamental feature determining the light climate of plant canopies is the overall pattern of the spatial dispersion of the phytoelements (Nilson 1971). In classical radiative transfer models, leaves have been assumed to be randomly dispersed in horizontally homogeneous layers. This assumption neglects the large variability in spatial patterns occurring in nature. More recently, the role of the spatial patterns has been considered both in ecological and remote-sensing applications (North 1996, Nilson and Ross 1997, Pinty et al. 2002). The effect of the nonrandom spatial dispersion of the leaves on light interception has been described using positive and negative binomial models, and conditional probability models (Markov chain) (see Eqs. 3.1 to 3.3 and Fig. 3.4) (Nilson 1971).

In nature, most plant canopies show a clustered architecture. Clustering may be related to different scales of spatial aggregation (e.g., plant spacing in crops and orchards, shoot clumping in conifers, petiole length and branch clumping in broad-leaved species) (Niinemets et al. 2004). Using observed profiles of leaf area density and canopy transmittance reported in Figure 3.21, the clumping coefficients (λ_0 in the Markov model) of five different canopies were computed assuming a spherical leaf angle distribution. Values are in the range of 0.45 to 0.70, exhibiting strongest clumping (lowest values of λ_0) in a coniferous canopy (*Picea sitchensis*) and lowest aggregation in a homogeneous maple forest (see Fig. 3.21).

The product of a given leaf area and clumping coefficient, defined as effective LAI (Chen, 1996), corresponds to the LAI of a random canopy with equivalent light interception capacity, and determines the fraction of intercepted radiation and the mean relative irradiance on the leaf area (Fig. 3.22). A spectrum of different combinations of LAI and clumping index may produce the same ef-

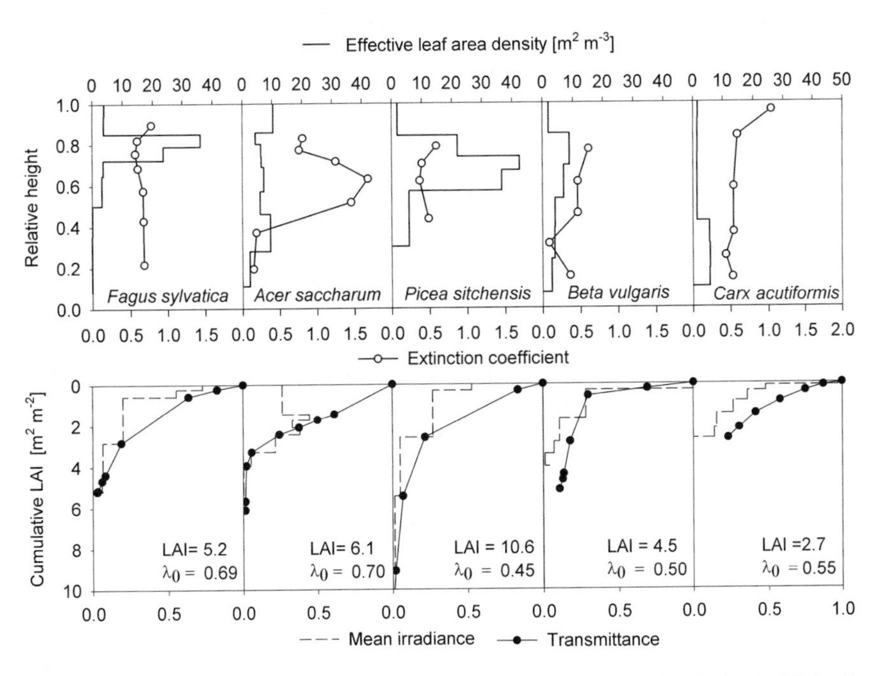

FIGURE 3.21. Experimental profiles of leaf area density (LAI per unit relative height), effective extinction coefficient for each canopy layer, light transmittance, and average absorptance of each layer in five canopies of contrasting architecture and total LAI. In the Markov model of radiative transfer (see Eq. 3.3 and Fig. 3.4), the effective extinction coefficient is equal to $k\lambda_0$, where k is the extinction coefficient and λ_0 is the coefficient of spatial clumping. Average values of λ_0 for the entire canopy were calculated assuming isotropic distribution of sky irradiance, and spherical distribution of leaf angles. All canopy characteristics were computed from profiles of LAI and light transmission. Total canopy height was 25.2 m for temperate broad-leaved tree *Fagus sylvatica* (data of Aussenac and Ducrey 1977), 18 m for *Acer saccharum* (Ellsworth and Reich 1993), 12 m for temperate conifer *Picea sitchensis* (Norman and Jarvis 1974, 1975), 0.65 m for the herb *Beta vulgaris* (Hodánová 1972), and 0.9 m for the grass *Carex acutiformis* (Hirose et al. 1989).

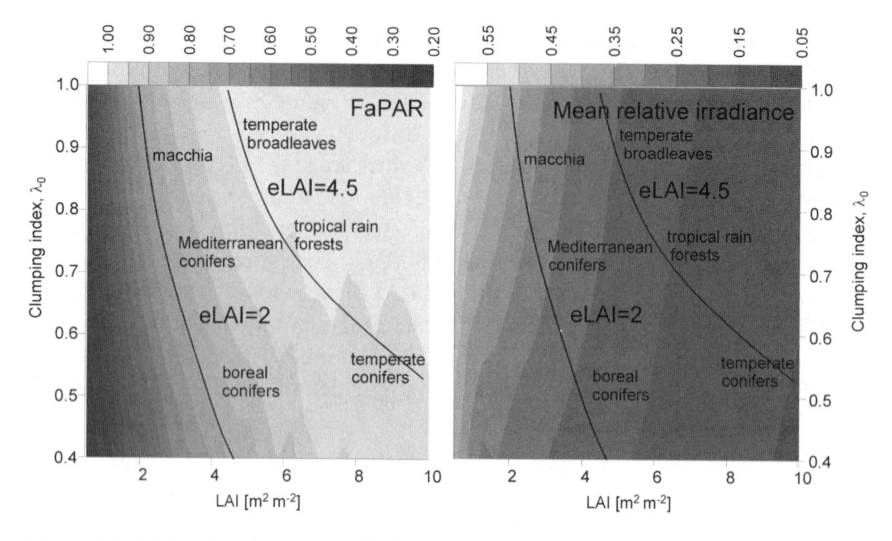

FIGURE 3.22. Fraction of intercepted photon flux density (F_{PPFD}) and mean relative irradiance on leaf area in dependence on LAI and spatial clumping coefficient (λ_0) in different plant communities. Canopy light interception is equal for the same effective leaf area index (λ_0LAI). The simulation was carried out with a 1-D Markov model (see Eq. 3.3).

fective LAI and therefore the same canopy-level efficiency in light interception, but with largely different mean relative irradiance on leaf area. Varying LAI and clumping, plant canopies can therefore generate different light regimes as a response to soil water and nutrient availabilities, competition and species physiological traits (evergreen vs. broadleaf, photosynthetic potentials, leaf anatomy, etc.). Different biomes can be tentatively located in this parameter gradient according to their characteristic architectural traits (see Fig. 3.22). Given that the leaf spatial distribution is so relevant for canopy-scale light interception, new methodologies are being developed to remotely derive the clumping index. In particular, the hot-dark spot index determined using the multi-angular remote-sensing techniques, seems to provide a promising tool to retrieve canopy-scale clumping index from satellite platforms (Lacaze et al. 2002).

Landscape

In previous sections, we examined the effects of different plant architectural and optical features on the canopy radiative regime. Apart from plant characteristics, atmospheric and topographic conditions play a major role on the canopy light climate at the landscape scale. The role of topography is particularly relevant in mountainous regions where differences in slope and aspects largely affect the distribution of the solar radiation on the ground (Revfeim 1978, Olseth and Skartveit 1997). The amount of direct sunlight incident on a slope depends on the cosine of the angle between the sun's beam and the normal to the slope and can be estimated with the topographic geometric projection.

$$R(i) = R_{\perp} \cos(i) \qquad \text{(Eq. 3.6)}$$
$$\cos(i) = \sin \alpha \cos \beta + \cos \alpha \sin \beta \cos \psi \qquad \text{(Eq. 3.7)}$$

where $R(i)$ is the radiation normal to an inclined surface, $R_{\perp}$ is the radiation flux density normal to the beam, α is the solar elevation angle, β is the inclination angle of the surface, ψ is the difference between the sun's azimuth angle and the azimuth angle of the normal to the surface (Campbell and Norman 1998). The flux of incoming diffuse radiation is also controlled by slope and aspect, since some fraction of the sky may be obscured by the ground (Tian et al. 2001). In addition, radiation reflected from proximal slopes may be important in mountainous regions, especially when the albedo is high, as in the case of snow or sand (Olseth and Skartveit 1997).

A number of models for estimating the distribution of solar radiation on complex terrain have been developed in last decade (Cooter and Dhakhwa 1995, Dubayah and Rich 1995, Olseth and Skartveit 1997, Thornton et al. 2000). Most models treat separately direct and diffuse components of solar radiation due to their different angular distribution in the sky hemisphere. Revfeim (1978) provided a general framework to estimate solar radiation starting from the incoming diffuse and direct components on a horizontal surface:

$$G_a = G_m[R_d(1 - K_r) + f_{\beta}K_r + 0.2(1 - f_{\beta})], \qquad \text{(Eq. 3.8)}$$

where G_a is the global radiation received on a surface with specific orientation; G_m is the global radiation on a flat surface; K_r is the ratio of diffuse radiation to global radiation for a horizontal surface; β is the slope and f_{β} is the "slope reduction factor," that is, the proportion of the sky hemisphere obscured by the slope, and is estimated as $f_{\beta} = (1 - \beta/180)$; R_d is the ratio of direct radiation on the slope to direct radiation on a horizontal surface, and can be expressed as a function of slope and aspect according to Equation 3.6 (Fig. 3.23). The term $G_mR_d(1 - K_r)$ is the contribution of direct irradiance from unobscured sun. The term $f_{\beta}K_r$ is the diffuse radiation component obtained by integrating the sky radiation for isotropic sky conditions. The diffuse component decreases as the surface becomes steeper, because of partial blocking of the sky hemisphere by the slope. The term $0.2(1 - f_{\beta})$ accounts for the reflection from the blocking land surface. In the latter term, albedo is set at 0.2, which is a typical value for grasslands.

More recently, the effect of topography on the distribution of solar radiation has been analyzed with digital elevation models (Olseth and Skartveit 1997) and finally implemented in GIS (Dubayah and Rich 1995). Nowadays, commercial GIS packages include specific routines to estimate the fraction of sky obscured by topographic shading and the cosine-corrected direct irradiance.

While the effect of topography on radiation distribution has been largely investigated and modeled, the interactions between radiation on slopes and canopy light climate have been rarely dealt with. In fact, the use of the topographic geometric projection to calculate the irradiance (R_i) on plant canopies growing in

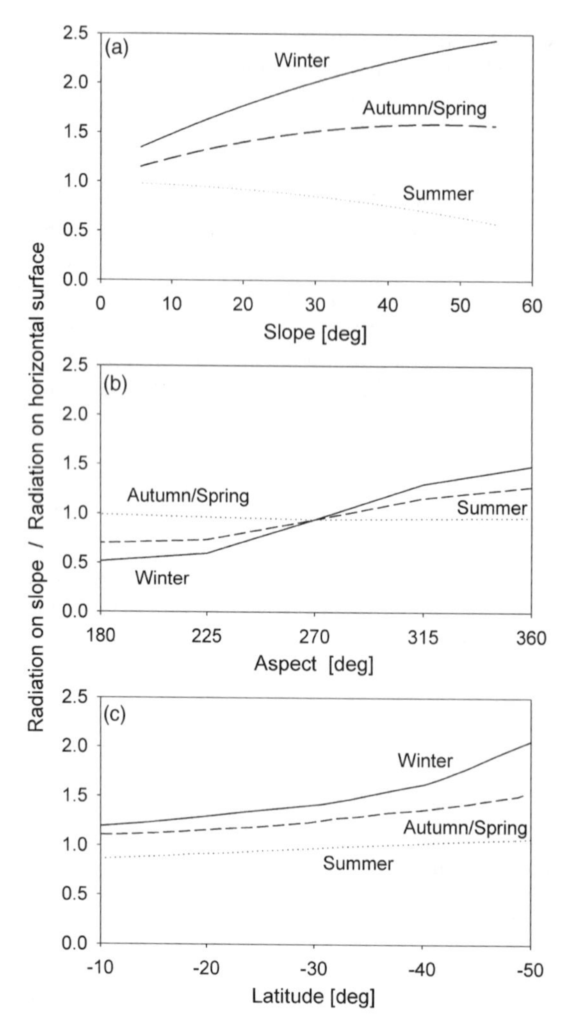

FIGURE 3.23. Radiation on a slope relative to radiation on a flat surface as a function of site (a) slope (N-facing aspect and latitude = 35°S), (b) aspect (south 180°; slope = 15° and latitude = 35°S), and (c) latitude (N-facing and slope = 15°) during different seasons (modified from Tian et al. 2001).

sloping terrains is not adequate, since it does not consider that trees grow upright and that the foliage angular distribution of the leaf surfaces is not directly affected by the slope (Wang et al. 2002). To include the effect of landscape topography on the canopy light climate, detailed 3-D canopy models are required (Cescatti 1997a,b, Courbaud et al. 2003) that account for the specific location of each tree in the landscape. As an alternative, Wang et al. (2002) presented a novel methodology that accounts for ground topography on the canopy light climate in 1-D canopy models, by changing the ratio of sunlit/shaded leaves as:

$$LAI_{sun} = \{1 - \exp[-k\, LAI/\cos(i)]\}\cos(i)/k \qquad \text{(Eq. 3.9)}$$
$$LAI_{shade} = LAI - LAI_{sun}$$

Equation 3.9 assumes that all LAI values referred to unit area of inclined surface.

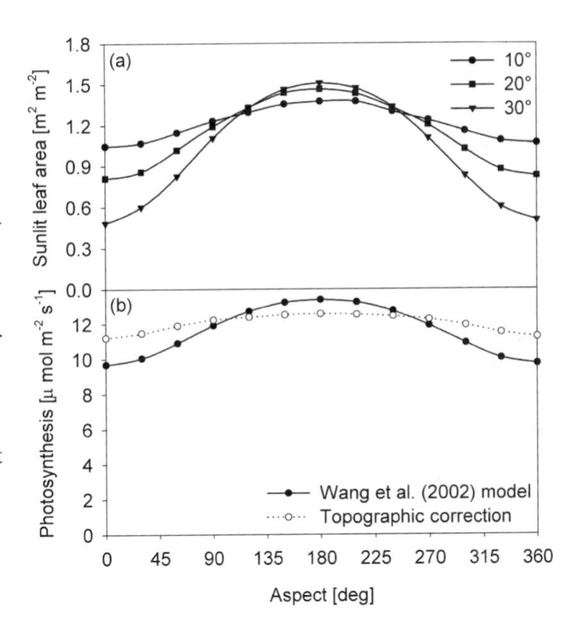

FIGURE 3.24. Effect of topographic aspect (south 180°) on the sunlit leaf area for three different slopes (a), and variations of light-limited canopy photosynthesis rate for a slope of 30° (b). In the latter simulations, photosynthesis rate was calculated using a simple geometric correction of light interception as in Eq. 3.8 (···) or considering the topographic effect on the fraction of sunlit area (see Eq. 3.9) (—). The simulations were carried out at a latitude of 47.3°N, day of year of 129, and for a canopy with LAI = 6.0 (see Eq. 3.9) (modified from Wang et al. 2002).

In this model framework, the beam pathlength in the canopy, cos(i), is determined by the sun-ground geometry (see Eq. 3.7) that in turn affects the distribution of leaf area between sunlit and shaded classes. This model predicts the same amount of total radiant energy interception as the geometric projection method, but it may lead to very different effect of topography on ecological processes due to nonlinearity of the photosynthetic response to light (Fig. 3.24).

Wang et al. (2002) proposed a methodology to calculate the diffuse fluxes as well, including the scattered radiation from land surface. Nevertheless, according to numerical analyses, the decrease of diffusive radiation incident on little or moderately inclined terrains is on the order of a few percent.

Summary and Conclusions

Simple radiative transfer models based on the turbid medium analogy have been previously applied to describe the stand light environment with varying degree of success. Nowadays, more advanced models based on Monte Carlo ray tracing allow the simulation of important canopy features, like leaf size, shape, and spatial distribution, and processes like radiation scattering and penumbra. In addition, current RTMs may deal with important architectural traits of plant canopies required to understand the influence of globally changing light conditions on the net primary productivity (Gu et al. 2003). As plants do not possess specific acclimation mechanisms to adjust to highly fluctuating light environments (Leakey et al. 2002), prediction of long-term changes in incoming radiation is highly relevant to gaining insight on long-term changes in the terrestrial carbon balance.

Our review demonstrates that leaf (leaf inclination and rolling in broad-leaved species; differences in needle projected to total area ratio), shoot (foliage aggregation), branch (aggregation of shoots), and canopy (crown length/width ratio) may importantly alter whole stand light interception capacity. Furthermore, plant traits at various hierarchical levels may even more strongly affect the mean irradiance on leaf surface than the total intercepted light.

Giving that the leaf photosynthetic response to light is typically a downward concave curve, canopy level responses simply determined from the mean leaf irradiances are systematically overestimated. In order to produce accurate predictions of the canopy-scale photosynthetic response, the actual light distribution within the canopy has to be described by measurements or modeling. In this regard, architectural features like the ratio of canopy height to leaf width, strongly affect the variability of the light distribution on the leaf area, due to the increasing importance of penumbra with increasing tree size and decreasing leaf width. Nowadays, the complex interplay between architectural and optical features of plant canopy, light climate, and primary productivity may be effectively investigated through the integration of geometrical canopy models and advanced RTMS.

Acknowledgments. This study was supported financially by the Fondazione Cassa di Risparmio Trento e Rovereto and by the Province of Trento (grant DGP 1060), the European Commission (CARBOMONT, EVK2-2002-0549), the Estonian Science Foundation (grant 4584), and the Estonian Ministry of Education and Science (grant 0182468As03).

References

Alcocer-Ruthling, M., Robberecht, R., and Thill, D. C. 1989. The response of *Bouteloua scorpioides* to water stress at two phenological stages. Bot. Gaz. 150:454–461.

Anderson, M. C. 1964. Studies of the woodland light climate. I. The photographic computation of light conditions. J. Ecol. 52:27–41.

Asner, G. P., and Wessman, C. A. 1997. Scaling PAR absorption from the leaf to landscape level in spatially heterogeneous ecosystems. Ecol. Model. 103:81–97.

Aussenac, G., and Ducrey, M. 1977. Etude bioclimatique d'une futaie feuillue (*Fagus sylvatica* L. et *Quercus sessiliflora* Salisb.) de l'Est de la France. I. Analyse des profils microclimatiques et des caractéristiques anatomiques et morphologiques se l'appareil foliaire. Ann. Sci. For. 34:265–284.

Baldocchi, D., and Collineau, S. 1994. The physical nature of solar radiation in heterogeneous canopies: Spatial and temporal attributes. In: Exploitation of Environmental Heterogeneity by Plants. Ecophysiological Processes Above- and Belowground. M. M. Caldwell and R. W. Pearcy (eds.), pp. 21–71. San Diego: Academic Press.

Baldocchi, D. D., and Harley, P. C. 1995. Scaling carbon dioxide and water vapour exchange from leaf to canopy in a deciduous forest. II. Model testing and application. Plant Cell Environ. 18:1157–1173.

Baldocchi, D. D., and Hutchison, B. A. 1986. On estimating canopy photosynthesis and stomatal conductance in a deciduous forest with clumped foliage. Tree Physiol. 2:155–168.

Baldocchi, D. D., and Wilson, K. B. 2001. Modeling CO_2 and water vapor exchange of a temperate broadleaved forest across hourly to decadal time scales. Ecol. Model. 142:155–184.
Campbell, G. S. 1986. Extinction coefficients for radiation in plant canopies calculated using an ellipsoidal inclination angle distribution. Agric. For. Meteorol. 36:317–321.
Campbell, G. S., and Norman, J. M. 1989. The description and measurement of plant canopy structure. In: Plant Canopies: Their Growth, Form and Function. G. Russell, B. Marshall, and P. G. Jarvis (eds.), pp. 1–19. Cambridge: Cambridge University Press.
Campbell, G. S., and Norman, J. M. 1998. An introduction to environmental biophysics, 2nd ed. New York: Springer-Verlag.
Carter, G. A., and Smith, W. K. 1985. Influence of shoot structure on light interception and photosynthesis in conifers. Plant Physiol. 79:1038–1043.
Casella, E., and Sinoquet, H. 2003. A method for describing the canopy architecture of coppice poplar with allometric relationships. Tree Physiol. 23:1153–1170.
Cescatti, A. 1997a. Modelling the radiative transfer in discontinuous canopies of asymmetric crowns. I. Model structure and algorithms. Ecol. Model. 101:263–274.
Cescatti, A. 1997b. Modelling the radiative transfer in discontinuous canopies of asymmetric crowns. II. Model testing and application in a Norway spruce stand. Ecol. Model. 101:275–284.
Cescatti, A. 1998. Effects of needle clumping in shoots and crowns on the radiative regime of a Norway spruce canopy. Ann. Sci. For. 55:89–102.
Cescatti, A., and Zorer, R. 2003. Structural acclimation and radiation regime of silver fir (*Abies alba* Mill.) shoots along a light gradient. Plant Cell Environ. 26:429–442.
Chen, J. M. 1996. Optically-based methods for measuring seasonal variation of leaf area index in boreal conifer stands. Agric. For. Meteorol. 80:135–163.
Chen, J. M., and Black, T. A. 1992. Defining leaf area index for non-flat leaves. Plant Cell Environ. 15:421–429.
Chen, S. G., Ceulemans, R., and Impens, I. 1994. A fractal-based *Populus* canopy structure model for the calculation of light interception. For. Ecol. Manage. 69:97–110.
Chen, S. G., Shao, B. Y., and Impens, I. 1993. A computerised numerical experimental study of average solar radiation penetration in plant stands. J. Quant. Spectrosc. Radiat. Transfer. 49:651–658.
Cooter, E. J., and Dhakhwa, G. B. 1995. A solar radiation model for use in biological applications in the south and southeast USA. Agric. For. Meteorol. 78:31–51.
Courbaud, B., de Coligny, F., and Cordonnier, T. 2003. Simulating radiation distribution in a heterogeneous Norway spruce forest on a slope. Agric. For. Meteorol. 116:1–18.
Cowan, I. R. 1968. The interception and absorption of radiation in plant stands. J. Appl. Ecol. 5:367–379.
Disney, M. I., Lewis, P., and North, P. R. J. 2000. Monte Carlo ray tracing in optical canopy reflectance modeling. Remote Sens, Rev, 18:163–196.
Dubayah, R., and Rich, P. M. 1995. Topographic solar radiation model for GIS. Int. J. Geograph. Informat. Sci. 9:405–419.
Ehleringer, J. R., and Björkman, O. 1978. Pubescence and leaf spectral characteristics in a desert shrub, *Encelia farinosa*. Oecologia 36:151–162.
Ellsworth, D. S., and Reich, P. B. 1993. Canopy structure and vertical patterns of photosynthesis and related leaf traits in a deciduous forest. Oecologia 96:169–178.
Ercoli, L., Mariotti, M., Masoni, A., and Massantini, F. 1993. Relationship between nitrogen and chlorophyll content and spectral properties in maize leaves. Eur. J. Agron. 2:113–117.

Evans, G. C., and Coombe, D. E. 1959. Hemispherical and woodland canopy photography and the light climate. J. Ecol. 47:103–113.

Farque, L., Sinoquet, H., and Colin, F. 2001. Canopy structure and light interception in *Quercus petraea* seedlings in relation to light regime and plant density. Tree Physiol. 21:1257–1267.

Fernandez, D., and Castrillo, M. 1999. Maize leaf rolling initiation. Photosynthetica 37:493–497.

Fleck, S., Niinemets, Ü., Cescatti, A., and Tenhunen, J. D. 2003. Three-dimensional lamina architecture alters light harvesting efficiency in *Fagus*: A leaf-scale analysis. Tree Physiol. 23:577–589.

Friday, J. B., and Fownes, J. H. 2001. A simulation model for hedgerow light interception and growth. Agric. For. Meteorol. 108:29–43.

Gastellu-Etchegorry, J. P., Demarez, V., Pinel, V., and Zagolski, F. 1996. Modeling radiative transfer in heterogeneous 3-D vegetation canopies. Remote Sens. Environ. 58:131–156.

Gijzen, H., and Goudriaan, J. 1989. A flexible and explanatory model of light distribution and photosynthesis in row crops. Agric. For. Meteorol. 48:1–20.

Gu, L., Baldocchi, D., Verma, S. B., Black, T. A., Vesala, T., Falge, E. M., and Dowty, P. R. 2002. Advantages of diffuse radiation for terrestrial ecosystem productivity. J. Geophys. Res. 107:doi:10.1029/2001JD001242.

Gu, L., Baldocchi, D. D., Wofsy, S. C., Munger, J. W., Michalsky, J. J., Urbanski, S. P., and Boden, T. A. 2003. Response of a deciduous forest to the Mount Pinatubo eruption: Enhanced photosynthesis. Science 299:2035–2038.

Gutschick, V. P. 1991. Joining leaf photosynthesis models and canopy photon-transport models. In: Photon-Vegetation Interaction: Applications in Optical Remote Sensing and Plant Ecology. R. B. Myneni and J. Ross (eds.), pp. 501–535. Berlin: Springer-Verlag.

Hale, S. E., and Edwards, C. 2002. Comparison of film and digital hemispherical photography across a wide range of canopy densities. Agric. For. Meteorol. 112:51–56.

Hikosaka, K., and Hirose, T. 1997. Leaf angle as a strategy for light competition: Optimal and evolutionary stable light extinction coefficient within a leaf canopy. Ecoscience 4:501–507.

Hirose, T., Werger, M. J. A., and van Rheenen, J. W. A. 1989. Canopy development and leaf nitrogen distribution in a stand of *Carex acutiformis*. Ecology 70:1610–1618.

Hodánová, D. 1972. Structure and development of sugar beet canopy. I. Leaf area–leaf angle relations. Photosynthetica 6:401–409.

Holmes, M. G., and Keiller, D. R. 2002. Effects of pubescence and waxes on the reflectance of leaves in the ultraviolet and photosynthetic wavebands: A comparison of a range of species. Plant Cell Environ. 25:85–93.

Jordan, D. N., and Smith, W. K. 1993. Simulated influence of leaf geometry on sunlight interception and photosynthesis in conifer needles. Tree Physiol. 13:29–39.

Kimes, D. S., and Kirchner, J. A. 1982. Radiative transfer model for heterogeneous 3-D scenes. Appl. Opt. 21:4119–4129.

Kimes, D., Gastellu-Etchegorry, J., and Estéve, P. 2002. Recovery of forest canopy characteristics through inversion of a complex 3D model. Remote Sens. Environ. 79:320–328.

Knyazikhin, Y., and Marshak, A. 1991. Fundamental equations of radiative transfer in leaf canopies and iterative methods for their solution. In: Photon-Vegetation Interactions: Applications in Optical Remote Sensing and Plant Ecology. R. B. Myneni and J. Ross (eds.), pp. 9–44. Berlin: Springer-Verlag.

Kursar, T. A., and Coley, P. D. 1992. The consequences of delayed greening during leaf development for light absorption and light use efficiency. Plant Cell Environ. 15: 901–909.

Kuuluvainen, T., and Pukkala, T. 1989. Simulation of within-tree and between-tree shading of direct radiation in a forest canopy: Effect of crown shape and sun elevation. Ecol. Model. 49:89–100.

Kuusk, A. 1992. Absorption profiles of shortwave radiation in a vegetation canopy. Agric. For. Meteorol. 62:191–204.

Kuusk, A., Nilson, T., and Paas, M. 2002. Angular distribution of radiation beneath forest canopies using a CCD-radiometer. Agric. For. Meteorol. 110:259–273.

Lacaze, R., Chen, J. M., Roujean, J.-L., and Leblanc, S. G. 2002. Retrieval of vegetation clumping index using hot spot signatures measured by POLDER instrument. Remote Sens. Environ. 79:84–95.

Lai, C. T., Katul, G. G., Ellsworth, D. S., and Oren, R. 2000. Modelling vegetation-atmosphere CO_2 exchange by a coupled Eulerian-Lagrangian approach. Bound.-Layer Meteorol. 95:91–122.

Lang, A. R. G. 1991. Application of some of Cauchy's theorems to estimation of surface areas of leaves, needles and branches of plants, and light transmittance. Agric. For. Meteorol. 55:191–212.

Law, B. E., Cescatti, A., and Baldocchi, D. D. 2001. Leaf area distribution and radiative transfer in open-canopy forests: Implications to mass and energy exchange. Tree Physiol. 21:777–787.

Leakey, A. D. B., Press, M. C., Scholes, J. D., and Watling, J. R. 2002. Relative enhancement of photosynthesis and growth at elevated CO_2 is greater under sunflecks than uniform irradiance in a tropical rain forest tree seedling. Plant Cell Environ. 25:1701–1714.

Leverenz, J. W., and Hinckley, T. M. 1990. Shoot structure, leaf area index and productivity of evergreen conifer stands. Tree Physiol. 6:135–149.

Leverenz, J. W., Whitehead, D., and Stewart, G. H. 2000. Quantitative analyses of shade-shoot architecture of conifers native to New Zealand. Trees. 15:42–49.

Macfarlane, C., Coote, M., White, D. A., and Adams, M. A. 2000. Photographic exposure affects indirect estimation of leaf area in plantations of *Eucalyptus globulus* Labill. Agric. For. Meteorol. 100:155–168.

Masoni, A., Ercoli, L., Mariotti, M., and Barberi, P. 1994. Changes in spectral properties of ageing and senescing maize and sunflower leaves. Physiol. Plant 91:334–338.

Miller, E. E., and Norman, J. M. 1971. A sunfleck theory for plant canopies. I. Length of sunlit segments along a transect. Agron. J. 63:735–739.

Monsi, M., and Saeki, T. 1953. Über den Lichtfaktor in den Pflanzengesellschaften und seine Bedeutung für die Stoffproduktion. Jap. J. Bot. 14:22–52.

Myneni, R. B. 1991. Modelling radiative transfer and photosynthesis in three-dimensional vegetation canopies. Agric. For. Meteorol. 55:323–344.

Myneni, R. B., and Impens, I. 1985. A procedural approach for studying the radiation regime of infinite and truncated foliage spaces. Part I. Theoretical considerations. Agric. For. Meteorol. 33:323–337.

Myneni, R. B., and Ross, J. (eds.) 1991. Photon-Vegetation Interactions. Applications in Optical Remote Sensing and Plant Ecology. Berlin: Springer-Verlag.

Nandy, P., Thome, K., and Biggar, S. 2001. Characterization and field use of a CCD camera system for retrieval of bidirectional reflectance distribution function. J. Geophys. Res. 106:957–966.

Niinemets, Ü. 1997. Role of foliar nitrogen in light harvesting and shade tolerance of four temperate deciduous woody species. Funct. Ecol. 11:518–531.

Niinemets, Ü., and Fleck, S. 2002. Petiole mechanics, leaf inclination, morphology, and investment in support in relation to light availability in the canopy of *Liriodendron tulipifera*. Oecologia 132:21–33.

Niinemets, Ü., and Kull, O. 1995a. Effects of light availability and tree size on the architecture of assimilative surface in the canopy of *Picea abies*: Variation in shoot structure. Tree Physiol. 15:791–798.

Niinemets, Ü., and Kull, O. 1995b. Effects of light availability and tree size on the architecture of assimilative surface in the canopy of *Picea abies*: Variation in needle morphology. Tree Physiol. 15:307–315.

Niinemets, Ü., Ellsworth, D. S., Lukjanova, A., and Tobias, M. 2001. Site fertility and the morphological and photosynthetic acclimation of *Pinus sylvestris* needles to light. Tree Physiol. 21:1231–1244.

Niinemets, Ü., Cescatti, A., Lukjanova, A., Tobias, M., and Truus, L. 2002a. Modification of light-acclimation of *Pinus sylvestris* shoot architecture by site fertility. Agric. For. Meteorol. 111:121–140.

Niinemets, Ü., Ellsworth, D. S., Lukjanova, A., and Tobias, M. 2002b. Dependence of needle architecture and chemical composition on canopy light availability in three North American *Pinus* species with contrasting needle length. Tree Physiol. 22:747–761.

Niinemets, Ü., Al Afas, N., Cescatti, A., Pellis, A., and Ceulemans, R. 2004a. Petiole length and biomass investments in support modify light-interception efficiency in dense poplar plantations. Tree Physiol. 24:141–154.

Niinemets, Ü., Cescatti, A., and Christian, R. 2004b. Constraints on light interception efficiency due to shoot architecture in broad-leaved *Nothofagus* species. Tree Physiol. (in press).

Niklas, K. J. 1988. The role of phyllotactic pattern as a "developmental constraint" on the interception of light by leaf surfaces. Evolution 42:1–16.

Niklas, K. J. 1989. The effect of leaf-lobing on the interception of direct solar radiation. Oecologia 80:59–64.

Niklas, K. J. 1999. Research review. A mechanical perspective on foliage leaf form and function. New Phytol. 143:19–31.

Nilson, T. 1971. A theoretical analysis of the frequency of gaps in plant stands. Agric. Meteorol. 8:25–38.

Nilson, T. 1992. Radiative transfer in nonhomogeneous plant canopies. Adv. Bioclimat. 1:59–88.

Nilson, T., and Ross, J. 1997. Modeling radiative transfer through forest canopies: Implications for canopy photosynthesis and remote sensing. In: The Use of Remote Sensing in the Modeling of Forest Productivity, H. L. Gholz, K. Nakane, and H. Shimoda (eds.), pp. 23–60. Dordrecht, The Netherlands: Kluwer Academic Publishers.

Norman, J. M., and Jarvis, P. G. 1974. Photosynthesis in Sitka spruce [*Picea sitchensis* (Bong.) Carr.]. III. Measurements of canopy structure and interception of radiation. J. Appl. Ecol. 11:375–398.

Norman, J. M., and Jarvis, P. G. 1975. Photosynthesis in Sitka spruce [*Picea sitchensis* (Bong.) Carr.]. V. Radiation penetration theory and a test case. J. Appl. Ecol. 12:839–878.

Norman, J. M., and Welles, J. M. 1983. Radiative transfer in an array of canopies. Agron. J. 75:481–488.

Norman, J. M., Miller, E. E., and Tanner, C. B. 1971. Light intensity and sunfleck-size distributions in plant canopies. Agron. J. 63:743–748.

North, P. R. J. 1996. Three-dimensional forest light interaction model using a Monte Carlo method. IEEE Transact. Geosci. Remote Sens. 34:946–955.

Oker-Blom, P. 1984. Penumbral effects of within-plant and between-plant shading on radiation distribution and leaf photosynthesis: A Monte-Carlo simulation. Photosynthetica 18:522–528.

Oker-Blom, P., and Kellomäki, S. 1982. Effect of angular distribution of foliage on light absorption and photosynthesis in the plant canopy: Theoretical computations. Agric. Meteorol. 26:105–116.

Oker-Blom, P., and Smolander, H. 1988. The ratio of shoot silhouette to total needle area in Scots pine. For. Sci. 34:894–906.

Olseth, J. A., and Skartveit, A. 1997. Spatial distribution of photosynthetically active radiation over complex topography. Agric. For. Meteorol. 86:205–214.

Palmer, J. W. 1989. The effects of row orientation, tree height, time of year and latitude on light interception and distribution in model apple hedgerow canopies. J. Hort. Sci. 64:137–145.

Palmroth, S., Palva, L., Stenberg, P., and Kotisaari, A. 1999. Fine scale measurement and simulation of penumbral radiation formed by a pine shoot. Agric. For. Meteorol. 95:15–25.

Palva, L., Markkanen, T., Siivola, E., Garam, E., Linnavuo, M., Nevas, S., Manoochehri, F., Palmroth, S., Rajala, K., Ruotoistenmäki, H., Vuorivirta, T., Seppälä, I., Vesala, T., Hari, P., and Sepponen, R. 2001. Tree scale distributed multipoint measuring system of photosynthetically active radiation. Agric. For. Meteorol. 106:71–80.

Pearcy, R. W., and Yang, W. 1996. A three-dimensional crown architecture model for assessment of light capture and carbon gain by understory plants. Oecologia 108:1–12.

Pearcy, R. W., and Yang, W. 1998. The functional morphology of light capture and carbon gain in the Redwood forest understorey plant, *Adenocaulon bicolor* Hook. Funct. Ecol. 12:543–552.

Pinty, B., Gobron, N., Widlowski, J.-L., Gerstl, S. A. W., Verstraete, M. M., Antunes, M., Bacour, C., Gascon, F., Gastellu, J.-P., Goel, N., Jacquemoud, S., North, P., Qin, W., and Thompson, R. 2001. Radiation transfer model intercomparison (RAMI) exercise. J. Geophys. Res. 106:11937–11956.

Pinty, B., Widlowski, J.-L., Gobron, N., Verstraete, M. M., and Diner, D. J. 2002. Uniqueness of multiangular measurements. Part 1. An Indicator of subpixel surface heterogeneity from MISR. IEEE Transact. Geosci. Remote Sens. 40:1560–1573.

Planchais, I., and Sinoquet, H. 1998. Foliage determinants of light interception in sunny and shaded branches of *Fagus sylvatica* L. Agric. For. Meteorol. 89:241–253.

Rakocevic, M., Sinoquet, H., Christophe, A., and Varlet-Grancher, C. 2000. Assessing the geometric structure of a white clover (*Trifolium repens* L.) canopy using 3-D digitising. Ann. Bot. 86:519–526.

Revfeim, K. J. A. 1978. A simple procedure for estimating global daily radiation on any surface. J. Appl. Meteorol. 17:1126–1131.

Roden, J. S., and Pearcy, R. W. 1993. Effect of leaf flutter on the light environment of poplars. Oecologia 93:201–207.

Roderick, M. L., Berry, S. L., and Noble, I. R. 2000. A framework for understanding the relationship between environment and vegetation based on the surface area to volume ratio of leaves. Funct. Ecol. 14:423–437.

Roderick, M. L., Farquhar, G. D., Berry, S. L., and Noble, I. R. 2001. On the direct effect of clouds and atmospheric particles on the productivity and structure of vegetation. Oecologia 129:21–30.

Ross, J. 1981. The Radiation Regime and Architecture of Plant Stands. The Hague: Dr. W. Junk.

Ross, J., Sulev, M., and Saarelaid, P. 1998. Statistical treatment of the PAR variability and its application to willow coppice. Agric. For. Meteorol. 91:1–21.

Shabanov, N. V., Knyazikhin, Y., Baret, F., and Myneni, R. B. 2000. Stochastic modeling of radiation regime in discontinuous vegetation canopies. Remote Sens. Environ. 74:125–144.

Sinclair, R., and Thomas, D. A. 1970. Optical properties of leaves of some species in arid South Australia. Am. J. Bot. 18:261–273.

Smith, W. K., and Carter, G. A. 1988. Shoot structural effects on needle temperatures and photosynthesis in conifers. Am. J. Bot. 75:496–500.

Smith, W. K., Knapp, A. K., and Reiners, W. A. 1989. Penumbral effects on sunlight penetration in plant communities. Ecology 70:1603–1609.

Smolander, H., Oker-Blom, P., Ross, J., Kellomäki, S., and Lahti, T. 1987. Photosynthesis of a Scots pine shoot: Test of a shoot photosynthesis model in a direct radiation field. Agric. For. Meteorol. 39:67–80.

Spitters, C. J. T. 1986. Separating the diffuse and direct component of global radiation and its implications for modeling canopy photosynthesis. Part II. Calculation of canopy photosynthesis. Agric. For. Meteorol. 38:231–242.

Stenberg, P. 1995. Penumbra in within-shoot and between-shoot shading in conifers and its significance for photosynthesis. Ecol. Model. 77:215–231.

Stenberg, P., Kuuluvainen, T., Kellomäki, S., Grace, J. C., Jokela, E. J., and Gholz, H. L. 1994. Crown structure, light interception and productivity of pine trees and stands. In: Environmental Constraints on the Structure and Productivity of Pine Forest Ecosystems: A Comparative Analysis. H. L. Gholz, S. Linder, and R. E. McMurtrie (eds.), pp. 20–34. Copenhagen: Munksgaard International Booksellers and Publishers.

Stenberg, P., Smolander, H., Sprugel, D. G., and Smolander, S. 1998. Shoot structure, light interception, and distribution of nitrogen in an *Abies amabilis* canopy. Tree Physiol. 18:759–767.

St-Jacques, C., Labrecque, M., and Bellefleur, P. 1991. Plasticity of leaf absorptance in some broadleaf tree seedlings. Bot. Gaz. 152:195–202.

Takenaka, A. 1994. Effects of leaf blade narrowness and petiole length on the light capture efficiency of a shoot. Ecol. Res. 9:109–114.

Tanner, V., and Eller, B. M. 1986. Veränderungen der spektralen Eigenschaften der Blätter der Buche (*Fagus sylvatica* L.) von Laubaustrieb bis Laubfall. Allg. Forst Jagdztg. 157:108–117.

Thomas, S. C., and Winner, W. E. 2000. A rotated ellipsoidal angle density function improves estimation of foliage inclination distributions in forest canopies. Agric. For. Meteorol. 100:19–24.

Thornton, P. E., Hasenauer, H., and White, M. A. 2000. Simultaneous estimation of daily solar radiation and humidity from observed temperature and precipitation: An application over complex terrain in Austria. Agric. For. Meteorol. 104:255–271.

Tian, Y. Q., Davies-Colley, R. J., Gong, P., and Thorrold, B. W. 2001. Estimating solar radiation on slopes of arbitrary aspect. Agric. For. Meteorol. 109:67–74.

Turner, D. P., Cohen, W. B., Kennedy, R. E., Fassnacht, K. S., and Briggs, J. M. 1999. Relationships between leaf area index and TM spectral vegetation indices across three temperate zone sites. Remote Sens. Environ. 70:52–68.

Utsugi, H. 1999. Angle distribution of foliage in individual *Chamaecyparis obtusa* canopies and effect of angle on diffuse light penetration. Trees 14:1–9.

Valladares, F., and Brites, D. 2004. Leaf phyllotaxis: Does it really affect light capture? Plant Ecol. (in press).

Valladares, F., and Pearcy, R. W. 1998. The functional ecology of shoot architecture in sun and shade plants of *Heteromeles arbutifolia* M. Roem., a Californian chaparral shrub. Oecologia 114:1–10.

Valladares, F., and Pearcy, R. W. 1999. The geometry of light interception by shoots of *Heteromeles arbutifolia*: Morphological and physiological consequences for individual leaves. Oecologia 121:171–182.

Valladares, F., and Pugnaire, F. I. 1999. Tradeoffs between irradiance capture and avoidance in semi-arid environments assessed with a crown architecture model. Ann. Bot. 83:459–469.

van Gardingen, P. R., Jackson, G. E., Hernandez-Daumas, S., Russell, G., and Sharp, L. 1999. Leaf area index estimates obtained for clumped canopies using hemispherical photography. Agric. For. Meteorol. 94:243–257.

Vesala, T., Markkanen, T., Palva, L., Siivola, E., Palmroth, S., and Hari, P. 2000. Effect of variations of PAR on CO_2 exchange estimation for Scots pine. Agric. For. Meteorol. 100:337–347.

Wagner, S. 1998. Calibration of grey values of hemispherical photographs for image analysis. Agric. For. Meteorol. 90:103–117.

Wagner, S. 2001. Relative radiance measurements and zenith angle dependent segmentation in hemispherical photography. Agric. For. Meteorol. 107:103–115.

Walter-Shea, E. A., and Norman, J. M. 1991. Leaf optical properties. In: Photon-Vegetation Interactions: Applications in Optical Remote Sensing and Plant Ecology. R. B. Myneni and J. Ross (eds.), pp. 229–252. Berlin: Springer-Verlag.

Wang, H., and Baldocchi, D. D. 1989. A numerical model for stimulating the radiation regime within a deciduous forest canopy. Agric. For. Meteorol. 46:313–337.

Wang, S., Chen, W. J., and Cihlar, J. 2002. New calculation methods of diurnal distribution of solar radiation and its interception by canopy over complex terrain. Ecol. Model. 155:191–204.

Wang, Y. P., and Jarvis, P. G. 1990. Influence of crown structural properties on PAR absorption, photosynthesis, and transpiration in Sitka spruce: Application of a model (MAESTRO). Tree Physiol. 7:297–316.

Wang, Y.-P., and Leuning, R. 1998. A two-leaf model for canopy conductance, photosynthesis and partitioning of available energy. I. Model description and comparison with a multi-layered model. Agric. For. Meteorol. 91:89–111.

Weiss, M., Baret F., Smith, G. J., Jonckheere, I., and Coppin, P. 2004. Review of methods for in situ leaf area index (LAI) determination. Part II. Estimation of LAI, errors and sampling. Agric. For. Meteorol. 121:37–53.

Welles, J. M. 1990. Some indirect methods for evaluating canopy structure. Remote Sens. Environ. 5:31–43.

Werner, C., Ryel, R. J., Correia, O., and Beyschlag, W. 2001. Structural and functional variability within the canopy and its relevance for carbon gain and stress avoidance. Acta Oecol. 22:129–138.

Part 3

Sunlight Processing

4
Chloroplast to Leaf

Neil R. Baker, Donald R. Ort, Jeremy Harbinson,
and John Whitmarsh

Introduction

A great deal is now understood about the photosynthetic reactions and processes occurring within chloroplasts that use light energy to convert inorganic molecules into organic compounds. From this knowledge base about chloroplast function is emerging an understanding of how chloroplasts operate within, and interact with, the environment of a leaf. Our goal in this chapter is to first overview energy capture and processing within the chloroplast. We then consider how chloroplast function is affected and regulated by different aspects of the environment within the leaf. We hope to provide a conceptual framework for scaling light energy processing from the chloroplast to the leaf level as well as highlight the gaps in understanding that will need to be filled to complete the task. Since scaling by its nature crosscuts a diverse set of disciplines over which none of us can be expert, we have tried to enhance the accessibility of our topic by preferentially referencing reviews and overviews that integrate the primary literature.

The Chloroplast

Photosynthetic Energy Capture and Conversion

Photosynthetic membranes use light energy to produce adenosine triphosphate (ATP) and reduced nicotinamide adenine dinucleotide phosphate (NADPH). Light is captured by an antennae array containing two classes of pigments, chlorophylls and carotenoids, which are responsible for the absorption of light that drives photosynthesis in higher plants. Chlorophyll is the dominant pigment and strongly absorbs red and blue light.

To convert the transient energy of light into stable chemical energy, the photosynthetic apparatus performs a series of energy-transforming reactions. The process is initiated by absorption of a photon by a chlorophyll or carotenoid molecule that converts light energy to an excited electronic state of the pigment molecule. These pigment molecules are arranged in the photosynthetic membranes of chloroplasts in groups of 250 to 300 bound to specialized proteins, which provide a scaffolding for the precise arrangement of each molecule within the antenna. Because of the proximity of other pigment molecules with the same or similar energy states, the excited state is rapidly transferred over the antenna system. Under optimum conditions, over 90% of the light captured by this antenna is successfully used to drive the next step in photosynthesis—primary charge separation.

Photosystems I (PSI) and II (PSII) of plants are the sites of the primary photochemical reactions of photosynthesis, which involve a separation of a positively charged molecule from a negatively charged one. Each photosynthetic membrane in a chloroplast contains thousands of these two photosystems, which are themselves complexes of multiple proteins embedded in the membranes. In PSII, the excited state energy is transferred to a specialized group of six chlorophyll molecules associated with proteins at the core of PSII. In the primary photochemical reaction a negatively charged electron is transferred from an excited chlorophyll molecule to an electron-accepting pigment creating positively and negatively charged molecules that are adjacent to each other. The negatively charged electron cannot flow back to the positively charge molecule and this separation of charge "captures" the energy and drives all subsequent electron transfer reactions in the photosynthetic membrane (Figure 4.1). In PS II the energy is used to remove electrons from water and to add electrons, as well as protons, to plastoquinone (PQ).

In PSI, the energy captured in the primary charge separation drives the oxidation of plastocyanin, a soluble copper-containing protein located in the thylakoid lumen, and the reduction of ferredoxin, a soluble iron-containing protein located in the stroma (Fig. 4.1). As with PSII, the reactions of PSI produce an electric potential across the photosynthetic membrane and generate a strong re-

FIGURE 4.1. Schematic drawing of the photosynthetic apparatus. The photosynthetic membrane contains the major protein complexes and pigments responsible for light absorption and photosynthetic electron and proton tranfer. Electron transport driven by excitation of PSI and PSII results in the reduction of NADP to NADPH and the accumulation of protons in the lumen of the thylakoid, which is used drive ATP synthesis by the ATP synthase. CO_2 assimilation is shown as a three-stage cycle. Carboxylation: Rubisco catalyses the covalent linkage of a molecule of CO_2 to ribulose 1,5-bisphosphate (RuBP) forming phosphoglyceric acid (PGA). Reduction: Energy in the form of ATP and NADPH is used to form triose phosphate. Regeneration: ATP is used to regenerate RuBP for carboxylation. Alternatively, in the presence of oxygen RuBP can be oxygenated by Rubisco to form PGA and phosphoglycolate, the first step in the process of photorespiration. Cyt bf, cytochrome bf complex; FNR, Ferredoxin-NADP reductase; LHCII, light-harvesting antennae complexes of PSII; PC, plastocyanin; PQ and PQH_2, plastoquinone and plastoquinol (reduced plastoquinone); PSII and PSI, photosystems II and I; P_{680}, reaction center chlorophyll of PSII; P_{700}, reaction center chlorophyll of PSI; Q_A, primary quinone electron acceptor of PSII; Q_B, secondary quinone acceptor of PSII (bound plastoquinone).

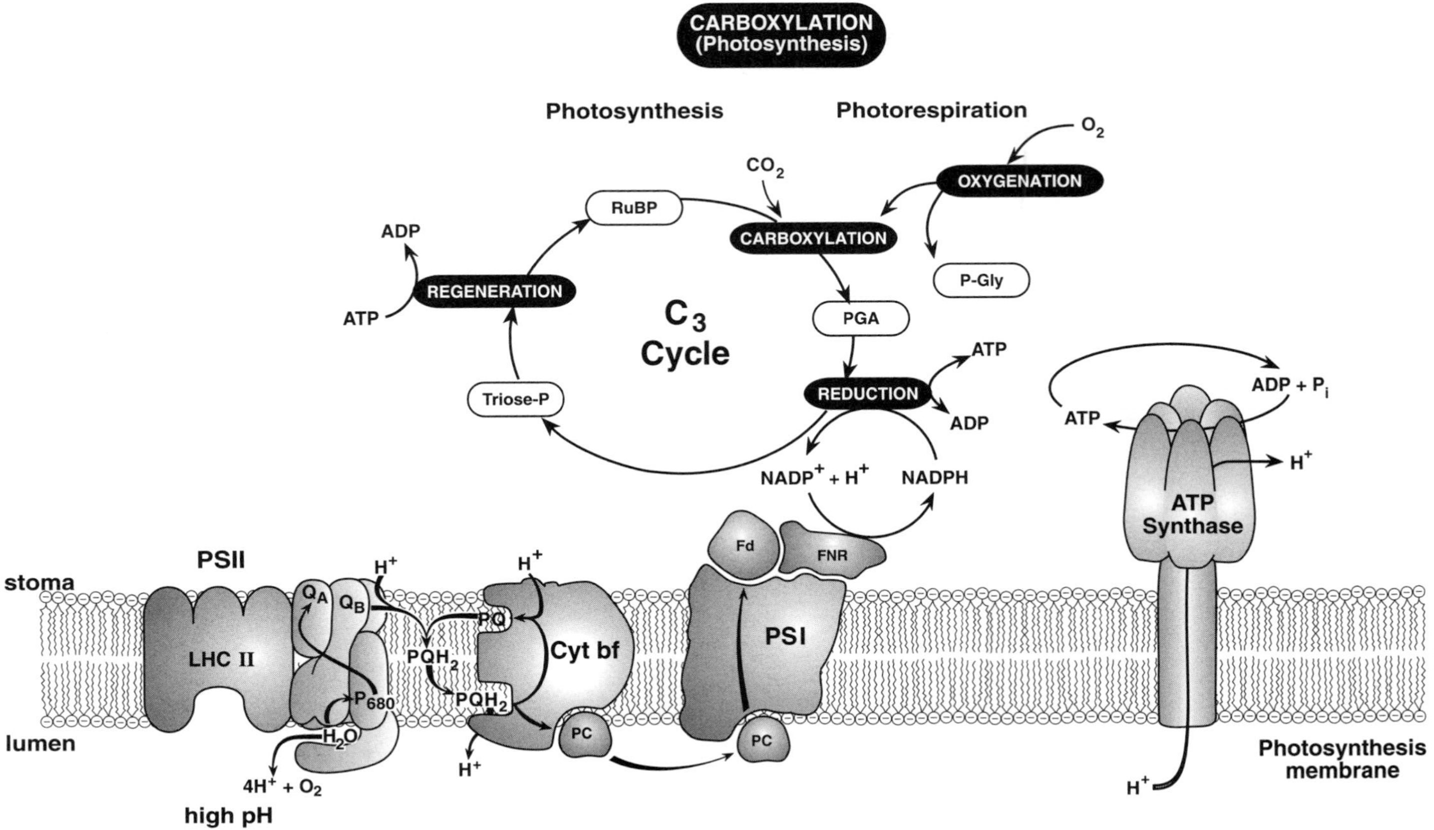

CARBOXYLATION
(Photosynthesis)
Photosynthesis
Photorespiration
O_2
OXYGENATION
P-Gly
CO_2
CARBOXYLATION
PGA
REDUCTION
ATP
ADP
C_3
Cycle
RuBP
REGENERATION
Triose-P
ADP
ATP
NADPH
$NADP^+ + H^+$
ATP
Synthase
ATP
$ADP + P_i$
H^+
H^+
Photosynthesis
membrane
FNR
PSI
Fd
PC
Cyt bf
PC
H^+
PQ
PQH_2
PQH_2
H^+
PSII
LHC II
Q_A
Q_B
H^+
P_{680}
H_2O
$4H^+ + O_2$
high pH
stoma
lumen

ductant. PSI differs from PSII in that it does not deposit protons into the lumen and uses its energy to reduce NADP producing NADPH.

The light-driven electron and proton transfer reactions of the two photosystems are interconnected through the activity of cytochrome bf complex, which catalyzes the energetically downhill reaction of reducing plastocyanin. Plastoquinone serves as a mobile hydrogen carrier within the membrane, transporting protons and electrons from PSII to the cytochrome bf complex, while plastocyanin serves as a mobile electron carrier, transporting electrons from the cytochrome bf complex to PSI. In addition to linking the activity of PSII and PSI, the cytochrome bf complex plays a central role in energy transformation and storage by converting energy available in reduced plastoquinone (PQH_2) into in a transmembrane pH difference as well as an electric potential difference. As shown in Figure 4.1, this is accomplished as cytochrome bf complex oxidizes PQH_2 at a site near the inside of the membrane vesicle, resulting in the release of protons into the vesicle lumen. In addition to oxidizing plastoquinone, the cytochrome bf complex also reduces plastoquinone at a second site that is near the outside of the membrane, acquiring the protons associated with its reduction from the stroma. The net result of these reactions, known as the cytochrome bf Q-cycle, is the transfer of protons from the stroma to the lumen, creating a transmembrane pH difference and, depending upon the extent of counter ion movement, an electric potential difference. The energy stored in the pH difference and electrical potential is used for the energy-requiring reaction of converting ADP to ATP by the addition of a phosphate group by the ATP-synthase inserted in the photosynthetic membrane (see Fig. 4.1).

Photosynthetic Energy is Used to Drive Carbon Metabolism

Although the energy stored in ATP and NADPH is chemically stable, plants do not accumulate high levels of these chemicals. Instead ATP and NADPH function as a rapidly turning over energy currency used in large part for the biosynthesis of carbohydrates from CO_2 and water. This intricate biosynthetic pathway, known as the C_3 photosynthetic carbon reduction cycle (C_3 cycle), takes place in the stroma of chloroplasts and involves more than a dozen different enzymes.

The C_3 cycle begins with a carboxylation reaction catalyzed by Rubisco (ribulose bisphosphate carboxylase/oxygenase) (see Fig. 4.1) in which CO_2 is attached to the 5-carbon-acceptor molecule RuBP. The reaction product is not stable and immediately splits into two 3-carbon molecules (PGA) giving the C_3 cycle its name. The next stage of the C_3 cycle requires energy in the form of both ATP and NADPH to form triose phosphate (glyceraldhyde-3-phosphate). Figure 4.1 illustrates that triose phosphate is the principal branch point within the C_3 cycle. Some of the triose phosphate is transported out of the chloroplast and used in the synthesis of sucrose. Triose phosphate is used within the chloroplast for starch synthesis and is reinvested into the C_3 cycle to regenerate RuBP, thereby completing the photosynthetic carbon reduction cycle.

In addition to the carboxylation of RuBP by CO_2, Rubisco will also catalyze its oxidation by atmospheric O_2 (oxygenation) to yield one molecule of PGA and a molecule of a 2 carbon compound. This oxygenation reaction creates a significant inefficiency in the photosynthetic process because the 2-carbon compound cannot enter the C_3 cycle. In a typical C_3 crop, the rate of Rubisco-catalyzed oxygenation is about 20% of the rate of Rubisco-catalyzed CO_2 fixation. To compensate for the oxygenation of RuBP by Rubisco, a metabolic pathway evolved in plants that recovers the 2C carbon skeletons that are diverted from the C_3 cycle by the oxygenation reaction. However, this scavenging operation, known as photorespiration, is energetically expensive. In the biochemical reactions of photorespiration, two of the 2-carbon molecules are converted into one 3-carbon molecule with the release of one CO_2 and at the expense of one ATP molecule. The cell thus succeeds in returning 75% of photorespiratory carbon to the C_3 cycle, with the remainder released as CO_2.

While photorespiration makes the best of a wasteful situation caused by the oxygenase activity of Rubisco, an alternative "strategy" taken by plants in the course of evolution is to prevent, or greatly reduce, Rubisco's oxygenase activity by exploiting the competition between CO_2 and O_2 as substrates of Rubisco. Plants such as maize and sugar cane greatly suppress or completely inhibit the oxygenation reaction of Rubisco by concentrating CO_2 in specialized leaf cells that contain Rubisco. These species are known as C_4 plants because the initial carboxylation reaction produces a 4-carbon acid. They have a unique leaf anatomy with two distinct photosynthetic cell types in which chloroplast-containing mesophyll cells surround chloroplast-containing bundle sheath cells, which in turn encircle the vascular bundles of leaf. A key feature of C_4 photosynthesis is that the initial fixation of CO_2 takes place in mesophyll cells and involves the carboxylation of the 3-carbon molecule PEP by PEP carboxylase. Unlike Rubisco, PEP carboxylase does not bind O_2 in competition with CO_2. The 4-carbon product of this reaction is transported to the bundle sheath cell and decarboxylated to release CO_2. The decarboxylation of the four carbon acids results in a significant elevation of the CO_2 concentration in the bundle sheath cell chloroplast where Rubisco and the other enzymes of the C_3 cycle are localized. The elevated concentration of CO_2 effectively competes with oxygen virtually eliminating photorespiration. The 3-carbon product of the decarboxylation reaction is transported back to the mesophyll cell so that the CO_2 acceptor PEP can be regenerated. Although the C_4 photosynthetic pathway effectively suppresses photorespiration, it is important to recognize that there are substantial energetic costs. In effect, the C_4 cycle is a light (i.e., energy)-driven CO_2 pump, concentrating CO_2 in the bundle sheath cells, the site of the C_3 cycle in C_4 plants.

Chloroplasts Use Photosynthetic Energy to Perform Other Important Processes

Although photosynthesis is most often considered in the context of CO_2 reduction to form sugars, a significant amount of reducing power and ATP are used within the chloroplasts for other processes that are essential to the growth and development of plants.

In many non-legumous herbaceous plants, most of the nitrate taken up by the roots is transported to the leaves where it is converted in the cytoplasm to nitrite and imported into the chloroplast. The reduction of nitrite to ammonia in the chloroplast consumes, per molecule, a third more reducing power than the reduction of CO_2 to carbohydrate. Nitrite reductase is located exclusively in the chloroplast and takes electrons directly from ferredoxin, thus competing with NADPH formation. The enzyme glutamine synthetase in the chloroplast transfers the newly formed NH_4^+ at the expense of ATP to glutamate forming glutamine, an important amino acid. From glutamine, the reduced N can be channeled into a host of other amino acids in the chloroplasts. The amino acids cysteine and methionine contain sulfur and the reduction of sulfate to sulfide in the chloroplast requires, on a per mole basis, 50% more energy than carbon assimilation. This multistep process also uses electrons that come directly from ferredoxin to produce sulfide, which is incorporated into cysteine. While sulfate reduction is very energy intensive the investment of photosynthetic energy is substantially greater for nitrate reduction.

Chloroplasts also synthesize proteins, lipids, photosynthetic pigments, various phytohormones and many other biochemicals essential to the functioning of the plant. These various reductive and biosynthetic processes have widely ranging energy requirements and use ATP and NADPH in different amounts and different ratios (Table 4.I). This fact implies that chloroplast energy transduction has mechanisms to adjust the ratio of ATP to NADPH production in order to meet different needs. Higher demands for ATP production could be meet by cyclic electron flow, electron flow to O_2, and perhaps by chlororespiration. If chloroplasts have mechanisms to disengage the extra proton pumping associated with Q-cycle operation of the cytochrome bf complex, this would enrich NADPH production relative to ATP synthesis. It is also important to consider that any tran-

TABLE 4.1. Requirements for ATP, NADPH, and electron transfer from ferredoxin in metabolic processes in leaves.

Metabolic Process	ATP Requirement (molecules)	NADPH Requirement (molecules)	Electrons Transferred from Ferredoxin
C_3 photosynthetic reduction cycle	3	2	
C_4 photosynthetic carbon cycle in mesophyll cells	2	0	
Assimilation of a CO_2 molecule in a C_4 leaf	5	2	
Photorespiratory cycle	3.3	2	
Reduction of nitrate to ammonia (via nitrite)		1	6
Reduction of sulfate to sulfide in the chloroplast	1		8

sient needs might also be met via exchange reactions with mitochondria that can liberate stored photosynthetic energy and deliver it to the chloroplast as either ATP or NADPH.

The Leaf Environment

Responses to the leaf light environment. The typical response of CO_2 assimilation in a leaf to increasing light and the changes in the efficiency with which absorbed photons are utilized for photosynthesis (quantum yield, Φ) are depicted in Figure 4.2. At low light photosynthesis increases linearly with increasing light absorption (region A in Figure 4.2) and the leaf is operating at its maximum quantum yield (Φ_{max}). As light absorption increases further the relationship between photosynthesis and absorbed light becomes non-linear (region B in Figure 4.2) and the quantum yield decreases. Eventually photosynthesis becomes light-saturated with no further increases in photosynthesis occurring with increasing light absorption (region C in Fig. 4.2) and consequently the quantum yield continues to decrease with increasing light absorption. However, the actual response of a leaf to increasing irradiance is more complex because factors that

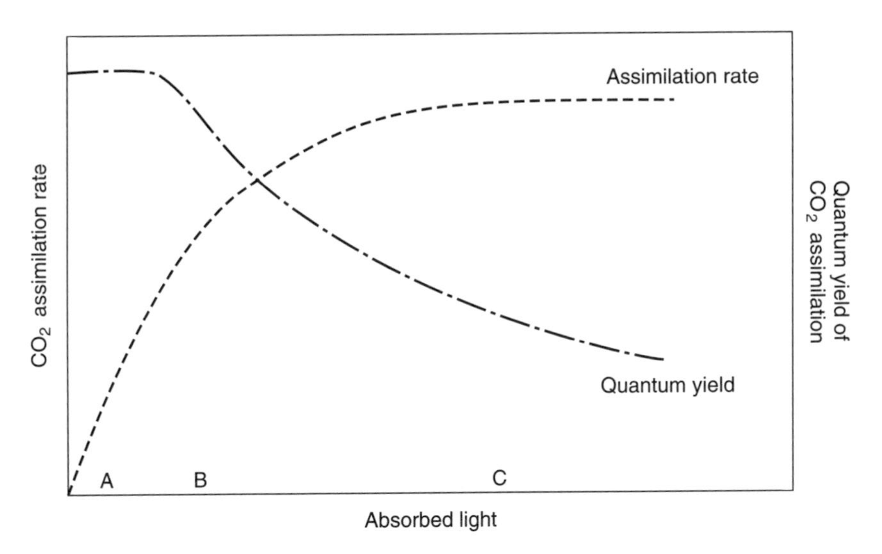

FIGURE 4.2. Response of leaf photosynthesis to increasing absorbed light. The changes in rate of CO_2 assimilation and the quantum yield of CO_2 assimilation are shown as a function of increasing light absorption. The response of CO_2 assimilation consists of 3 phases. A: CO_2 assimilation increases linearly with increasing light absorption and the quantum yield is maintained constant. B: CO_2 assimilation continues to increase, but nonlinearly with light absorption and quantum yield decreases. C: CO_2 assimilation has become light-saturated and remains constant with increasing light absorption, although the quantum yield continues to decrease.

determine the effective use of light for photosynthesis change as light absorption increases.

Quantum Efficiency Under Limiting Light

When leaves are operating at Φ_{max}, the rate of photosynthetic electron transport is dependent upon the appropriate delivery of excitation from absorbed photons to the reaction centers of PSI and PSII. Theoretically, in C_3 leaves the upper limit of Φ_{max} is 0.125, meaning that 8 moles of photons are required to reduce 1 mole of CO_2. This assumes that photochemistry at both PSI and PSII has a perfect quantum yield and that all photons absorbed by the photosynthetic pigments have only one fate, the generation of irreversible charge separation that results in linear electron transport through both photosystems. Empirical estimations of Φ_{max} for both CO_2 assimilation and O_2 evolution (which has an expected 1:1 stoichiometric relationship with CO_2 assimilation in the absence of photorespiration) are always below the theoretical maximum of 0.125 (Genty and Harbinson 1996, Singsaas et al. 2001). In an extensive study of Φ_{max} for O_2 evolution in leaves of 42 species from contrasting habitats, a highly conserved value of 0.106 was found (Björkman and Demmig 1987) suggesting that leaves can operate quite close to their theoretical Φ_{max} and are capable of maintaining a closely balanced distribution of absorbed light between PSI and PSII. Such observed high quantum yields also imply that, at low light in the absence of photorespiration, CO_2 assimilation is the major sink for ATP and NADPH in mature leaves. Clearly, consumption of ATP, NADPH and electrons from ferredoxin by other metabolic activities, such as nitrate assimilation and sulfate reduction as well as electron transfer from PSI directly to oxygen must be less than 15% of the total flux involved in CO_2 assimilation under limiting light. It may be that these other energy-requiring processes are preferentially activated at higher, nonlimiting light levels.

Quantum Efficiency Under Partially Limiting Light

The decrease in quantum yield of photosynthesis from Φ_{max} as light absorption by the leaf increases indicates that delivery of excitation energy to PSI and PSII reaction centers is then not the sole factor limiting electron transport. As light intensity increases, the quantum yield of electron transport decreases, which can be attributed to two processes (Fig. 4.3); one is a direct consequence of electron transport components becoming reduced, and the second is due to the activation of a mechanism that diverts excitation energy in the light-harvesting antennae from photochemistry and converts it to heat. The first mechanism operates to reduce the quantum yield of both PSI and PSII electron transport. In PSII this is due to an increase in the proportion of quinone electron acceptors of PSII that are in the reduced state. Reduction of the primary quinone acceptor of PSII, Q_A, effectively prevents further primary charge separation at the PSII reaction center. As light intensity increases, the proportion of PSI reaction centers, P_{700}, in

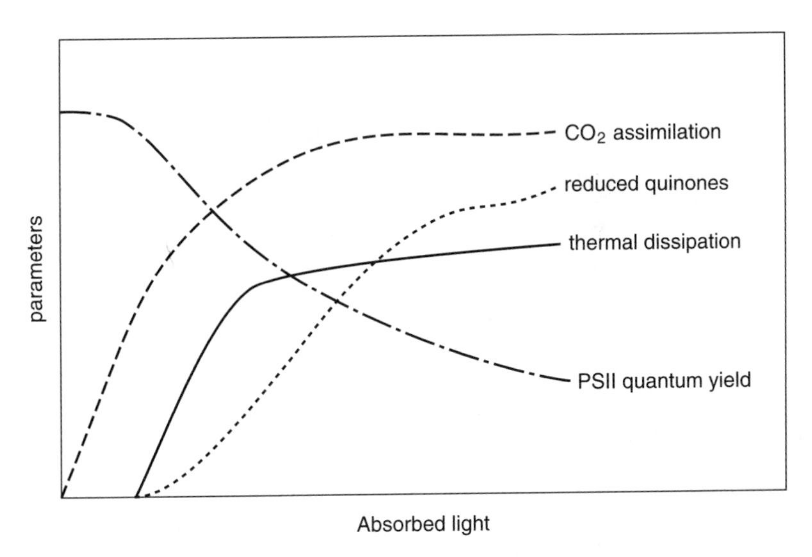

FIGURE 4.3. Thermal dissipation of excitation energy and reduction of quinone electron acceptors by PSII in a leaf as light absorption increases. Changes in the efficiency of dissipating absorbed light as heat (thermal dissipation), the reduction of PSII quinone electron acceptors (reduced quinones), the quantum yield of PSII electron transport, and the rate of CO_2 assimilation in a leaf as light absorption increases.

the oxidized state increases. When P_{700} is oxidized, PSI cannot participate in electron transfer reactions. Both the reduction of Q_A in PSII and the oxidation of P_{700} in PSI are consequences of a restriction on electron transfer between PSII and PSI imposed at the cytochrome bf complex. Initially with increasing light, the limiting step in electron transport appears to be the oxidation of plastoquinone by the cytochrome bf complex. Restriction of electron flux through the cytochrome bf complex is under physiological control mediated via a light-induced decrease in the pH of the thylakoid lumen. However, with further light increases and as photosynthesis moves toward light-saturation, CO_2 assimilation becomes increasingly limited by the metabolic capacity to consume ATP and NADPH. This results in a further increase in the proportion of reduced electron transport components thereby decreasing the probability of PSII photochemistry. The second process causing a decrease in the quantum yield in partially limiting light is the increase in the amount and proportion of excitation energy absorbed by the PSII antennae that is diverted directly to heat through a thermal dissipation mechanism. This thermal dissipation decreases the rate at which photons are delivered to PSII reaction centers helping prevent overreduction of quinone acceptors.

The light-dependent increases in thermal dissipation of excitation energy are triggered by a decrease in the pH of the thylakoid lumen (Ort 2001). With increasing absorbed light, the rate of proton transfer from the stroma to the lumen increases and the lumen becomes increasingly acidified. This results in the activation of a membrane-bound violaxanthin de-epoxidase, which converts violax-

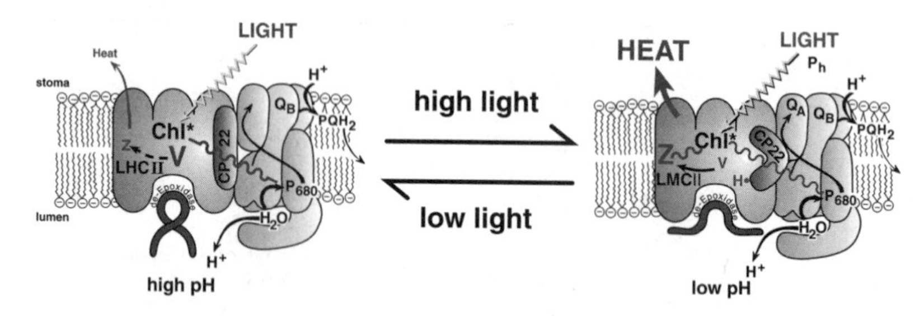

FIGURE 4.4. Mechanism for pH-dependent thermal dissipation of light absorbed by PSII. At low light, the rate of proton accumulation in the lumen is low and the thermal dissipation of excitation in the PSII antennae via zeaxanthin is low. As light absorption increases, proton transfer into the thyalkoid lumen increases and induces a conformational change in the light-harvesting chlorophyll protein, CP22 and an increase in the zeaxanthin (Z) content of the antennae due to activation of a de-epoxidase in the thylakoid lumen that converts violaxanthin to zeaxanthin. This results in a large increase in the rate of thermal dissipation of excitation energy by zeaxanthin thereby lowering the rate of proton accumulation. LHCII, light-harvesting antennae complexes of PSII; PQH_2, plastoquinol (reduced plastoquinone); P_{680}, reaction center chlorophyll of PSII; Q_A, primary quinone electron acceptor of PSII; Q_B, secondary quinone acceptor of PSII (bound plastoquinone).

anthin, a xanthophyll pigment bound to the PSII light-harvesting antennae, to zeaxanthin (Fig. 4.4). Excited chlorophyll molecules can transfer energy to zeaxanthin, which will then dissipate this excitation energy as heat. Accumulation of zeaxanthin has also been proposed to result in aggregation of the light-harvesting chlorophyll protein complexes associated with PSII (LHCII), which increases the rate of thermal dissipation of excitation energy. Protonation of a minor light-harvesting chlorophyll protein associated with PSII, called CP22 also appears to have a regulatory role in switching on the pH-dependent quenching of excitation in the PSII antenna.

Energy dissipation in PSI is much less studied than in PSII. However, it is clear that oxidized P_{700}, P_{700}^+, is a strong quencher of excited states in the PSI light-harvesting antennae. Although the photophysical mechanism of this quenching of chlorophyll excited states remains a matter of debate, it does suggest a reasonable mechanism to balance PSI light energy utilization with zeaxanthin/ΔpH-dependent energy dissipation in PSII. Thus, when PSI absorbs more light quanta than it receives electrons from PSII, P_{700} becomes oxidized and stays oxidized until an electron comes along from PSII. In this way thermal energy dissipation in PSI by P_{700}^+ quenching tracks the ΔpH-dependent regulation of PSII thermal energy dissipation.

Quantum Efficiency at Light Saturation and Beyond

Although thermal dissipation of absorbed light is an important photoprotective mechanism in helping prevent the overreduction of quinone acceptors and increases with increasing light intensity, the rate of increase in the proportion of absorbed light released through thermal dissipation decreases as the photosynthetic rate approaches light saturation (see Fig. 4.3). In healthy, mature leaves exposed to full sunlight, which is generally well in excess of the light required to saturate leaf photosynthesis, the quinone acceptors rarely become fully reduced, and in many cases remain substantially oxidized. The maintenance of the quinone acceptors in a partially oxidized state is important to prevent photoinactivation and photodamage to the PSII reaction centers. When PSII is excited, there is a probability that damage to the D1 protein of the reaction center will occur. This probability is low at low excitation levels and increases with increasing light absorption and a decreasing ability to perform photochemistry. Under normal physiological conditions, the rate of photodamage does not exceed the capacity to repair the damage. However, under unusual conditions, when the PSII quinone acceptors do become highly reduced, the rate of damage can exceed considerably the rate of repair and consequently there is a loss of functional PSII reaction centers (see Chapter 9).

An additional protection to help prevent the PSII quinone acceptors from becoming reduced under saturating light are alternative sinks for electrons, other than CO_2 assimilation. For instance, it has been suggested that in saturating light up to 30% of the total photosynthetic electron flux can be involved in the non-photorespiratory reduction of O_2 (Asada 1999). PSI can directly reduce O_2 to produce superoxide radicals that are rapidly converted within the chloroplast to hydrogen peroxide by superoxide dismutase. The hydrogen peroxide is then detoxified to produce water by ascorbate peroxidase, which oxidizes ascorbate to monodehydroascorbate (MDA) (Fig. 4.5). This photosynthetic electron flow from water at PSII through PSI to produce the ascorbate peroxidase-generated water is often termed the water-water cycle. For continuous electron flux through this cycle, MDA must be converted back to ascorbate. This is achieved by an electron flux from ferredoxin (see Fig. 4.5). Consequently, operation of the water-water cycle will serve as a sink for electrons, which are utilized for reduction of oxygen and MDA. Although there is not yet definitive evidence for the operation of the water-water cycle at high rates in leaves, a substantial body of circumstantial support exists in the literature (Ort and Baker 2002).

If the water-water cycle does play an important role in protecting PSII reaction centers at high light intensities, it would appear that the process must be induced as light intensity increases. As discussed above, at limiting light levels when leaves are operating at Φ_{max}, there is little possibility of significant electron flux to electron acceptors other than CO_2 since Φ_{max} is often close to the theoretical maximum quantum yield of CO_2 assimilation.

Responses of Photosynthetic Light-Use Efficiency to Environmentally-Imposed Limitations

In field situations, leaves are frequently exposed to conditions that reduce the capacity for photosynthesis and consequently decrease the quantum efficiency of

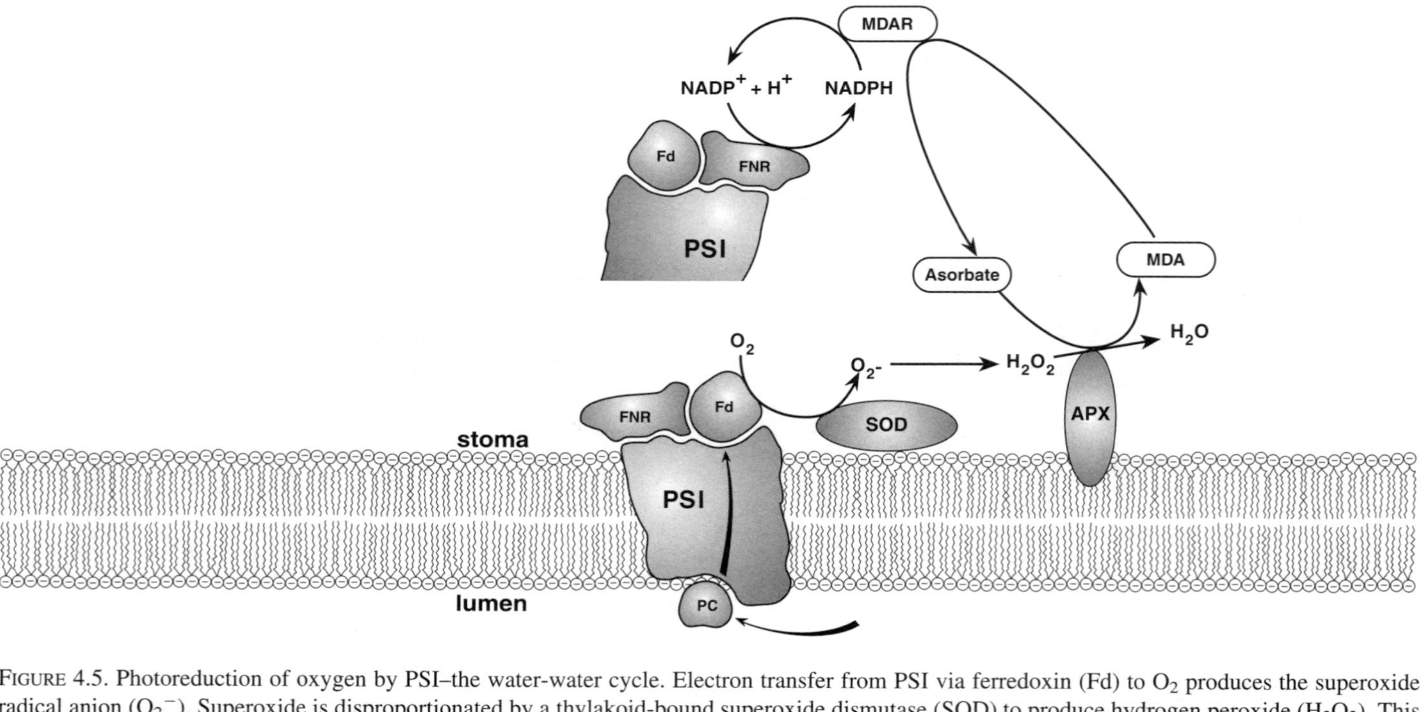

FIGURE 4.5. Photoreduction of oxygen by PSI–the water-water cycle. Electron transfer from PSI via ferredoxin (Fd) to O_2 produces the superoxide radical anion (O_2^-). Superoxide is disproportionated by a thylakoid-bound superoxide dismutase (SOD) to produce hydrogen peroxide (H_2O_2). This H_2O_2 is reduced by ascorbate to form water in a reaction catalysed by ascorbate peroxidase (APX). Monodehydroascorbate (MDA) is also produced in this reaction, which provides an additional sink for electrons as ascorbate is regenerated by electron transfer from ferredoxin in a reaction mediated by monodehydroascorbate reductase (MDAR). Electrons used for the reduction of O_2 by PSI originate from the photooxidation of water by PSII (see Fig. 4.1) and consequently this process is often termed the water-water cycle.

photosynthesis. Many crops, even when irrigated, exhibit maximum photosynthetic capacity during mid-morning and then photosynthesis declines at mid-day and the early afternoon as stomates partially close in response to declining water status. Even under conditions that may not generally be considered stressful, partial stomata closure can substantially restrict CO_2 entry into leaves, which results in a decrease in CO_2 assimilation, an increased reduction of the PSII quinone acceptors and an increase in thylakoid lumen pH. The acidification of the lumen thus triggers an increase in the rate of thermal dissipation of PSII excitation energy at a lower intensity than had stomata been fully open. However, in some environmental conditions, photosynthetic and photorespiratory metabolism cannot utilize fully the excitation energy reaching the PSII reaction centers and the potential for photodamage is high. Many plants that regularly experience such environmental stresses appear to have the ability to increase both the capacity for thermal dissipation of excitation energy (Demmig-Adams and Adams 1992) as well as their capacity to scavenge active oxygen species, suggesting that they increase the rate of photosynthetic electron flux to oxygen (Ort and Baker 2002).

Acclimation to the Light Environment

Adjustments to the efficiencies of electron transfer and CO_2 assimilation are occurring continually in leaves under natural conditions. This continuous tuning of photosynthesis, however, occurs in a background of slower and less dynamic changes to the structure of the photosynthetic apparatus driven by acclimatory and developmental responses of leaves to long-term changes in the growth light environment. The short-term adjustments of photosynthesis discussed in the preceding sections of this chapter must operate within the outcome of the long-term developmental and structural changes. Thus, in circumstances where the long-term adjustments are insufficient or incapable of responding quickly enough, a greater role must be borne by short-term responses. If these are incapable of responding to a sufficient degree, then damage to the photosynthetic apparatus and long-term, possibly permanent, depression of the quantum yield with result.

Stable changes in the natural growth light environment most often involve simultaneous changes in both the intensity and spectral distribution of light because stable shade is usually the result of over reaching vegetation. The maximum capacity for photosynthesis is highly responsive to changes in growth irradiance and differs greatly between plant species that evolved to be primarily adapted to sun versus shade environments. In contrast, Φ_{max} in unstressed C_3 leaves remains high and largely invariant whatever the growth environment and origin. Thus, as changes are made to photosynthetic apparatus to optimize performance at partially limiting or nonlimiting irradiance levels, the ability to achieve a highly balanced distribution of excitation energy to PSI and PSII is retained. It is worth noting that substantial gradients in the light environment in terms of both intensity and spectral distribution exist across leaves from the upper to lower surface. This gradient across the leaf induces much of the same response of the photosynthetic apparatus as a function of depth into the leaf as are seen between leaves at different depths within a leaf canopy.

Substantial variations in the relative amounts of the major thylakoid components in responses to changes in growth irradiance have been widely observed (Evans 1996, Genty and Harbinson 1996). With increasing irradiance the most consistent and major change in the thylakoid membrane organization is a decrease in the size of the chlorophyll antennae serving each reaction center. Importantly, this change represents a significant change in the way that the chloroplast within the leaf environment is allocating resources between light harvesting and the processing of absorbed light. Despite these changes in chlorophyll content at the reaction center level in response to changes in irradiance, there are no reliable indicators of this at the leaf level as most leaves contain enough chlorophyll that leaf absorbance is only weakly dependent on pigment concentration.

Decreases in growth irradiance are generally associated with lower chlorophyll a to b ratios and an increase in the proportion of appressed or stacked regions of the thylakoid membranes. Measurements of an increased titer of PSII relative to PSI in shade grown leaves forms a coherent picture of the photosynthetic apparatus responding to a light environment depleted in PSII excitation. Other components of the thylakoid apparatus also adjust to the growth light environment in seemingly consistent ways. Overall it appears that the growth irradiance-driven adjustments in stoichiometries among the photosystems and electron transfer components are consistent in purpose with the better-understood short-term functional responses of photosynthesis.

Summary

The foundation for understanding light processing by leaves is a detailed understanding of how chloroplasts operate within, and interact with, the environment of leaf. To convert the transient energy of light into stable chemical energy, the photosynthetic apparatus performs a series of energy-transforming reactions. The energy of absorbed light is converted to charge separations within each of the two photosystems. All subsequent reactions of photosynthesis leading to the production of ATP and NADPH and thereafter to the reduction of CO_2, as well as lesser amounts of other inorganic molecules, is driven by the light energy captured by the photosystems. As irradiance on a leaf increases, the relationship between photosynthesis and absorbed light transitions from linear to nonlinear and eventual reaches saturation where no further increases in photosynthesis accompany increased light absorption. Within the linear region of the light response, the quantum yield of CO_2 assimilation operations within 15% of its theoretical maximum. This remarkable efficiency requires both that the distribution of absorbed light between the two photosystems is very closely balanced and that under these limiting light conditions nearly all ATP and NADPH production is used for CO_2 assimilation.

The quantum yield of leaf photosynthesis decreases with increasing irradiance because the two processes that determine the effective use of light for photosynthesis change as light absorption increases. As irradiance increases, the proportion of PSII quinone acceptors in the reduced state increases, thus lowering the utilization efficiency of absorbed light by PSII. Similarly, the proportion of oxidized P_{700} increases and reduces the efficiency of light utilization by PSI. Also accompanying increased irradiance is the increase in the amount and proportion of excitation energy absorbed by the PSII antennae that is diverted directly to heat through a thermal dissipation mechanism. This thermal dissipation decreases the rate at which photons are delivered to PSII reaction centers and is photoprotective by helping prevent the overreduction of quinone acceptors. As irradiance levels reach saturation, alternative electron sinks, that is, acceptors other than CO_2, become increasingly more important in preventing overreduction of the PSII quinone acceptors and providing photoprotection. Exactly what constitutes saturating or excess light for a leaf depends on its instantaneous environmental conditions and can vary over quite a wide range of light levels. Thus, conditions external to the leaf, such as conditions that cause stomatal closure, frequently cause quantum yield to decline more rapidly with increasing irradance than would otherwise be the case. The continuous tuning of efficiencies of electron transfer and CO_2 assimilation occurs in a background of slower and less-dynamic changes to structure of the photosynthetic apparatus driven by acclamatory and developmental responses of leaves to permanent changes in the growth light environment. These growth irradiance-driven adjustments in structure appear to be consistent in purpose with the better-understood short-term functional responses of photosynthesis.

References

Asada, K. 1999. The water-water cycle in chloroplasts: Scavenging of active oxygens and dissipation of excess photons. Annu. Rev. Plant Physiol. Plant Molec. Biol. 50:601–639.

Björkman, O., Demmig, B. 1987. Photon yield of O_2 evolution and chlorophyll fluorescence characteristics at 77K among vascular plants of diverse origins. Planta 170: 489–504.

Demmig-Adams, B., Adams, W. W. 1992. Photoprotection and other responses of plants to high light. Annu. Rev. Plant Physiol. Plant Molec. Biol. 43:599–626.

Evans, J. R. 1996. Developmental constraints on photosynthesis: Effects of light and nutrition. In: Photosynthesis and the Environment. N. R. Baker (ed.), pp. 281–304. Dordrecht, The Netherlands: Kluwer Academic Publishers.

Genty, B., Harbinson, J. 1996. Regulation of light utilisation for photosynthetic electron transport. In: Photosynthesis and the Environment. N. R. Baker (ed.), pp. 67–99. Dordrecht, The Netherlands: Kluwer Academic Publishers.

Li, X.-P., Björkman, O., Shih, C., Grossman, A. R., Rosenquist, M., Jansson, S., and Niyogi, K. K. 2000. A pigment-binding protein essential for regulation of photosynthetic light-harvesting. Nature 403:391–395.

Ort, D. R. 2001. When there is too much light. Plant Physiol. 125:29–32.

Ort, D. R., Baker, N. R. 2002. A photoprotective role for O_2 as an alternative electron sink in photosynthesis? Curr. Opin. Plant Biol. 5:193–198.

Ort, D. R., Long, S. P. 2003. Converting solar energy into crop production. In: Plants, Genes, and Crop Biotechnology. M. J. Chrispeels and D. E. Sadava (eds.), pp. 240–269. Boston: Jones and Bartlett Publishers.

Singsaas, E. L., Ort, D. R., DeLucia, E. H. 2001. Variation in measured values of photosynthetic quantum yield in ecophysiological studies. Oecologia 128:15–23.

Part 4

CO_2 Capture

5
Chloroplast to Leaf

John R. Evans, Ichiro Terashima, Yuko Hanba,
and Francesco Loreto

Introduction

Leaf shapes and morphologies are very diverse. Obviously there are many successful solutions to the challenge of constructing an organ that intercepts light, enables CO_2 uptake, restricts water loss, and withstands or avoids temperature extremes, herbivory, and disease. Tradeoffs exist among capturing CO_2 and light, water loss, or construction cost. So, to some extent, the various leaf structures represent different compromises that are reached in different environments.

Carbon dioxide is captured during photosynthesis, either in the chloroplast or cytoplasm, depending on the metabolic pathway. The removal of a CO_2 molecule within the cell creates a gradient in partial pressure between the cell and surrounding air. This drives the diffusion of other CO_2 molecules from the surrounding air into the cell, according to Fick's law. While there may be a net flux of CO_2 into or out from the leaf, at the same time, the flux of a particular isotope of CO_2 may be in the opposite direction. The difference between isotopes then provides a signal that can be used to interpret leaf physiology, for example, water use efficiency or leaf temperature.

The diffusion pathway is typically divided into several components, namely, the boundary layer, stomata, intercellular airspace, and mesophyll cell. These can be seen in Figure 5.1. The surface view of a wheat leaf shows the spacing of stomata across the epidermis as well as the presence of trichomes (hairs), which would slow down the movement of air across the leaf surface. The cross-sectional view is approximately alligned to show the diffusion path for CO_2 as it enters the leaf through stomata, diffuses across the substomatal cavity and into further intercellular airspaces before entering mesophyll cells lined with chloroplasts. Issues associated with each of these components are considered in turn.

Boundary Layer

Surrounding any object is a layer of still air called the boundary layer. CO_2 and water diffusion and heat conduction are all influenced by the boundary layer.

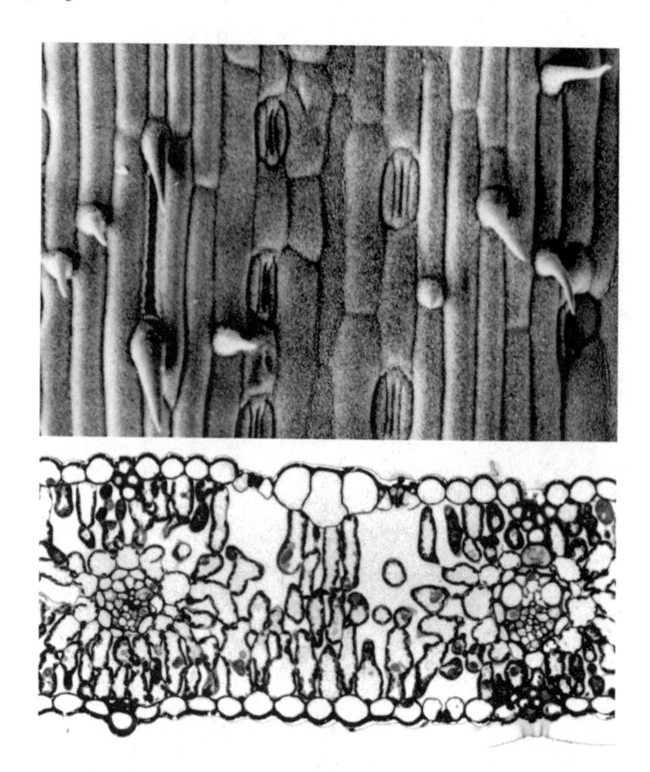

FIGURE 5.1. Wheat leaf. A scanning electron micrograph of the upper surface is alligned with a light micrograph of a transverse section below. Note the trichomes that increase the thickness of the boundary layer and the two files of epidermal cells with stomata which have large bulliform cells in between to allow the leaf to roll when under water stress. In the transverse section, the large substomatal cavities are evident with chloroplasts covering most of the mesophyll cell walls exposed to airspace. The distance between the files of stomata is about 100 μm.

The thickness of the boundary layer depends primarily on the length of the surface and the windspeed:

$$g_b\ (\text{mol H}_2\text{O m}^{-2}\ \text{s}^{-1}) = 0.25\sqrt{[\text{windspeed (m s}^{-1})/\text{leaf length (m)}]} \quad \text{(Eq. 5.1)}$$

at 25°C (Nobel 1999).

Larger leaves have smaller boundary layer conductances. The formation of compound leaves, lobing, or shredding can increase the boundary layer conductance for a given leaf area. At the finer scale, grooves, hairs, and waxes frequently decorate leaf surfaces. These would influence the boundary layer conductance and have consequences for leaf reflectance, wettability, and fungal or insect deterrence. Because CO_2 and water have different relative diffusivities through the boundary layer and stomata, CO_2 diffusion is slightly favored when the boundary layer dominates the diffusion limitation (Evans and Loreto 2000).

Stomata

The water contained in an average leaf would evaporate within 10 min if the leaf had no cuticle. To restrict such rapid water loss, leaf cuticles are highly impermeable to water. Such a barrier also prevents CO_2 diffusion into the leaf. Consequently, the epidermis is perforated with many pores, called stomata, which provide a variable channel for CO_2 entry into the leaf. Although the pores cover only 1% of the leaf surface, transpiration in the light is reduced to only one sixth of that of a wet surface because of the boundary layer. A typical leaf has a boundary layer thickness that is much greater than the distance separating stomata. The net result is that a larger flux passes through stomatal pores than would be expected simply on the basis of the proportion of leaf surface they cover.

Stomatal guard cells change their turgor and hence aperture in response to various environmental variables. The most important variables are light, water status, and CO_2 concentration.

Light

Stomata in well-watered plants open in the light. This is important for photosynthetic CO_2 fixation as it enables CO_2 to diffuse into the leaf. The role of light is threefold. Firstly, there is a blue light reaction that activates the H^+-ATPase. Secondly, photosynthesis in the mesophyll cells decreases CO_2 concentration inside the leaf (see below) but may also increase sugar concentration in the apoplast. Thirdly, the guard cells themselves have photosynthetic activity.

The action spectra of stomatal opening show a notable blue peak that is superimposed on that of photosynthetically active radiation (Hsiao and Allaway 1973, Ogawa et al. 1978, Sharkey and Raschke 1981, Karlsson 1986). A recent study with *Arabidopsis thaliana* revealed engagement of photropin 1 (phot1) and phototropin 2 (phot2) flavin proteins in photosensing (Kinoshita et al. 2001). The mutant lacking both of these proteins did not open stomata in response to blue light illumination. However, mutants that lacked only one of these proteins still opened their stomata in response to blue light, albeit with reduced sensitivity. These results indicate that the effects of these photoreceptors are additive.

Recent progress in understanding molecular mechanisms of blue light effects relies on studies with guard cell protoplasts. Blue light induces swelling of guard cell protoplasts and a decrease in pH of the bathing medium (Zeiger and Hepler, 1977). This acidification is due to the activation of the plasmalemma H^+-ATPase by blue light (Shimazaki et al. 1986, Kinoshita and Shimazaki 1999). Plasmalemma H^+-ATPase has an autocatalytic site on its C-terminus that is activated by phosphorylation of Ser and Thr residues (Kinoshita and Shimazaki 1999). A 14-3-3 protein, first described in the animal brain, associates with the phosphorylated H^+-ATPase (Kinoshita and Shimazaki 1999). For activation of H^+-ATPase, both phosphorylation and association of the 14-3-3 protein are needed. Fusicoccin, which has been known to induce stomatal opening for a long time (Schnabl 1978), stabilizes H^+-ATPase in the activated state (Sze et al. 1999).

Patch clamp techniques revealed that the activated H^+-ATPase causes hyperpolarization of the membrane potential of guard cells. When the membrane potential is hyperpolarized and falls below -120 mV, an inward K^+ channel opens (Schroeder et al. 1994). The inward flux of K^+ enhances malate^{2-} formation (Ogawa et al. 1978) and the influx of Cl^- (Outlaw 1983). K^+ concentrations in the cell sap can reach as high as 500 mM. Accumulation of K^+ and anions leads to water influx into guard cells, guard cell swelling, and stomatal opening.

K^+ may not always be the most abundant osmotic substance in the guard cells. There is evidence showing that, in *Vicia faba*, the main osmotic substance is sucrose in the afternoon (Talbott and Zeiger 1998). The origin of this sucrose is unknown. In plants employing apoplastic phloem loading, the concentration of sugar in the apoplast increases in the daytime and the transpiration stream from the veins to stomata carries such sugars to the sites of evaporation, which are around the stomatal guard cells (Ewert et al. 2000). Thus, sugar could be available in the apoplast of the guard cells or be transported symplastically. It would be interesting to compare the osmotic substance near the guard cells between the apoplastic and symplastic loaders.

Photosynthetic activities of guard cells have been studied in some species. For example, CO_2 fixation rate is much lower than the O_2 evolution rate in *Vicia faba* guard cells (Gotow et al. 1988), indicating that ATP and NADPH produced in thylakoids are preferentially used in reactions other than in the Calvin cycle and photorespiration. There is considerable species diversity in photosynthesis by guard cells, and further investigations using a variety plants are needed. Fluorescence imagery of chloroplasts now enables measurements of photosynthetic electron transport in stomatal guard cells (Lawson et al. 2002, 2003).

Water Status

When water stress is imposed on plants, stomata close. The stomatal closure occurs when the turgor in guard cells decreases. However, except under very severe drought conditions, stomatal closure in response to water stress occurs well before any appreciable dehydration of leaf tissues. Such sensitive stomatal closure occurs through the involvement of the hormone, abscisic acid (ABA). Abscisic acid is synthesized in the leaf and/or in the roots and transported to the leaf. The role of ABA that is synthesized in the root has been clarified by a series of experiments employing the split root technique (Davies and Zhang 1991). In these experiments, the roots of a seedling, such as *Zea mays*, were split into two pots, and one pot was amply watered while another pot was subject to drying. Stomata closed in response to an increase in the ABA concentration in the xylem sap, without any change in the water potential of the shoot system.

ABA is also synthesized when a leaf is subject to desiccation and this induces stomatal closure. Rapid closure via ABA signaling is advantageous in avoiding severe damage to cells by delaying dehydration and avoiding the damage that would otherwise occur if dehydration had to occur before stomata closed (Zeevaart and Creelman 1988).

Responses of guard cells to ABA have been studied with epidermal strips and guard cell protoplasts. When ABA is applied to guard cells, the concentration of Ca^{2+} inside the guard cells increases (Schroeder et al. 1994). This stimulates an efflux of K^+ and anions (Blatt 1999, Leung and Giraudat 1998). Ca^{2+} also inactivates the H^+-ATPase (Kinoshita et al. 1995). Effects of ABA are summarized in Figure 5.2. In mutants where ABA-induced stomatal closure does not occur, various ABA effects are absent (Leung and Giraudat 1998). Thus, the effects of ABA on stomatal closure occur upstream of ABA signal transduction.

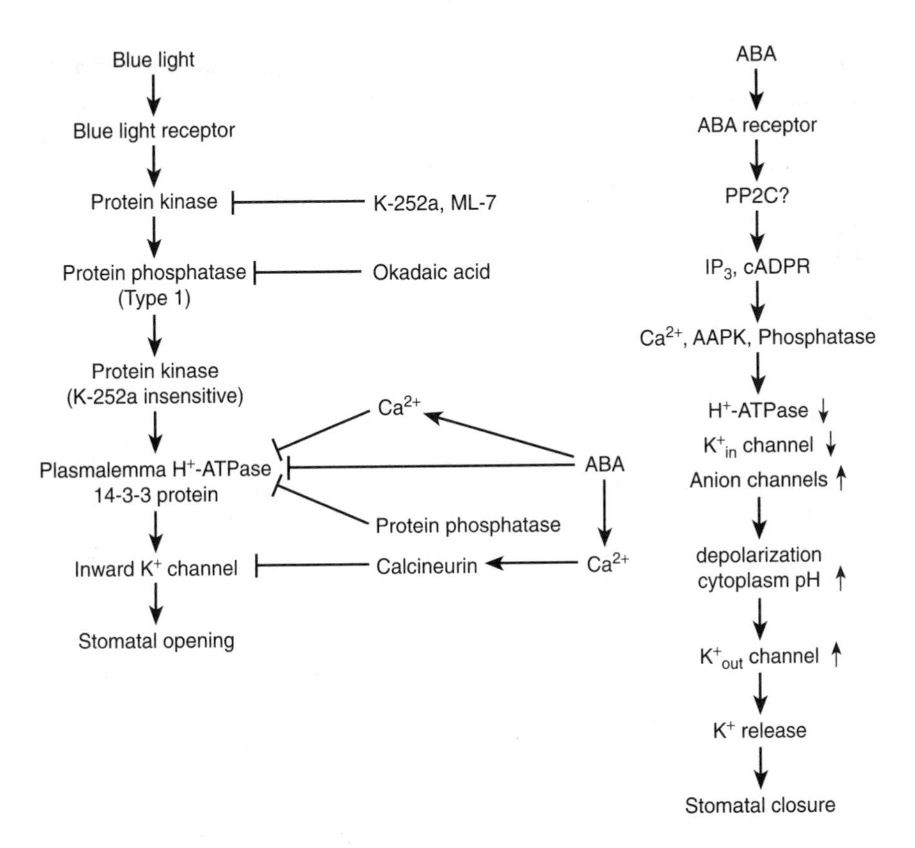

FIGURE 5.2. Hyopthetical signal transduction systems of blue light (left) and ABA (right). Blue light: there are two protein kinases that are sensitive and insensitive to k-252a (kinase inhibitor). Calcineurin is a protein phosphatase that is activated by Ca^{2+} and specifically removes phosphates from Ser and Thr residues (also called protein posphatase 2B). ABA: introduction of inositol-3 phosphate (IP_3) or cyclic-ADP-ribose (cADPR) increases cytosolic Ca^{2+} concentration, and thereby induces stomatal closure. Ca^{2+} influx into stomatal guard cells is increased by ABA. ABA increases cytosolic concentration of Ca^{2+} via phosphatidyl inositol and ADP-ribosyl cyclase systems. These are mostly English versions of figures in Shimazaki and Kinoshita (2001) and Shimazaki (2001). See also Blatt (1999) and Leung and Giraudat (1998).

A single stoma can respond to air humidity. When dry air was blown onto a single stoma with a fine capillary, the stoma opened for a while (Mott and Franks 2001) but eventually closed (Lange et al. 1971). Application of humid air could reverse the sequence. Such dynamic responses as well as many steady-state responses indicate that stomata respond to air humidity. However, it is unclear whether stomata sense air humidity per se or the transpiration rate. A series of experiments using Helox (air in which helium replaces nitrogen) clarified that stomata can measure their transpiration rate rather than the air humidity per se (Mott and Parkhurst 1991). Because diffusion of H_2O is 2.33 times faster in Helox than in ordinary air, Helox results in 2.33 times greater transpiration for a given stomatal aperture than air with the same water vapor concentration. When the leaf was switched from air to Helox with the same humidity, stomata closed. On the other hand, when the water vapor concentration in Helox was increased so that transpiration rate remained the same following the switch from air to Helox, stomatal aperture did not change. These results indicate that stomata can sense their transpiration rate. Perhaps, peristomatal transpiration (cuticular transpiration from the vicinity of the guard cells) is responsible for monitoring the transpiration rate using a negative-feedback loop (Maier-Maercker 1999).

CO_2

Stomata open when the CO_2 concentration in the air is low. As CO_2 concentration increases, stomata close. This applies to illuminated as well as to darkened leaves, and to both etiolated and green leaves (Meidner and Mansfield 1968). However, when stomata tightly close in the dark, a decrease in CO_2 concentration in the ambient air does not induce stomatal opening (Heath 1950). Using leaves of *Allium cepa*, it was clearly shown that the concentration of CO_2 in the substomatal cavity rather than that in the ambient air is sensed. In the dark, CO_2 concentration in the substomatal cavity of the leaf with tightly closed stomata is very high because of respiratory activity. In the light, photosynthetic CO_2 uptake decreases CO_2 concentration in the leaf, thereby inducing stomatal opening. The molecular identity of the CO_2 sensor is unknown (Assmann 1999, Zeiger et al. 2002).

There is evidence that the magnitude of the closing response to CO_2 varies among species. Moreover, it depends on the previous history. Sensitivity of stomata to CO_2 in unstressed plants is very low. The response increases after mild water or cold stress has been experienced (Raschke 1979). While plants acclimate to environmental variables through adjusting responses of their stomata, these acclimation mechanisms have not yet been addressed.

In addition to stomatal aperture changing in response to sensing CO_2, the density of stomata in the epidermis also responds to atmospheric CO_2 concentration. The atmospheric CO_2 is sensed by mature leaves and a signal sent to the expanding leaves where it alters the frequency of stomatal cells relative to other epidermal cells (Lake et al. 2001). An analogous signal exists that influences the development of palisade tissue by altering the plane of cell division. The light environment of neighboring mature leaves determines whether a sun or shade leaf is formed at the apex, regardless of the light environment of the developing leaf (Yano and Terashima 2001).

Stomatal Regulation in CAM Plants

In Crassulacean acid metabolism (CAM) plants, stomata close by day and open at night. In *Mesembryanthemum crystallinum*, an inducible CAM plant, CAM is induced by desiccation or salt stress (Cushman et al. 2000). During the transition from C_3 to CAM, stomatal properties change drastically. In epidermal strips from well-watered *M. crystallinum* plants, blue light was effective in opening stomata. When in CAM mode, however, illumination of the epidermal strip did not induce stomatal opening, suggesting that the blue light receptors became inert. Treatment of epidermal tissues from the CAM plants with fusicoccin did not induce stomatal opening. On the other hand, ABA-induced stomatal closure was still functional. When guard-cell photoreceptors are inactivated, other factors including the CO_2 sensor determine the stomatal movement (Mawson and Zaugg 1994).

The fixation of CO_2 by PEPCase at night reduces the CO_2 concentration in the intercellular airspaces, and stomata open. In the decarboxylating phase, CO_2 concentration in the leaf increases, which induces stomatal closure. In phase IV that occurs in the late afternoon, malate in the vacuole is exhausted, decarboxylation is suppressed and fixation of CO_2 by Rubisco decreases CO_2 concentration in the intercellular spaces, which induces stomatal opening.

Hydropassive Regulation

In leaves, shrinkage of epidermal cells can stretch the stomatal complex outwards, thereby opening the stomata. On the other hand, when epidermal cells swell, stomata may be forced shut even when their turgor is considerable. When turgor is increased equally in both epidermal and guard cells, the pore will close slightly (Franks et al. 1995). These stomatal behaviors affected by turgor of surrounding epidermal cells are called hydropassive movements (Stålfelt 1955). When stomatal opening/closure occurs due to changes in the turgor of guard cells, these movements are called hydroactive. Hydropassive stomatal movements are best illustrated when a transpiring leaf is suddenly excised. Stomata transiently close due to the swelling of other epidermal cells that have been under tension but are now supplied with apoplastic water at a high water potential. Subsequent dehydration of other epidermal cells then causes stomatal opening. Further general dehydration causes stomatal closure (Meidner and Mansfield 1968).

Patchy Stomatal Closure

CO_2 concentration in the substomatal cavity (C_i) is an important parameter that can be calculated with conventional gas exchange techniques (Farquhar and Sharkey 1981). When a stress factor induces stomatal closure but does not alter the relationship between the rate of photosynthesis (A) and calculated C_i (A-C_i

relationship), one can assume that this stress factor induces only stomatal closure but does not suppress photosynthetic capacity of the mesophyll. A stress factor is judged to affect mesophyll photosynthesis when the A-C_i relationship is changed by the stress factor. The calculation of C_i assumes uniform stomatal behavior over the leaf area, but this is not always true. In 1988, patchy photosynthesis in leaves treated with ABA was demonstrated by a starch iodine test (Terashima et al. 1988) and autoradiography of leaves (Downton et al. 1988). Subsequently, the imaging of the chlorophyll fluorescence parameter NPQ (Daley et al. 1989) or the photochemical yield of photosystem II (Genty and Meyer 1995) can also be used to visualize patchiness. Although there is no direct evidence showing that nonuniform photosynthesis is solely due to nonuniform stomatal closure, there are several lines of circumstantial evidence strongly supporting the causal relationship. One such example is the measurement of photosynthesis using a gas phase oxygen electrode system at very high CO_2 concentration (10%). This CO_2 concentration can overcome any resistance to CO_2 diffusion imposed by closed stomata. Leaves of *Helianthus annuus* and *Vicia faba* that had been treated with ABA showed lower rates of photosynthesis than untreated control leaves at any C_i value up to 1000 μmol mol^{-1}. However, they did not show any depression of photosynthetic O_2 evolution rate when measured under 10% CO_2 concentrations (Terashima et al. 1988). These results clearly showed that the diagnosis of stress effects based on the changes in the A-Ci relationship could be spurious. As will be seen below, if photosynthesis is nonuniform due to patchy stomatal closure over the leaf and if mesophyll photosynthesis is not impaired, C_i will be overestimated. Several reviews provide detailed analyses of this problem (Terashima 1992, Pospisilova and Santrucek 1994, Weyers and Lawson 1987, Beyschlag and Eckstein 1998, Mott and Buckley 2000a). Here, some general points will be discussed.

The relationship between net photosynthetic rate and stomatal conductance is typically convex (see Fig. 5.3). For simplicity, we initially neglect cuticular transpiration. Assume that the leaf area under consideration is separated into two parts (patches) of the same area and these two parts have different values of stomatal conductance, g_1 and g_2 ($g_1 < g_2$). Then, the averaged value of C_i for these areas is,

$$C_i = C_a = 1.6 \cdot (A(g_1)/g_1 + A(g_2)/g_2)/2, \qquad \text{(Eq. 5.2)}$$

where C_a is CO_2 concentration in the ambient air, and $A(g_1)$ and $A(g_2)$ are the photosynthetic rates corresponding to g_1 and g_2, respectively. On the other hand, C_i calculated assuming the uniform conductance over the leaf area, C_i*, is,

$$C_i^* = C_a = 1.6 \cdot (A(g_1) + A(g_2))/(g_1 + g_2) \qquad \text{(Eq. 5.3)}$$

The difference between C_i* and C_i is

$$C_i^* - C_i = 0.8 \cdot \frac{(g_2 - g_1) \cdot (A(g_1)/g_1 - A(g_2)/g_2)}{(g_1 + g_2)} > 0 \qquad \text{(Eq. 5.4)}$$

because $A(g_1)/g_1 > A(g_2)/g_2$ (see Fig. 5.3).

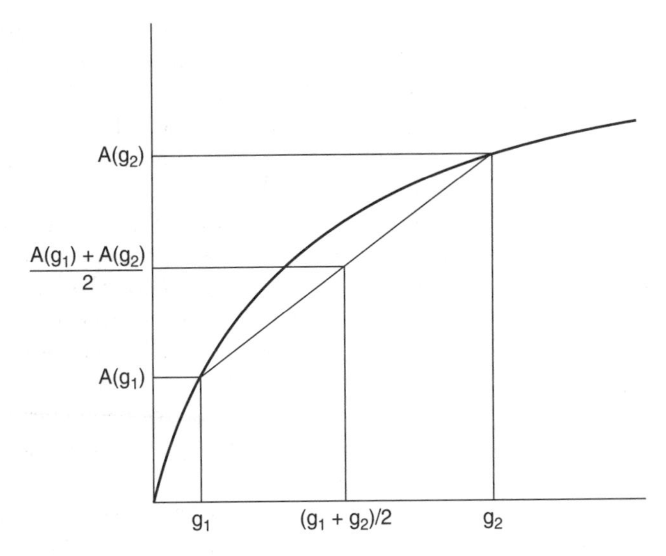

FIGURE 5.3. Relationship between assimilation rate (A) and stomatal conductance (g). Stomatal conductances of the patches having the same areas are assumed to be g_1 and g_2, respectively. If there is no lateral diffusion of gas between these patches, the averaged assimilation rate for these patches is less than it would be if these patches had the same conductance, $(g_1 + g_2)/2$. Redrawn from Terashima et al. (1988).

Thus, when there are differences in stomatal conductance over the leaf, C_i* is always overestimated. The effect of nonuniform stomatal conductance on the calculation of C_i is most pronounced when g_1 and g_2 differ in a bimodal manner (e.g., $g_1 >> g_2$). However, when g_1 and g_2 are small and in the range where the relationship between A and g is almost linear, C_i* will not be very different from C_i. This is clearly shown with a sophisticated computer simulation model called PATCHY (Cheeseman 1991, see also Buckley et al. 1997). A further problem of cuticular transpiration needs to be included. When stomatal conductance decreases, the relative contribution of cuticular transpiration to the total transpiration increases, which results in the overestimation of C_i. As already mentioned above, C_i is calculated as,

$$C_i = C_a - 1.6\ A/(g_{leaf}), \qquad \text{(Eq. 5.5)}$$

where g_{leaf} is the leaf conductance to water vapor. Because CO_2 diffuses almost exclusively through stomata but not across the epidermis, stomatal conductance (g_s) rather than g_{leaf} should be used. However, due to the technical difficulty, g_{leaf} is conventionally used instead of g_s. g_{leaf} is expressed as

$$g_{leaf} = g_s + g_c, \qquad \text{(Eq. 5.6)}$$

where g_c is cuticular conductance. When g_c is much smaller than g_s, g_{leaf} can be used for g_s. Since g_c is relatively constant irrespective of g_s values, the contribution of g_c may become significant when g_s is small. Thus, when stomatal closure is induced by some stress factor, calculated C_i using g_{leaf} tends to be overestimated (Kirschbaum and Pearcy 1988). When stomatal conductance values are low over the leaf area, overestimation of C_i may be mainly due to neglecting cuticular conductance (Boyer et al. 1997). Meyer and Genty (1998) clearly showed this in leaves of *Rosa rubiginosa* treated with ABA.

It has been pointed out that patchy photosynthesis and transpiration would be more pronounced in leaves with bundle sheath extension compartments, called heterobaric leaves, than in homobaric leaves, which do not have bundle sheath extensions (Terashima et al. 1988, Terashima 1992). In heterobaric leaves, photosynthetic rates can differ from compartment to compartment, because lateral diffusion of CO_2 across the bundle sheath extension does not occur. In homobaric leaves, lateral diffusion can moderate the effect of nonuniform stomatal conductance. It is likely that a compartment bounded by bundle sheath extensions is a unit for water relations.

Mott and coworkers analyzed the time course of photosynthetic oscillations using fluorescence imaging techniques. In response to sudden changes in environmental variables such as humidity or irradiance, photosynthetic oscillations can be induced (Mott and Buckley 2000b). Moreover, the phase of oscillation may differ in the respective spaces over the leaf. The mechanisms of local oscillation may be due to hydropassive movements of stomata. Once a group of stomata open widely due to some perturbation, they transpire rapidly and dehydrate themselves and neighboring epidermal cells, which induces neighboring stomata to open, while they close. Such dynamic changes would underlie the oscillation (Mott and Buckley 2000b, Mott and Franks 2001). Oscillations may occur at the scale of several compartments or over a much greater area (Siebke et al. 1995a,b) and further studies at this mesoscopic level are needed.

Internal Conductance

Internal conductance (g_i), is defined as the conductance for CO_2 diffusion from the substomatal cavities to the chloroplast stroma. This pathway involves diffusion in gas and liquid phases, both of which are influenced by leaf anatomy (Fig. 5.4). Internal CO_2 conductance is positively correlated to CO_2 assimilation rate (von Caemmerer and Evans 1991, Evans and Loreto 2000). Internal conductance is of a similar magnitude to and quite closely correlated with stomatal CO_2 conductance across C_3 species from different functional types (Fig. 5.5). Annual herbs generally have high internal CO_2 conductance compared with deciduous or evergreen trees (Epron et al. 1995, Evans 1999), but the drawdown in CO_2 between C_i and C_c is similar for both herbs and trees. For comparison, the drawdown in CO_2 across stomata ($C_a - C_i$) is 121 $\pm$ 5 c.f. 90 $\pm$ 3 μmol mol^{-1} in mesophytic versus sclerophytic C_3 leaves measured under high irradiance, whereas for $C_i - C_c$ it is 77 $\pm$ 3 c.f. 81 $\pm$ 4 μmol mol^{-1}, respectively (Evans 1999).

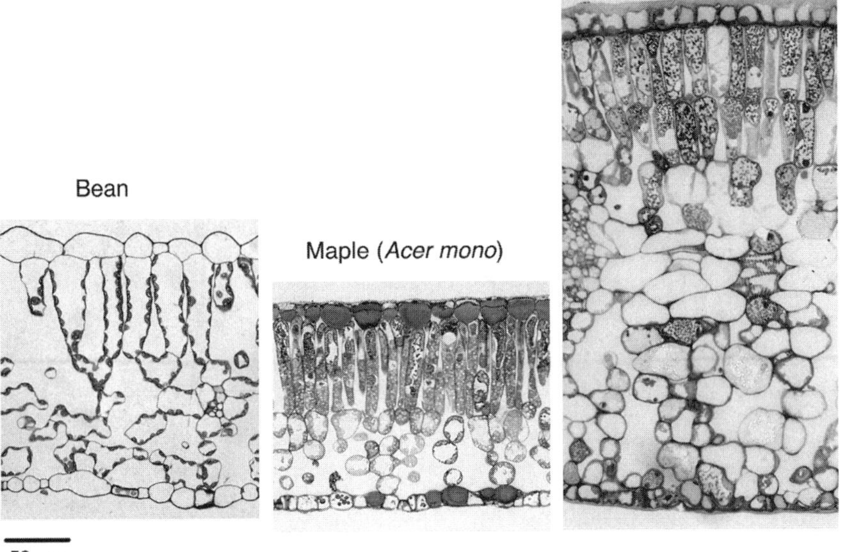

FIGURE 5.4. Variation in leaf mesophyll anatomy between bean (annual herb species, Hanba unpublished), maple (deciduous tree species, Hanba et al. 2002) and camellia (evergreen tree species, Hanba et al. 1999). All plants were grown in high light. Note the considerable difference in mesophyll cell packing, mesophyll thickness, and chloroplast distribution between species.

Gas Phase

Diffusion in the gas phase depends on leaf anatomy. As experimental approaches to measuring gaseous diffusion within the airspaces of a leaf are limited, the problem has been approached by developing various models. Diffusion is significantly affected by stomatal distribution, mesophyll porosity and thickness (Parkhurst 1994, Pachepsky and Acock 1996, Terashima et al. 2001) as well as by bundle sheath extensions (Terashima et al. 1988). Plants with amphistomatous leaves or leaves with high mesophyll porosity (e.g., bean, see Fig. 5.4), have a high CO_2 conductance in the gas phase. For amphistomatous leaves, the difference between C_i measured on the two surfaces was less than 10 μmol mol^{-1} (Parkhurst et al. 1988). Experiments with Helox increased photosynthetic rates by 2 and 12% for amphistomatous and hypostomatous leaves, respectively (Parkhurst and Mott 1990), suggesting that mesophyll thickness could play a significant role in restricting CO_2 diffusion through the airspaces. However, more recent Helox experiments where C_c was calculated from chlorophyll fluorescence have also shown that sclerophyllous leaves of oak have little diffusive resistance to CO_2 diffusion through the intercellular airspaces (Genty et al. 1998). Other experimental data and

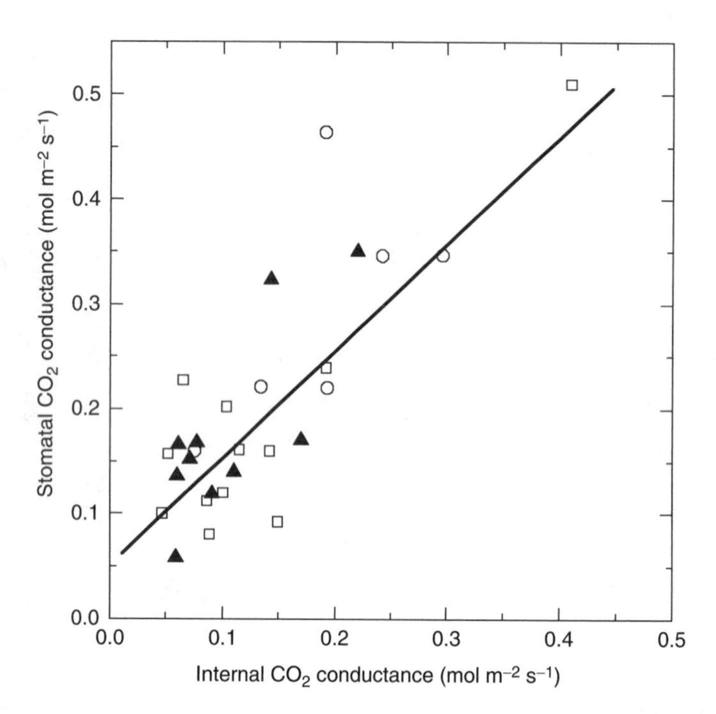

FIGURE 5.5. Relationship between stomatal conductance to CO_2 measured under high irradiance and internal conductance for a variety of species: annual herbs (○), deciduous trees (□) and evergreen trees (▲). Each point represents a different species. Data are from Lloyd et al. (1992), Loreto et al. (1992), Epron et al. (1995), Lauteri et al. (1997), Hanba et al. (1999, 2001, 2002), Kogami et al. (2001), Piel et al. (2002) and unpublished data by Hanba et al.

theoretical analyses suggest that diffusion through intercellular airspaces of leaves generally contributes very little to the drawdown in CO_2 partial pressure between C_i and C_c (Aalto and Juurola 2002, Cordell et al. 1999).

There is huge variation in leaf anatomy with leaf thickness ranging from 55 to 1960 μm and mesophyll porosity ranging from 10 to 36% (Niinemets 1999). While this affects the profiles of light absorption through leaves by enabling different distributions of pigment and Rubisco, the consequences for gaseous diffusion seem considerably smaller. Thicker leaves with higher photosynthetic capacity tend to be amphistomatous, thereby halving the diffusion distance. Little evidence exists supporting any significant limitation by mesophyll porosity. Consequently, it is fair to assume that most of the internal resistance is associated with the liquid phase.

Liquid Phase

Conductance through the liquid phase depends on anatomical properties of the mesophyll. To date, three components have been studied: (1) chloroplast surface area facing intercellular airspaces, S_c (Evans et al. 1994), (2) thickness and/or composition of cell walls (Kogami et al. 2001), and (3) other factors related to membrane permeability (Terashima and Ono 2002). The factor most responsible for variation in internal conductance may depend on species and growth environments compared.

There are increasing data available where both internal conductance and S_c have been measured (Fig. 5.6). An upper bound for the conductance of 24 mmol CO_2 m_c^{-2} s^{-1} is evident for amphistomatous annual herbs. This limit is also achieved by peach and citrus. However, for poplar, alder and macadamia trees, the conductance is considerably less, at around 6.5 mmol CO_2 m_c^{-2} s^{-1}. Internal conductances for the hypostomatous deciduous maple tree species were low and independent of S_c. In wheat, g_i decreases over the course of senescence (Loreto et al. 1994), but this is not accompanied by changes in S_c (see Fig. 5.6, dashed arrow) (Evans and Vellen 1996). Therefore it is clear that the conductance per unit chloroplast surface exposed to intercellular airspace is not a unique value.

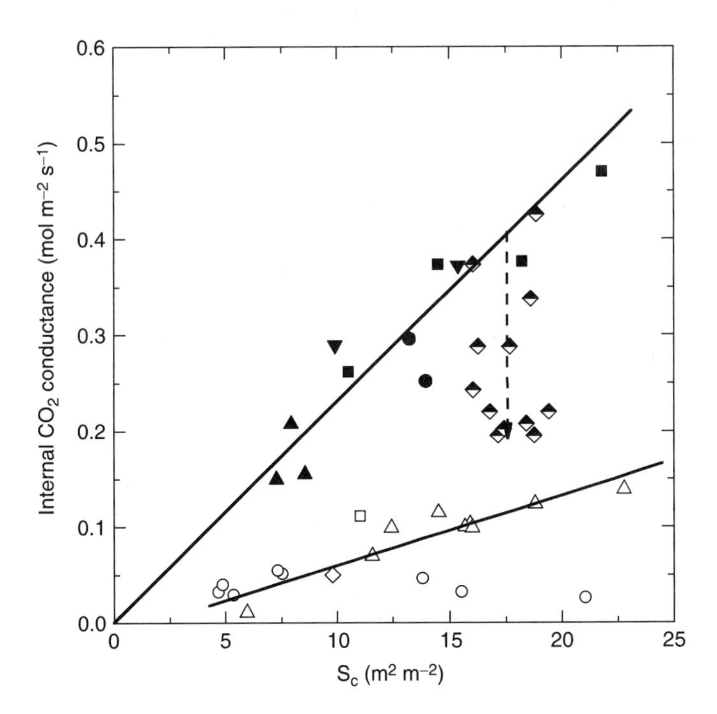

FIGURE 5.6. Dependence of internal conductance on the surface area of chloroplasts facing intercellular airspaces (S_c). Data are for mature leaves grown in high light, measured under high irradiance and ambient CO_2 (350 μmol mol^{-1}); tobacco (■) (Evans et al. 1994); peach (▼), citrus (▲, macadamia (□) (Syvertsen et al. 1995); wheat (◇) (Evans and Vellen 1996); Kalanchoe (◇) (Maxwell et al. 1997); poplar and alder (△) (Hanba et al. 2001); maple (○) (Hanba et al. 2002); bean (●) (unpublished data by Hanba et al.).

In mesophyll cells in C_3 leaves, chloroplasts are appressed against the cell membrane/cell walls that directly face the intercellular spaces. For leaves with high photosynthetic capacities, they generally occupy most of the exposed surface, leaving only the surfaces adjacent to other cells empty. This is probably a good indication of the limitation to CO_2 diffusion into chloroplasts not directly adjacent to exposed cell surfaces. Chloroplasts are anchored to such positions by actin filaments. In the dark, actin filaments appear to be dissociated (Takagi 2000). Given that chloroplast anchorage occurs in response to environmental signals, changes in g_i may be related to chloroplast anchorage. In senescing wheat leaves, association of chloroplasts to the cell membranes becomes weaker and considerable cytoplasm intrusion can be seen between chloroplast envelopes and cell membranes (Ono, K., unpublished data). A tobacco mutant with excess phytochrome shows similar intrusion of cytoplasm and low internal conductance for CO_2 (Sharkey et al. 1991). Those wanting a more detailed explanation should see the review by Takagi (2000).

It appears that chloroplasts attach to cell membranes that directly face intercellular spaces but not to the wall attached to neighboring cells. This is an important feature in CO_2 diffusion. Although Senn discussed chloroplast positioning with respect to CO_2 diffusion as early as 1908, this problem has seldom been addressed for nearly one century. Photoacoustic analysis of *Alocasia* leaves suggested chloroplast movement did not alter internal conductance to CO_2 (Gorton et al. 2003), but the mesophyll cell surfaces of these leaves are only partly covered by chloroplasts. Mechanisms of chloroplast positioning should be studied and its role in leaves with more tightly packed chloroplasts examined. It is interesting to point out that, in C_4 plants, mesophyll chloroplasts are not necessarily attached to cell walls directly facing the intercellular spaces and can be seen along cell walls between the cells (Tazoe, Y., unpublished data). In bundle sheath cells, chloroplast position also differs depending on C_4 decarboxylation type (e.g., Evans and Loreto 2000).

Aquaporins

Aquaporins, one of the most abundant proteins in plant plasma membranes, mainly transfer water molecules according to the gradient of water potential (Kjellbom et al. 1999). In plant cells, more than 90% of the water that moves across cell membranes would be through aquaporins rather than the lipid bilayer. It is also known these proteins are sensitive to mercurial reagents. Recently, it was shown that animal aquaporin 1 transports CO_2 as well as water. In these studies, *Xenopus* oocytes and/or liposomes were used and the CO_2 permeability was monitored as changes in pH (Cooper and Boron 1998, Nakhoul et al. 1998, Ramesh Prasad et al. 1998). However, Yang et al. (2000) criticized the interpretation of these results, suggesting that instead they were due to inhibitory effects of $HgCl_2$ on carbonic anhydrase rather than the aquaporins. Expression of the tobacco plasma membrane aquaporin NtAQP1 in *Xenopus* oocytes has also shown that it is a pore for CO_2 (Uehlein et al. 2003). Attempts to examine the possibility that aquaporin facilitates CO_2 diffusion across the cell membranes in mesophyll cells have begun.

Terashima and Ono (2002) examined whether mercury-sensitive aquaporins are involved in photosynthetic CO_2 diffusion across the plasma membrane in higher plants. $HgCl_2$ was fed to the leaf of *Phaseolus vulgaris* via the transpiration stream. After the leaf was fed with 1 mM $HgCl_2$, the rate of photosynthesis on leaf area basis (A) at a given C_i decreased. However, when A was plotted against CO_2 concentration in the chloroplast stroma, which was calculated from fluorescence data and Rubisco kinetics, no change was found in the relationship between A and C_c. More directly, *Vicia faba* leaflets without an abaxial epidermis were treated with $HgCl_2$ solution. After $HgCl_2$ treatment, A was reduced at any given ambient CO_2 concentration, similar to the way in which A-C_i relationships had changed. If the primary site of $HgCl_2$ inhibition is the aquaporin, then these results may indicate that CO_2 diffusion is greatly facilitated by aquaporins. It has been argued that the temperature dependence of internal conductance is also consistent with a significant involvement of aquaporins (Bernacchi et al 2002). Thus, not only are CO_2 and water fluxes linked through stomata, they may also be linked across the mesophyll cell wall interface.

Stress

It was suggested that growth temperature could alter internal conductance in rice. Photosynthesis per unit Rubisco at 25 °C and 200 μbar C_i was almost 40% lower for plants grown under 18/15 °C day/night temperatures (a cold stress for tropical rice) compared with a 30/23 °C regime (Makino et al. 1994). The expression of rice aquaporin genes decreased under a combined water stress/chilling treatment (Li et al. 2000), which is significant, given the potential role they may play in membrane CO_2 permeability.

The comparison of leaves sampled from plants grown at different altitudes reveals an intriguing difference. It is likely that leaves from higher altitude experience both greater variation and extremes (such as freezing) in temperature. The internal conductance of *Polygonum cuspidatum* (synonymous with *Reynoutria japonica*) leaves from high altitude was only 40% of that from low altitude, despite having similar exposed chloroplast surface area (Kogami et al. 2001) (Table 5.1). While the mesophyll cell walls were thicker, it was suggested that membrane composition may have contributed to variation in internal conductance. A similar relative difference in internal conductance between leaves sampled at low and high altitude was observed by Sakata and Yokoi (2002). During freezing and thawing, there is redistribution of cellular water (Ball et al. 2002). The ability to withstand freezing stress may involve altered expression of aquaporins to be able to deal with this water movement. If aquaporins are also the major pathway for CO_2 diffusion across the plasmalemma, then any change in aquaporins associated with freezing tolerance would also influence internal conductance to CO_2.

There is also evidence that other stresses can reduce internal conductance more than stomatal conductance. Water stress (Renou et al. 1990, Roupsard et al. 1996,

TABLE 5.1. Internal conductance and leaf properties from plants grown at low and high altitude. *Polygonum cuspidatum* plants were collected from their natural habitat in a lowland (10 m a.s.l.) or highland (2500 m a.s.l.) (Kogami et al. 2001).

	10 m	2500 m
Internal conductance (mol m^{-2} s^{-1})	0.193 ± 0.015	± 0.026
LMA (g m^{-2})	42 ± 2	88 ± 5
S_c (m^2 m^{-2})	21 ± 2	21 ± 1
Palisade cell wall thickness (μm)	0.22 ± 0.03	0.35 ± 0.03
Spongy cell wall thickness (μm)	0.29 ± 0.03	0.42 ± 0.06

Ridolfi and Dreyer 1997, Brugnoli et al. 1997, Scartazza et al. 1998) and salt stress (Bongi and Loreto 1989, Delfine et al. 1998, 1999) both led to greater reductions in the CO_2 drawdown through the mesophyll ($C_i - C_c$) than through stomata ($C_a - C_i$). In the case of spinach, the reduction in internal conductance due to the imposition of salt stress could be partially reversed following removal of the stress (Delfine et al. 1999).

Respiratory Efflux and Refixation

In illuminated leaves, CO_2 uptake by photosynthesis is counteracted by CO_2 release by photorespiration and (mitochondrial) respiration. Photorespiration in C_3 plants can be calculated from photosynthetic electron transport rate if one knows the kinetic properties of Rubisco, the CO_2 concentration at the sites of carboxylation, and the rate of CO_2 release not associated with photorespiration. However, accurate measurements of the actual photorespiratory CO_2 release in the atmosphere are difficult because some unknown fraction of the formed CO_2 is re-assimilated by photosynthesis before it can escape the leaf. Because photorespiratory CO_2 release occurs near the sites of fixation, this CO_2 is more likely to be assimilated than CO_2 outside the leaf, which encounters additional physical resistances mentioned in other sections of this chapter.

Mitochondrial respiration at night in the dark (R_n) is similar in C_3 and C_4 plants and can be simply measured by gas exchange. It is preferable to enclose the whole leaf in a chamber, otherwise errors due to respiration underneath the gasket of clamp-on chambers can be encountered (Pons and Welschen 2002). A considerable fraction of the CO_2 fixed by photosynthesis during the day can be released back to the atmosphere by R_n (Van der Werf et al. 1994). As with photorespiration, the rate of mitochondrial respiration in daylight (R_d) is difficult to measure by gas exchange, mainly because of the overwhelming simultaneous carbon fixation by photosynthesis. In C_3 plants, R_d has been indirectly estimated by linear extrapolation to zero light of the relationship between photosynthesis and light intensity (Kok 1948), or from the intersection of relationships between CO_2 assimilation and intercellular CO_2 obtained at different light intensities, where CO_2 assimilation rate exactly matches the rate of photorespiratory CO_2 release (Laisk 1977). Both methods suggest that R_d is lower than R_n, though the estimated reduction is generally higher with the Kok method than with the Laisk method. In general, R_d is 50% or less of R_n (Atkin et al. 1998, 2000). In C_4 plants, R_d is conveniently estimated as the CO_2 flux missing at the intercept in the linear relationship between electron transport rate and photosynthesis (Edwards and Baker 1993). R_d may apparently reach 75% of R_n in C_4 plants.

Direct measurements of CO_2 refixation are rare and methodologically complex. Pulse and chase experiments with radiolabeled carbon suggested a photorespiratory CO_2 emission from 35 to 100% of the actual rate of photorespiratory CO_2 production (Ludwig and Canvin 1971, Gaudillere 1981). The insensitivity of infrared gas analyzers to $^{13}CO_2$ has been recently exploited to conveniently and rapidly measure photorespiratory and respiratory CO_2 efflux from leaves (Loreto et al. 1999, 2001). This is detected as $^{12}CO_2$ emission in a $^{13}CO_2$ atmosphere. The CO_2 formed by mitochondrial respiration is labeled more slowly than that formed by photorespiration, and this delay is used to detect the CO_2 emission attributable to each of the two processes. A refinement of this technique, simultaneously detecting both isotopes (^{12}C and ^{13}C), has been made possible by the use of mass spectrometry (Haupt-Herting et al. 2001), but this measurement is expensive and cannot be made routinely. Direct measurements indicate that the CO_2 emissions by photorespiration and mitochondrial respiration are similar, and close to those indirectly estimated. This confirms the validity of the assumptions on which the indirect estimates of emissions are based.

Why is R_d lower than R_n? There are reasons to believe that respiration can be inhibited in the light. However, part of the CO_2 produced by mitochondrial respiration can also be refixed before exiting the leaf. Perhaps the most convincing evidence for considerable refixation of CO_2 produced in the light is based on the discrepancy between the calculated amount of internal $^{12}CO_2$ produced by respiration and the actual $^{12}CO_2$ emission rate (Loreto et al. 2001). Refixation of CO_2 produced during respiration is generally 50 to 70% of R_n, a percentage similar to that of refixation of CO_2 produced during photorespiration. The proportion of CO_2 that is refixed can range from almost complete, particularly in plant species with elevated rates of photosynthesis, or may be very low when photosynthetic rate is limited by low light, elevated temperatures, low CO_2, and other environmental stresses (Fig. 5.7). Measurements of photorespiratory CO_2 refixation under stress conditions are missing, but a reduction similar to that observed for respiratory CO_2 is conceivable. It is clear that the proportion of photorespiratory and respiratory CO_2 production that is refixed is related to the rate of CO_2 assimilation. CO_2 is produced in the mitochondria from either the citric acid cycle or glycine decarboxylation and therefore faces the same diffusion pathway to escape from the leaf. Exceptions to this would be C_3-C_4 intermediates and C_4 species where glycine decarboxylation is confined to certain mesophyll cells.

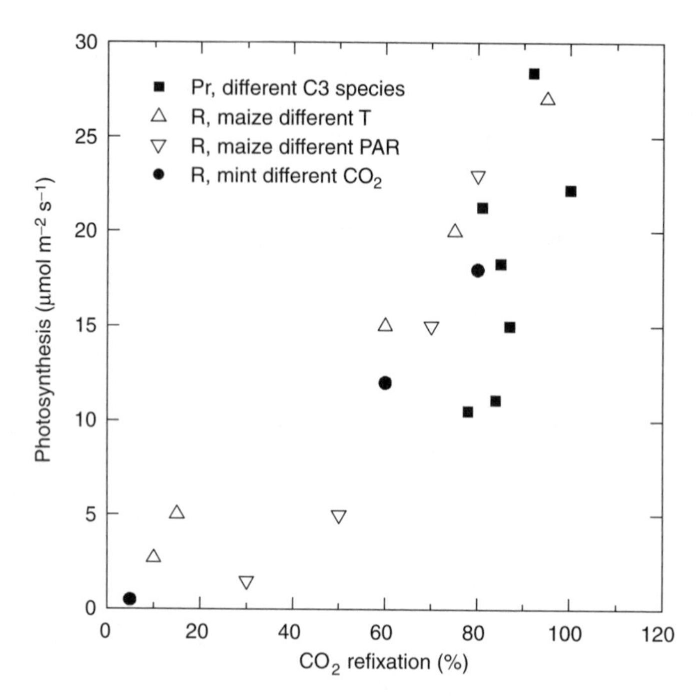

FIGURE 5.7. Relationship between photosynthetic rate and percentage of refixation of photorespiratory and respiratory CO_2 in leaves of C_3 and C_4 plants: ●, photorespiration (Pr) refixation in leaves of different C_3 plants (Loreto et al. 1999); upward and downward triangles show respiration (R) refixation in leaves of maize exposed to different temperatures and photosynthetic active radiation (PAR), respectively (Loreto et al. 2001); closed circles show respiration refixation in leaves of mint exposed to different CO_2 concentrations (Pinelli and Loreto 2003).

Summary and Conclusions

The compromises facing leaves in enabling CO_2 entry while controlling water loss, intercepting light without gaining too much heat, and opting for a particular lifespan combine to result in great morphological diversity. However, the physiological diversity in CO_2 capture by leaves is quite limited even without setting aside the pathways used for CO_2 fixation. This suggests that in the first instance, it is easier to change cellular arrangements to alter leaf properties than to alter the biochemistry.

Stomata have been the focus of much research because they are accessible, responsive to many environmental parameters and dominate the CO_2 diffusion pathway between the atmosphere and the chloroplast. While outside the scope of this chapter, the role of stomata in the compromise between CO_2 gain versus water loss is obviously crucial to plant survival and performance. By unravelling the signaling pathways and mechanisms used by stomata, one hopes that it may be possible to engineer or select for variation that may result in improved plant performance under certain environments. Progress has already been possible by direct selection for altered water-use efficiency (dry matter gain per unit water transpired) without knowing the exact mechanism. Since it is the major limiting resource for plant growth in many environments, this area will continue to receive attention.

Once CO_2 reaches the substomatal cavity, the next major barrier appears to be close to the mesophyll cell wall interface. Both stomatal and internal conductance covary with photosynthetic capacity, showing no systematic variation with respect to leaf type. It is clear that the conductance varies per unit of chloroplast surface area exposed to intercellular airspace, with an upper bound of 23 mmol CO_2 m_c^{-2} s^{-1}. Leaves with lower conductance per unit chloroplast area must also have reduced photosynthetic capacity per unit chloroplast area. This correlation cannot be via direct sensing of CO_2 because antisense reduction of Rubisco did not alter internal conductance when both are expressed per unit exposed chloroplast area (Evans et al. 1994). The recent emergence of the role of aquaporins in CO_2 diffusion provides a target for progress. If internal conductance were to be increased in those species with low conductances per unit chloroplast area by increasing aquaporin expression, it is not clear what the hydraulic consequences would be. For gas exchange calculations, it is assumed that the intercellular airspaces are effectively water saturated. Therefore, additional aquaporins should not increase the flux of water into the cell walls, but could increase the conductance to CO_2.

The compromises faced by leaves can be deduced from careful observation of their anatomy. As Haberlandt (1914) stated: "the peripheral location [of the chloroplasts] facilitates the gaseous interchanges associated with the photosynthetic function. . . . In the photosynthetic tissues of higher plants, the chloroplasts adhere exclusively, or in great part, to those walls that abut upon airspaces; by this means they evidently obtain the most favorable conditions for the absorption of carbon dioxide." We now have instruments and techniques that will enable us to obtain a better understanding of the mechanisms involved, which may lead to engineering plants with better performance under particular conditions.

References

Aalto, T., and Juurola, E. 2002. A three-dimensional model of CO_2 transport in airspaces and mesophyll cells of a silver birch leaf. Plant Cell Environ. 25:1399–1409.

Assmann, S. M. 1999. The cellular basis of guard cell sensing of rising CO_2. Plant Cell Environ. 22:629–637.

Atkin, O. K., Evans, J. R., and Siebke, K. 1998. Relationship between the inhibition of leaf respiration by light and enhancement of leaf dark respiration following light treatment. Aust. J. Plant Physiol. 25:437–443.

Atkin, O. K., Holly, C., and Ball, M. C. 2000. Acclimation of snow gum (*Eucalyptus pauciflora*) leaf respiration to seasonal and diurnal variations in temperature: The importance of changes in the capacity and temperature sensitivity of respiration. Plant Cell Environ. 23:15–26.

Ball, M. C., Wolfe, J., Canny, M., Hofmann, M., Nicotra, A. B., and Hughes, D. 2002. Space and time dependence of temperature and freezing in evergreen leaves. Funct. Plant Biol. 29:1259–1272.

Bernacchi, C. J., Portis, A. R., Nakano, H., von Caemmerer, S., and Long, S. P. 2002. Temperature response of mesophyll conductance. Implications for the determination of Rubisco enzyme kinetics and for limitations to photosynthesis in vivo. Plant Physiol. 130:1992–1998.

Beyschlag, W., and Eckstein, J. 1998. Stomatal patchiness. In: Progress in Botany. K. Behnke et al. (eds.), pp. 283–298, Berlin: Springer-Verlag.

Blatt, M. R. 1999. Reassessing roles for Ca^{2+} in guard cell signaling. J. Exp. Bot. 50:989–999.

Boyer, J. S., Wong, S. C., and Farquhar, G. D. 1997. CO_2 and water vapor exchange across leaf cuticle (epidermis) at various water potentials. Plant Physiol. 114:185–191.

Brugnoli, E., Scartazza, A., Lauteri, M., Monteverdi, M. C., and Máguas, C. 1997. Carbon isotope discrimination in structural and non-structural carbohydrates in relation to productivity and adaptation to unfavourable conditions. In: Stable Isotopes: Integration of Biological, Ecological and Geochemical Processes. H. Griffiths (ed.), pp. 203–222. Oxford: Bios Scientific Publishers Limited.

Bongi, G., and Loreto, F. 1989. Gas-exchange properties of salt-stressed olive (*Olea europea* L.) leaves. Plant Physiol. 90:1408–1416.

Buckley, T. N., Farquhar, G. D., and Mott, K. A. 1997. Qualitative effects of patchy stomatal conductance distribution features on gas-exchange calculations. Plant Cell Environ. 20:867–880.

Cheeseman, J. M. 1991. PATCHY—Simulating and visualizing the effects of stomatal patchiness on photosynthetic CO_2 exchange studies. Plant Cell Environ. 14:593–599.

Cooper, G. J., and Boron, W. F. 1998. Effects of PCMBS on CO_2 permeability of *Xenopus* oocytes expressing aquaporin 1 or its 189S mutant. Am. J. Physiol. 275C: 1481–1486.

Cordell, S., Goldstein, G., Meinzer, F. C., and Handley, L. L. 1999. Allocation of nitrogen and carbon in leaves of *Metrosideros polymorpha* regulates carboxylation capacity and delta ^{13}C along an altitudinal gradient. Funct. Ecol. 13:811–818.

Cushman, J. C., Taybi, T., and Bohnert, H. J. 2000. Induction of Crassulacean acid metabolism—Molecular aspects. In: Photosynthesis: Physiology and Metabolism. R. C. Leegood, T. D. Sharkey, and S. von Caemmerer (eds.), pp. 551–582. Dordrecht, The Netherlands: Kluwer Academic Publishers.

Daley, P. F., Raschke, L., Ball, J. T., and Berry. J. A. 1989. Topography of photosynthetic activity of leaves obtained from video images of chlorophyll fluorescence. Plant Physiol. 90:1233–1238.

Davies, X. J., and Zhang, J. 1991. Root signals and the regulation of growth and development of plants in drying soil. Annu. Rev. Plant Physiol. Plant Mol. Biol. 42:55–76.

Delfine, S., Alvino, A., Zacchini, M., and Loreto, F. 1998. Consequences of salt stress on conductance to CO_2 diffusion, Rubisco characteristics and anatomy of spinach leaves. Aust. J. Plant Physiol. 25:395–402.

Delfine, S., Alvino, A., Villani, M. C., and Loreto, F. 1999. Restrictions to CO_2 conductance and photosynthesis in spinach leaves recovering from salt stress. Plant Physiol. 119:1101–1106.

Downton, W. J. S., Loveys, B. R., and Grant, W. J. R. 1988. Stomatal closure fully accounts for the inhibition of photosynthesis by abscisic acid. New Phytol. 108:263–266.

Edwards G. E., and Baker, N. R. 1993. Can CO_2 assimilation in maize leaves be predicted accurately from chlorophyll fluorescence analysis? Photosyn. Res. 37:89–102.

Epron D., Godard D., Cornic G., and Genty B. 1995. Limitation of net CO_2 assimilation rate by internal resistances to CO_2 transfer in the leaves of two tree species (*Fagus sylvatica* L. and *Castanea sativa* Mill.). Plant Cell Environ. 18:43–51.

Evans, J. R. 1999. Leaf anatomy enables more equal access to light and CO_2 between chloroplasts. New Phytol. 143:93–104.

Evans, J. R., and Loreto, F. 2000. Acquisition and diffusion of CO_2 in higher plant leaves. In: Photosynthesis: Physiology and Metabolism. R. C. Leegood, T. D. Sharkey, and S. von Caemmerer (eds.), pp. 321–351. Dordrecht, The Netherlands: Kluwer Academic Publishers.

Evans, J. R., and Vellen, L. 1996. Wheat cultivars differ in transpiration efficiency and CO2 diffusion inside their leaves. In: Crop Research in Asia: Achievements and Perspective. R. Ishii and T. Horie (eds.), pp. 326–329. Tokyo: Asian Crop Science Association.

Evans, J. R., von Caemmerer, S., Setchell, B. A., and Hudson, G. S. 1994. The relationship between CO_2 transfer conductance and leaf anatomy in transgenic tobacco with a reduced content of Rubisco. Aust. J. Plant Physiol. 21:475–495.

Ewert, M. S., Outlaw, W. H. Jr., Zhang, S., Aghoram, K., and Riddle, K. A. 2000. Accumulation of an apoplastic solute in the guard-cell wall is sufficient to exert a significant effect on transpiration in *Vicia faba* leaflets. Plant Cell Environ. 23:195–203.

Farquhar, G. D., and Sharkey, T. D. 1982. Stomatal conductance and photosynthesis. Annu. Rev. Plant Physiol. 33:317–345.

Franks, P. J., Cowan, I. R., and Farquhar, G. D. 1995. A study of stomatal mechanics using the cell pressure probe. Plant Cell Environ. 18:795–800.

Gaudillere, J. P. 1981. CO_2 evolution outside leaves during photosynthesis. In: Photosynthesis 4. Regulation of Carbon Metabolism. G. Akoyunoglou (ed.), pp. 661–666. Philadelphia: Balaban International Science.

Genty, B., and S. Meyer. 1995. Quantitative mapping of leaf photosynthesis using chlorophyll fluorescence imaging. Aust. J. Plant Physiol. 22:277–284.

Genty, B., Meyer, S., Piel, C., Badeck, F., and Liozon, R. 1998. CO_2 diffusion inside leaf mesophyll of ligneous plants. In: Photosynthesis: Mechanisms and Effects. G. Garab (ed.), pp. 3961–3966. Dorecht, The Netherlands: Kluwer Academic Publishers.

Gorton, H. L., Herbert, S. K., and T.C. Vogelmann. 2003. Photoacoustic analysis indicates that chloroplast movement does not alter liquid-phase CO_2 diffusion in leaves of *Alocasia brisbanensis*. Plant Physiol. 132:1529–1539.

Gotow, K., Taylor, S., and Zeiger, E. 1988. Photosynthetic carbon fixation in guard cell protoplasts of *Vicia faba* L.: Evidence from radiolabel experiments. Plant Physiol. 86:700–705.

Haberlandt, G. 1914. Physiological Plant Anatomy, 4th German ed. Leipzig: W. Engelmann.

Hanba, Y. T., Miyazawa, S.-I., and Terashima, I. 1999. The influence of leaf thickness on the CO_2 transfer conductance and leaf stable carbon isotope ratio for some evergreen tree species in Japanese warm temperate forests. Funct. Ecol. 13:632–639.

Hanba, Y. T., Miyazawa, S., Kogami, H., and Terashima, I. 2001. Effects of leaf age on internal CO_2 transfer conductance and photosynthesis in tree species having different types of shoot phenology. Aust. J. Plant Physiol. 28:1–9.

Hanba, Y. T., Kogami, H., and Terashima, I. 2002. The effect of growth irradiance on leaf anatomy and photosynthesis in *Acer* species differing in light adaptation. Plant Cell Environ. 25:1021–1030.

Haupt-Herting, S., Klug, K., Fock, H. P. 2001. A new approach to measure gross CO_2 fluxes in leaves. Gross CO_2 assimilation, photorespiration, and mitochondrial respiration in the light in tomato under drought stress. Plant Physiol. 126:388–396.

Heath, O. V. S. 1950. The role of carbon dioxide in the light response of stomata. J. Exp. Bot. 1:29–62.

Hsiao, T. C., and Allaway, W. G. 1973. Action spectra for guard cell Rb^+ uptake and stomatal opening in *Vicia faba*. Plant Physiol. 51:82–88.

Karlsson, P. E. 1986. Blue light regulation of stomata in wheat seedlings. II. Action spectrum and search for action dichroism. Physiol. Plant 66:207–210.

Kinoshita T., and Shimazaki, K.-I. 1999. Blue light activates the plasma membrane H^+-ATPase by phosphorylation of the C-terminus in stomatal guard cells. EMBO J. 20:5548–5558.

Kinoshita, T., Nishimura, M., and Shimazaki, K.-I. 1995. Cytosolic concentration of Ca^{2+} regulates the plasma membrane H^+-ATPase in guard cells of faba bean. Plant Cell 7:1333–1342.

Kinoshita, T., Doi, M., Suetsugu, N., Kagawa, T., Wada, M., and Shimazaki, K.-I. 2001. phot1 and phot2 mediate blue light regulation of stomatal opening. Nature 414:656–660.

Kirschbaum, M. U. F., and Pearcy R. W., 1988. Gas exchange analysis of the relative importance of stomatal and biochemical factors in photosynthetic induction in *Alocasia macrorrhiza*. Plant Physiol. 86:782–785.

Kjellbom, P., Karlsson, C., Johansson, I., Karlsson, M., and Johanson, U. 1999. Aquaporins and water homeostasis in plants. Trends Plant Sci. 4:308–314.

Kogami H., Hanba Y. T., Kibe T., Terashima I., and Masuzawa T. 2001. CO_2 transfer conductance, leaf structure and carbon isotope discrimination of *Polygonum cuspidatum* leaves from low and high altitude. Plant Cell Environ. 24:529–538.

Kok, B. 1948. A critical consideration of the quantum yield of Chlorella-photosynthesis. Enzymologia 13:1–56.

Laisk, A. K. 1977. Kinetics of photosynthesis and photorespiration in C_3-plants. Nauka, Moscow (in Russian).

Lange, O. L., Losch, R., Schulze, E. -D., and Kappen, L. 1971. Responses of stomata to changes in humidity. Planta 100:76–86.

Lauteri, M., Scartazza, A., Guido, M. C., and Brugnoli, E. 1997. Genetic variation in photosynthetic capacity, carbon isotope discrimination and mesophyll conductance in provenances of *Castanea sativa* adapted to different environments. Funct. Ecol. 11:675–683.

Lawson, T., Oxborough, K., Morison, J. I. L., and Baker, N. R. 2002. Responses of photosynthetic electron transport in stomatal guard cells and mesophyll cells in intact leaves to light, CO2, and humidity. Plant Physiol. 128:52–62.

Lawson, T., Oxborough, K., Morison, J. I. L., and Baker, N. R. 2003. The responses of guard and mesophyll cell photosynthesis to CO2, O-2, light, and water stress in a range of species are similar. J. Exp. Bot. 54:1743–1752.

Leung, J., and Giraudat, J. 1998. Abscisic acid signal transduction. Annu. Rev. Plant Physiol. Plant Mol. Biol. 49:1999–222.

Li, L. G., Li, S. F., Tao, Y., and Kitagawa, Y. 2000. Molecular cloning of a novel water channel from rice: Its products expression in Xenopus oocytes and involvement in chilling tolerance. Plant Sci. 154:43–51.

Lloyd, J., Syvertsen, J. P., Kriedemann, P. E., and Farquhar, G. D. 1992. Low conductances for CO_2 diffusion from stomata to the site of carboxylation in leaves of woody species. Plant Cell Environ. 15:873–899.

Loreto, F., Harley, P. C., Di, M. M., and Sharkey, T. D. 1992. Estimation of mesophyll conductance to Co2 flux by three different methods. Plant Physiol. 98:1437–1433.

Loreto, F., DiMarco, G., Tricoli, D., and Sharkey, T. D. 1994. Measurements of mesophyll conductance, photosynthetic electron transport and alternative electron sinks of field-grown wheat leaves. Photosynth. Res. 41:397–403.

Loreto, F., Delfine, S., Di Marco, G. 1999. Estimation of photorespiratory carbon dioxide recycling during photosynthesis. Aust. J. Plant Physiol. 26:733–736.

Loreto, F., Velikova, V., Di Marco, G. 2001. Respiration in the light measured by $^{12}CO_2$ emission in $^{13}CO_2$ atmosphere in maize leaves. Aust. J. Plant Physiol. 28:1103–1108.

Ludwig, L. J., Canvin, D. T. 1971. The rate of photorespiration during photosynthesis and the relationship of the substrate of light respiration to the products of photosynthesis in sunflower leaves. Plant Physiol. 48:712–719.

Maier-Maercker, U. 1999. New light on the importance of peristomatal transpiration. J. Exp. Bot. 26:9–16.

Makino A., Nakano H., and Mae T. 1994. Effects of growth temperature on the responses of ribulose-1,5-bisphosphate carboxylase, electron transport components, and sucrose synthesis enzymes to leaf nitrogen in rice, and their relationships to photosynthesis. Plant Physiol. 105:1231–1238.

Mawson, B. T., and Zaugg, M. W. 1994. Modulation of light-dependent stomatal opening in isolated epidermis following induction of Crassulacean acid metabolism in *Mesembryanthemum crystallinum*. J. Plant Physiol. 144:740–746.

Maxwell, K., Von Caemmerer, S., and Evans, J. R.. 1997. Is a low internal conductance to CO_2 diffusion a consequence of succulence in plants with crassulacean acid metabolism? Aust. J. Plant Physiol. 24:777–786.

Meidner, H., and Mansfield, T. A. 1968. Physiology of Stomata. Maidenhead: McGraw-Hill.

Meyer, S., and Genty, B. 1998. Mapping intercellular CO_2 mole fraction (C_i) in *Rosa rubiginosa* leaves fed with abscisic acid by using chlorophyll fluorescence imaging. Significance of C_i estimation from leaf gas exchange. Plant Physiol. 116:947–957.

Mott, K. A., and Buckley, T. N. 2000a. Stomatal heterogeneity. J. Exp. Bot. 49:407–417.

Mott, K. A., and Buckley, T. N. 2000b. Patchy stomatal conductance: Emergent collective behaviour of stomata. Trend Plant Sci. 5:258–262.

Mott, K. A., and Franks, P. J. 2001. The role of epidermal turgor in stomatal interactions following a local perturbationin himidity. Plant Cell Environ. 24:657–662.

Mott, K. A., and Parkhurst, D. 1991. Stomatal responses to humidity in air and helox. Plant Cell Environ. 18:431–438.

Nakhoul, N. L., Davis, B. A. Romero, M. F., and Boron, W. F. 1998. Effect of expressing the water channel aquaporin-1 on the CO_2 permeability of *Xenopus* oocytes. Am. J. Physiol. 274C:543–548.

Niinemets, Ü. 1999. Components of leaf dry mass per area—thickness and density—alter leaf photosynthetic capacity in reverse directions in woody plants. New Phytologist 144:35–47.

Nobel, P. S. 1999. Physicochemical and environmental plant physiology, 2nd edition. San Diego: Academic Press.

Ogawa, T., Ishikawa, H., Shimada, K., and Shibata, K. 1978. Synergistic action of red and blue light and action spectrum for malate formation in guard cells of *Vicia faba* L. Planta 142:61–65.

Outlaw, W. H. Jr. 1983. Current concepts on the role of potassium in stomatal movements. Physiol. Plant 59:302–311.

Pachepsky, L. B., and Acock, B. 1996. A model 2DLEAF of leaf gas exchange: Development, validation, and ecological application. Ecol. Model 93:1–18.

Parkhurst, D. F. 1994. Diffusion of CO_2 and other gases inside leaves. New Phytologist 126:449–479.

Parkhurst, D. F., and Mott, K. A. 1990. Intercellular diffusion limits to CO_2 uptake in leaves. Plant Physiol. 94:1024–1032.

Parkhurst, D. F., Wong S. C., Farquhar, G. D.,, and Cowan, I. R.. 1988. Gradients of intercellular CO_2 levels across the leaf mesophyll. Plant Physiol. 86:1032–1037.

Piel, C., Frak, E., Le Roux, X., and Genty, B. 2002. Effect of local irradiance on CO_2 transfer conductance of mesophyll in walnut. J. Exp. Bot. 53:2423–2430.

Pinelli, P., and Loreto, F. 2003. $^{12}CO_2$ emission from different metabolic pathways measured in illuminated and darkened C_3 and C_4 leaves at low, atmospheric and elevated CO_2 concentration. J. Exp. Bot. 54:1761–1769.

Pons, T. L., and Welschen, R. A. M. 2002. Overestimation of respiration rates in commercially available clamp-on leaf chambers. Complications with measurement of net photosynthesis. Plant Cell Environ. 25:1367–1372.

Pospisilova, J., and Santrucek, J. 1994. Stomatal patchiness. Biol. Plant. 36:481–510.

Ramesh Prasad, G. V., Coury, L. A. Finn, F., and Zeidel, M. L. 1998. Reconstituted aquaporin 1 water channels transport CO_2 across membranes. J. Biol. Chem. 273:33123–33126.

Raschke, K. 1979. Movement of stomata. In: Encyclopedia of Plant Physiology, New Series, Vol. 7. W. Haupt and M. E. Feinleib (eds.), pp. 383–441. Berlin: Springer-Verlag.

Renou, J. L., Gerbaud, A., Just, D., and Andre, M. 1990. Differing substomatal and chloroplastic CO_2 concentrations in water-stressed wheat. Planta 182:415–419.

Roupsard, O., Gross, P., and Dreyer, E. 1996. Limitation of photosynthetic activity by CO_2 availability in the chloroplasts of oak leaves from different species and during drought. Ann. Sci. Forestieres 53:243–254.

Ridolfi, M., and Dreyer, E. 1997. Responses to water stress in an ABA-unresponsive hybrid poplar (*Populus koreana* × *trichocarpa* cv peace). 3. Consequences for photosynthetic carbon assimilation. New Phytologist 135:31–40.

Sakata, T., and Yokoi, Y. 2002. Analysis of the O-2 dependency in leaf-level photosynthesis of two Reynoutria japonica populations growing at different altitudes. Plant Cell Environ. 25:65–74.

Scartazza, A., Lauteri, M., Guido, M. C., and Brugnoli, E. 1998. Carbon isotope discrimination in leaf and stem sugars, water-use efficiency and mesophyll conductance during different developmental stages in rice subjected to drought. Aust. J. Plant Physiol. 25:489–498.

Schnabl, H. 1978. The effect of Cl^- upon the sensitivity of starch-containing and starch-deficient stomata and guard cell protoplasts towards potassium ions, fusicoccin and abscisic acid. Planta 144:95–100.

Schroeder, J. I., Ward, J. M., and Gassmann, W. 1994. Perspectives on the physiology and structure of inward-rectifying K+ channels in higher plants: Biophysical implications for K^+ uptake. Annu. Rev. Biophys. Biomol. Struct. 23:441–471.

Senn, G. 1908. Die Gestalts-und Lägeveranderung der Pflanzen-Chromatophoren. Leipzig: Verlag von Wilheim Engelmann.

Sharkey, T. D., and Raschke, K. 1981. Effects of light quality on stomatal opening in leaves *Xanthium strumarium* L. Plant Physiol. 68:1170–1174.

Sharkey, T. D., Vassey, T. L., Vancerveer, P. J., and Vierstra, R. D. 1991. Carbon metabolism enzyme and photosynthesis in transgenic tobacco (*Nicotiana tabacum* L.) having excess phytochrome. Planta 185:287–296.

Shimazaki, K.-I. 2001. Stomatal movemement. In: Responses to Environmental Variables in Plants. I. Terashima (ed.), pp. 40–48, Tokyo: Asakura (in Japanese).

Shimazaki, K.-I., and Kinoshita, T. 2001. Signal transduction mediated by blue light receptor: Blue light response in stomatal guard cells. In: Photosensing in Plants. M. Wada, S. Tokutomi, A. Ngatani, and M. Hasebe (eds.), pp. 114–122. Tokyo: Shujyunsha (in Japanese).

Shimazaki, K.-I., Iino, M., and Zeiger, E. 1986. Blue light-dependent proton extrusion by guard cell protoplasts of *Vicia faba*. Nature 319:324–326.

Siebke, K., and Weis, E. 1995a. Imaging of chlorophyll-a-fluorescence in leaves—Topography of photosynthetic oscillations in leaves of *Glechoma hederacea*. Photosynth. Res. 45:225–237.

Siebke, K., and Weis, E. 1995b. Assimilation images of leaves of *Glechoma hederacea*—Analysis of non-synchronous stomata related oscillations. Planta 196:155–165.

Stålfelt, M. G. 1955. The stoma as a hydrophotic regulator of the water deficit of the plant. Physiol. Plant 8:572–593.

Syvertsen, J. P., Lloyd, J., McConchie, C., Kriedemann P. E., and Farquhar, G. D. 1995. On the relationship between leaf anatomy and CO_2 diffusion through the mesophyll of hypostomatous leaves. Plant Cell Environ. 18:149–157.

Sze, H., Li, X., and Palmgren, M. G. 1999. Energization of plant cell membranes by H^+-pumping ATPases: Regulation and biosynthesis. Plant Cell 11:677–689.

Takagi, S. 2000. Roles for actin filaments in chloroplast motility and anchoring. In: Actin: A Dynamic Framework for Multiple Plant Cell Functions. C. J. Staiger, F. Baluska, D. Volkmann, and P. W. Barlow (eds.), pp. 203–212. Dordrecht, The Netherlands: Kluwer Academic Publishers.

Talbott, L. D., and Zeiger, E. 1998. The role of sucrose in guard cell osmoregulation. J. Exp. Bot. 49:329–337.

Terashima, I. 1992. Anatomy of non-uniform leaf photosynthesis. Photosynth. Res. 31:195–212.

Terashima, I., and Ono, K. 2002. Effects of $HgCl_2$ on CO_2 dependence of leaf photosynthesis: Evidence indicating involvement of aquaporins in CO_2 diffusion across the plasma membrane. Plant Cell Physiol. 43:70–78.

Terashima, I., Wong, S.-C., Osmond, C. B., and Farquhar, G. D. 1988. Characterisation of non-uniform photosynthesis induced by abscisic acid in leaves having different mesophyll anatomies. Plant Cell Physiol. 29:385–394.

Terashima, I., Miyazawa, S. I., and Hanba, Y. T. 2001. Why are sun leaves thicker than shade leaves? Consideration based on analyses of CO2 diffusion in the leaf. J. Plant Res. 114:93–105.

Uehlein, N., Lovisolo, C., Siefritz, F., and Kaldenhoff, R. 2003. The tobacco aquaporin NtAQP1 is a membrane CO2 pore with physiological functions. Nature 425:734–737.

Van der Werf, A., Poorter, H., and Lambers, H. 1994. Respiration is dependent on a species' inherent growth rate and on the nitrogen supply to the plants. In: A Whole Plant Perspective on Carbon-nitrogen Interactions. J. Roy and E. Garnier (eds.), pp. 83–103. The Hague: SPB Academic Publishing BV.

von Caemmerer, S., and Evans, J. R. 1991. Determination of the average partial pressure of CO_2 in chloroplasts from leaves of several C_3 species. Aust. J. Plant Physiol. 18:287–305.

Weyers, J. D. B., and Lawson, T. 1997. Heterogeneity on stomatal characteristics. Adv. Bot. Res. 26:317–351.

Yang, B., Fukuda, N., van Hoek, A., Matthay, M. A., Ma, T., and Verkman, A. S. 2000. Carbon dioxide permeability of aquaporin-1 measured in erythrocytes and lung of aquaporin-1 nul mice and in reconstituted proteoliposomes. J. Biol. Chem. 275:2686.

Yano, S., and Terashima. I. 2001. Separate localization of light signal perception for sun or shade type chloroplast and palisade tissue differentiation in *Chenopodium album*. Plant Cell Physiol. 42:1303–1310.

Zeevaart, J. A. D., and Creelman, R. A. 1988. Metabolism and physiology of abscisic acid. Annu. Rev. Plant Physiol. Plant Mol. Biol. 39:439–473.

Zeiger, E., and Hepler, P. K. 1977. Light and stomatal function: Blue light stimulates swelling of guard cell protoplasts. Science 196:887–889.

Zeiger, E., Talbott, L. D., Frechilla, S., Srivastava, A., and Zhu, J. X. 2002. The guard cell chloroplast: A perspective for the twenty-first century. [review] New Phytologist 153:415–424.

6
Leaf to Landscape

MATHEW WILLIAMS, F. IAN WOODWARD, DENNIS D. BALDOCCHI,
AND DAVID S. ELLSWORTH

Introduction

This chapter focuses on the diffusion of CO_2 from the atmosphere to the leaf interior. CO_2 diffusion occurs along a concentration gradient from the bulk atmosphere to the sub-stomatal cavity. Along this gradient there are "bottlenecks," or resistors, which control the rate of flow. The most important of these resistors occur at the leaf and stomatal level. This chapter explores the functioning of these resistors and their importance in constraining CO_2 capture.

Starting with the resistance that operates at the finest space and time scales, we examine arrangements of stomata on leaves and their relationship to stomatal conductance. Differences in stomatal structures are explored across key functional groups, including their phenotypic plasticity and variations within the canopy and across biomes. Models that have been constructed to analyse and predict stomatal conductance are described and compared. These models are often key components of current efforts to explore issues ranging from crop and forest production under global change, to plant-water relations, and the global carbon cycle. We then examine how leaf size and shape affects boundary-layer conductance, and, at the canopy scale, explore how stomatal and leaf-level constraints on CO_2 capture manifest themselves in aggregate. This aggregation is carried up to the global scale to describe patterns of global CO_2 capture, and the chapter closes with a discussion of the effects of atmospheric CO_2 enrichment on CO_2 capture.

Stomatal Control of Diffusion

Stomata present a variable resistance to the diffusion of CO_2 into the leaf during photosynthesis and the simultaneous outward diffusion of water vapor during transpiration. The total pore area of stomata is, at a maximum, only about 5% of leaf area, but rates of diffusion from the leaf could reach as high as 70% of a leaf with no stomatal resistance to diffusion. Field and laboratory observations of the resistance or conductance of stomata to diffusion are primarily concerned with measuring the combined, parallel opening of all of the stomata on a leaf. It is also possible to calculate stomatal conductance from that of an individual stoma to the whole leaf using theoretical considerations of stomatal size and diffusion. This is an important consideration as it provides a method of integrating the understanding of genetic, ontogenetic, and environmental controls on the development of stomata, with short-term measurements of conductance.

The diffusion conductance of stomatal pores can be defined as follows:

$$g_s = \frac{n_s DaP}{RT_K d} \qquad \text{(Eq. 6.1)}$$

where g_s is the stomatal conductance (units of mol m^{-2} s^{-1}) of the leaf surface s, n_s is stomatal density (mm^{-2}), D is the diffusivity of water vapor in air (m^2 s^{-1}), a is the stomatal pore area (m^2), P is atmospheric pressure (Pa), R the gas constant, T_K absolute temperature and d is the depth of the stomatal pore). The total stomatal pore area is a small fraction of the total leaf area, so the lines of gaseous flux converge when entering the pore, reducing diffusion. This effect is accounted for by adding an end correction term to Equation 6.1:

$$g_s = \frac{n_s DaP}{RT_K\left(d + \frac{\pi}{4}\sqrt{\frac{a}{\pi}}\right)} \qquad \text{(Eq. 6.2)}$$

Van Gardingen et al. (1989) modified the relationship determined by Brown and Escombe (1900) and others, on the basis of observations of low-temperature scanning microscopy of rapidly fixed leaves:

$$g_s = \frac{n_s DP}{RT_K\left(\frac{d}{\pi lw} + \frac{\log_e\left(\frac{4l}{w}\right)}{\pi l}\right)} \qquad \text{(Eq. 6.3)}$$

In Equation 6.3, stomatal pore area has been replaced by its two components of w, pore width, and l, length (m).

A comparison of the three models (Fig. 6.1) indicates close agreements between Equations 6.2 and 6.3 by Brown and Escombe and by Van Gardingen and coworkers. It is clear that without end-correction (Eq. 6.1) the estimates of stomatal conductance are much higher. Data presented in Franks and Farquhar (2001) indicate close agreement between observations of stomatal conductance and calculations by both Equations 6.2 and 6.3 (Table 6.1). Equations 6.2 and 6.3 indicate that stomatal conductance increases linearly with stomatal density (Fig. 6.2).

Stomatal Variability over the Leaf Surface

Weyers and Lawson (1997) describe many examples of heterogeneity in stomatal characteristics over the surface of a leaf and indicate that this is an area in need of much more research in order to understand the causes of this variability. However, there are interesting conclusions to be drawn and illustrated about

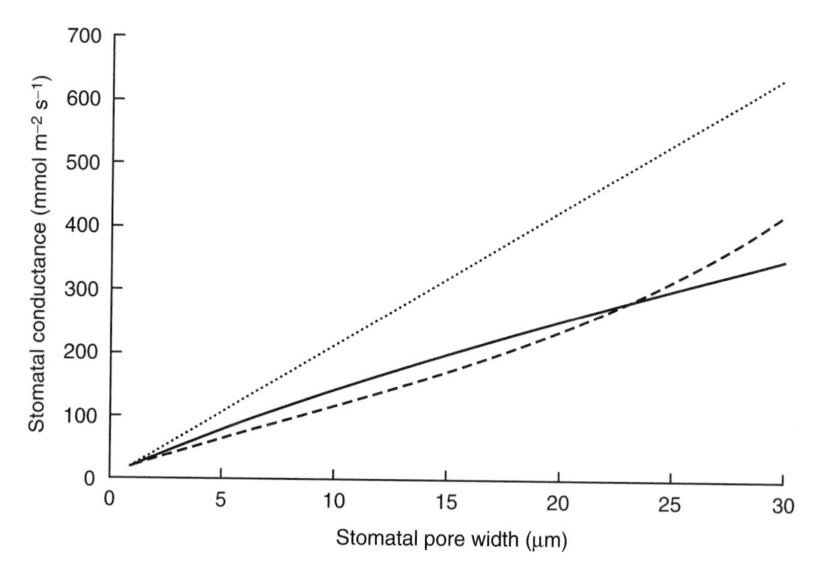

FIGURE 6.1. Predicted changes in stomatal conductance with stomatal pore width for the abaxial leaf surface of *Tradescantia virginiana*, using data presented in Franks and Farquhar (2001), and with the characteristics indicated on Table 6.1. Three models are used, Brown and Escombe (1900), without end correction (· · · · ·) and with end correction (———), and Van Gardingen et al. (1989) (- - - - -).

this variability. Laisk and coworkers (1980) measured marked variability of stomatal aperture over a leaf surface, using in situ microscopy. This variability also changed markedly with irradiance (Fig. 6.3). One feature of the calculations of stomatal conductance using theoretical considerations (Eqs. 6.2 and 6.3) is that predicted conductances are often much higher than observations. Variability of stomatal opening over the leaf surface is likely, therefore, to be an important component in determining the actual stomatal conductance.

TABLE 6.1. Relationship between observed maximum whole leaf stomatal conductance and predicted stomatal conductance for *Tradescantia virginiana*, using data presented in Franks and Farquhar (2001). For the control, adaxial stomatal density is 4.2 mm^{-2}, abaxial is 11.4 mm^{-2}, maximum stomatal width is 28 μm, and depth of the stomatal pore is 18 μm. For the treatment, adaxial stomatal density is 4.5 mm^{-2}, abaxial is 19.7 mm^{-2}, maximum stomatal width is 13 μm, and depth of the stomatal pore is 18 μm. The two model formulations for stomatal conductance are BE (Brown and Escombe 1900) and VG (Van Gardingen et al. 1989).

	Control	Treatment
Observed range maximum conductance (mmol m^{-2} s^{-1})	460–660	150–225
Predicted conductance, BE	480	210
Predicted conductance, VG	548	205

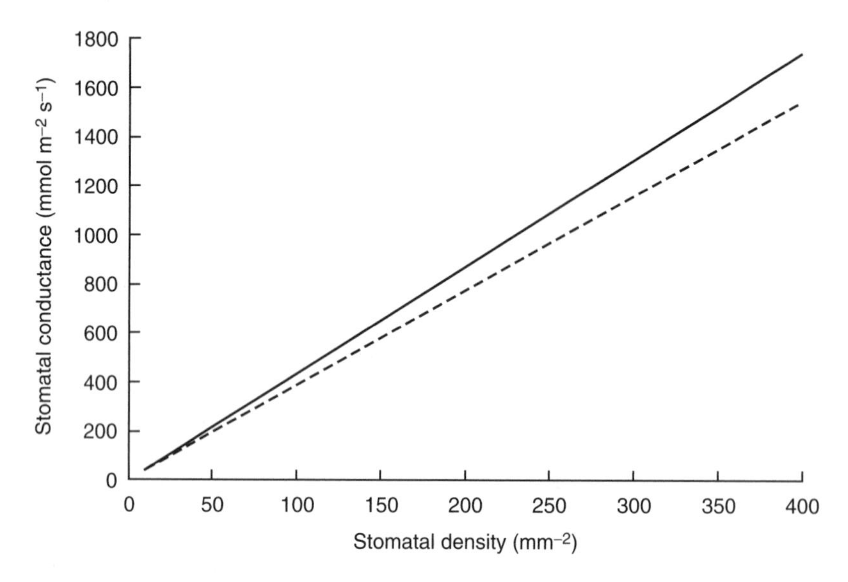

FIGURE 6.2. Predicted changes in stomatal conductance with stomatal density for the abaxial leaf surface of *Tradescantia virginiana*. Stomatal pore width 3 μm, stomatal pore length 10 μm. Models of Brown and Escombe (1900) with end correction (———) and Van Gardingen et al. (1989) (- - - - -).

Groups of stomata may respond differentially to a range of environmental conditions, leading to patches differing in stomatal conductance. This response has been observed for a range of environmental conditions (Buckley et al. 1999) and has led to a question of the impact on whole leaf processes. Buckley and coworkers have addressed this theoretically and indicate that patchiness exerts little impact on the plant carbon/water balance. The effect on water-use efficiency may be slightly positive for large leaves, but negative for small leaves. The packing of cells in leaves means that diffusion of CO_2 within a leaf is slow (Farquhar 1989), and this means that any variability in stomatal opening results in heterogeneity of CO_2 concentration within leaves. A mean internal CO_2 concentration (C_i) can be calculated in such a case, based on a weighting by either conductance ($C_{i,g}$), carboxylation efficiency ($C_{i,k}$), or by local assimilation rate ($C_{i,A}$). Farhquar (1989) shows that $C_{i,k} < C_{i,A} < C_{i,g}$, and concludes that sometimes the value of C_i obtained from gas exchange measurements will exceed that required for modeling photosynthesis.

A further interesting feature of stomatal development is that there may be considerable variation in stomatal density over the leaf surface (Weyers and Lawson 1997). Miranda and coworkers (1981) quantified this variability along a leaf of *Zea mays* (Fig. 6.4). The first 5 cm of the leaf, from the base, are located in the leaf sheath and 5 cm is the position of the ligule. Developmental and ontogenetic differences in stomatal characteristics were observed for this first 5 cm

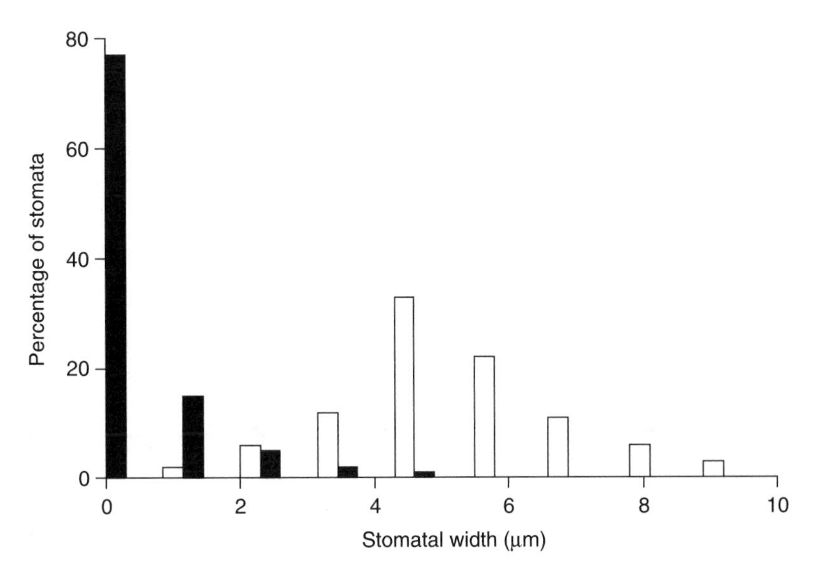

FIGURE 6.3. Fraction of stomata with different widths for *Vicia faba* at a low irradiance (black bars) and a higher irradiance (open bars).

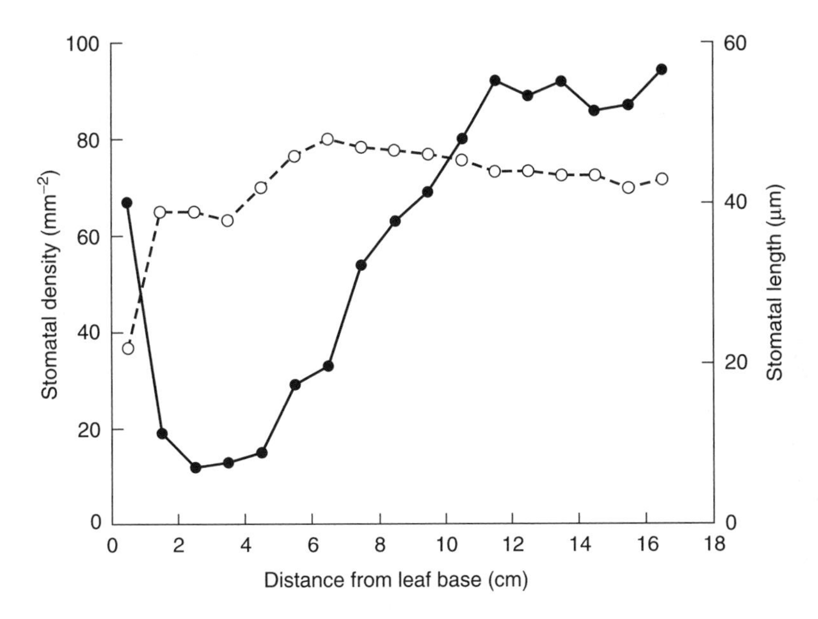

FIGURE 6.4. Variations in stomatal density (●) and stomatal length (○) along a leaf of *Zea mays* (from Miranda et al. 1981).

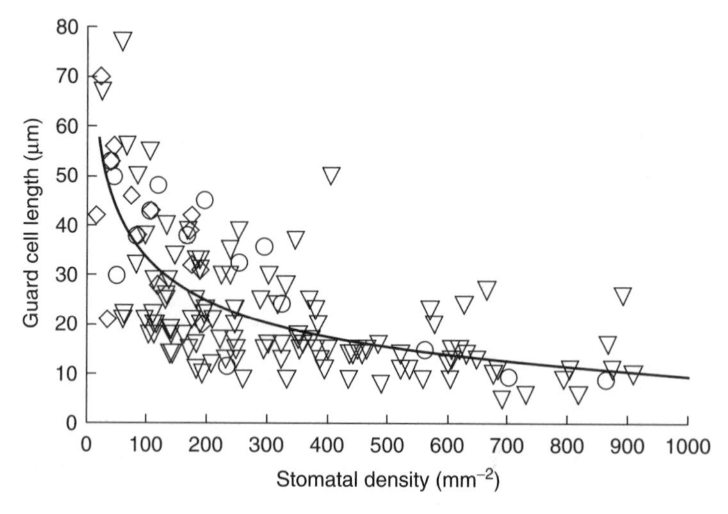

FIGURE 6.5. Relationship between stomatal width and density in amphistomatous species (◇), hypostomatous species (▽), and grasses (○). Curve is log-normal, $y = -28.75 + 162.x^{-0.2086}$, $r^2 = 0.5$ (data from Willmer and Fricker 1996, Sugden 1985, Tanner and Kapos 1982, Knapp 1993, and personal observations of F. I. Woodward).

of leaf. However stomatal density but not stomatal length varies for some considerable distance along the leaf. No explanation was offered for this change, particularly as the plants had been grown in constant environmental conditions.

The relationship between stomatal density and size, as seen for maize (see Fig. 6.4) is reflected more generally in a survey of 143 species (Fig. 6.5). A general increase in stomatal size occurs as stomatal density decreases, particularly below a stomatal density of about 200 mm^{-2}. This relationship holds for amphistomatous and hypostomatous leaves and for the leaves of grasses and suggests a general and global relationship between size and density. This response ensures the possibility of high stomatal conductances for plants with low stomatal densities, and therefore large stomata, such as during periods of abundant water supply, therefore minimizing the diffusion resistance to CO_2 uptake during photosynthesis. However, there is very little, if any, information on the evolutionary advantages of large versus small stomata; this intriguing topic remains understudied.

Stomatal Characteristics at the Plant Level

Ticha (1982) reviewed stomatal characteristics during plant development. She observed marked changes in stomatal development and density with plant age. Stomatal density and frequency of amphistomaty increased markedly with plant age in tomato (Gay and Hurd 1975) and this was ascribed to increasing irradiance during development. The decrease in irradiance through a plant canopy is associated with reductions in leaf nitrogen, photosynthesis, and stomatal conductance (Abrams and Kubiske 1990, Woodward et al. 1995), and at least some of these reductions in conductance may be caused by reductions in stomatal density. Smith and coworkers (1998) noted increases in amphistomaty for plant communities exposed to increasing irradiances. Support for this response has been provided by Peat and Fitter (1994) and Mott and coworkers (1982) have shown that amphistomaty is positively correlated with leaf thickness.

Amphistomaty may maximize photosynthesis by the parallel conductances of the two leaf surfaces to the larger surface areas of photosynthetic cells within the leaf (Parkhurst 1978). However Foster and Smith (1986) indicated that water-use efficiency is likely to be lower for amphistomatous leaves. Amphistomatous plants can behave as functionally hypostomatous, closing or nearly closing the adaxial stomata, independently of the abaxial stomata (Reich 1984). This will enhance water-use efficiency but photosynthesis may be reduced more than for a true hypostomatous species.

It seems that the impacts of natural selection on the occurrence of different distributions of stomata are rather unclear in some environments, and this may account for mixtures of all distributions in vegetation (Salisbury 1927). However, in more extreme environments the impacts are much clearer. Figure 6.6 shows the distribution of amphistomaty and hypostomaty with leaf thickness for 301 species of warm desert plants (Gibson 1996). In this environment, hypostomatous species only account for 10% of the species. Most species of both hypostomatous and amphistomatous species have leaf thicknesses between 200 and 400 μm, however on average, the hypostomatous species have relatively thinner leaves than the amphistomatous species.

It is very rare, in the literature, to find experimental investigations on the interactions between leaf structure, environment, and stomatal distribution. One

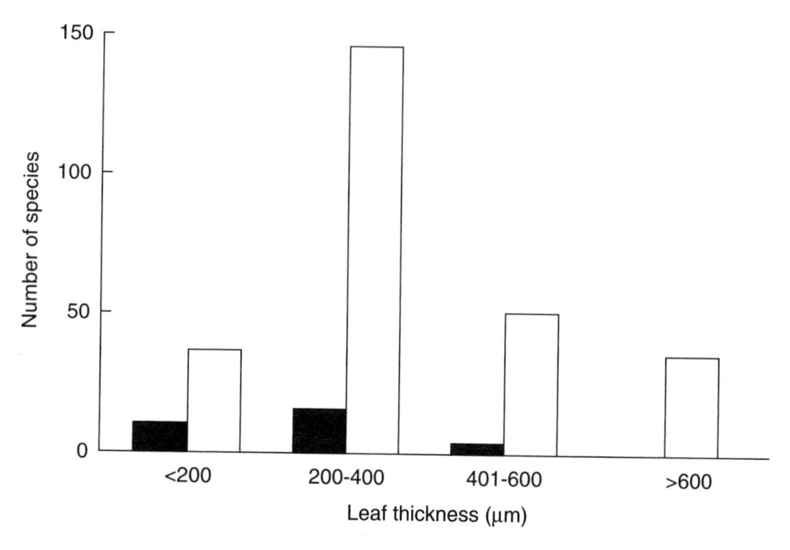

FIGURE 6.6. Number of warm desert plant species with either amphistomatous (open bars) or hypostomatous (black bars) leaves in relation to leaf thickness (data from Gibson 1996).

particular case which nicely describes the relationship between ontogeny, irradiance, and leaf thickness and stomatal development is that by James and Bell (2000). They grew saplings of *Eucalyptus globulus* ssp. *globulus* provenances under 100, 50, or 10% sunlight for 2 mo. They then analyzed a range of leaf structure characteristics from the three different light treatments. In this species, leaf form is heteroblastic and exists as either the juvenile, transitional, or adult form. The juvenile leaf form is thin and horizontally oriented, while the adult leaf is vertically oriented and isobilateral with both adaxial and abaxial palisade mesophyll. The transitional form is closer to the adult form in structure (James and Bell 2001). The experiment demonstrated (Fig. 6.7) that leaf thickness increased with irradiance and there was a trend to more equal amphistomaty. This trend was caused by an increase in adaxial stomatal density and a decrease in abaxial density with irradiance. The juvenile leaves were effectively hypostomatous under all treatments, indicating that the dorsiventral development of leaf structure and the horizontal orientation of the leaf itself appear to favor hypostomaty, rather in keeping with the observations of Smith and coworkers (1998).

Figure 6.8 and Table 6.2 indicate stomatal densities for different functional types of plant species. In all cases, there is skewed distribution with a long tail at high densities but most densities at about 200 to 400 mm^{-2}, except for succulent species with significantly lower stomatal densities. Among broad-leaved tree species there is significant (see Table 6.2) support for the observation (Mott

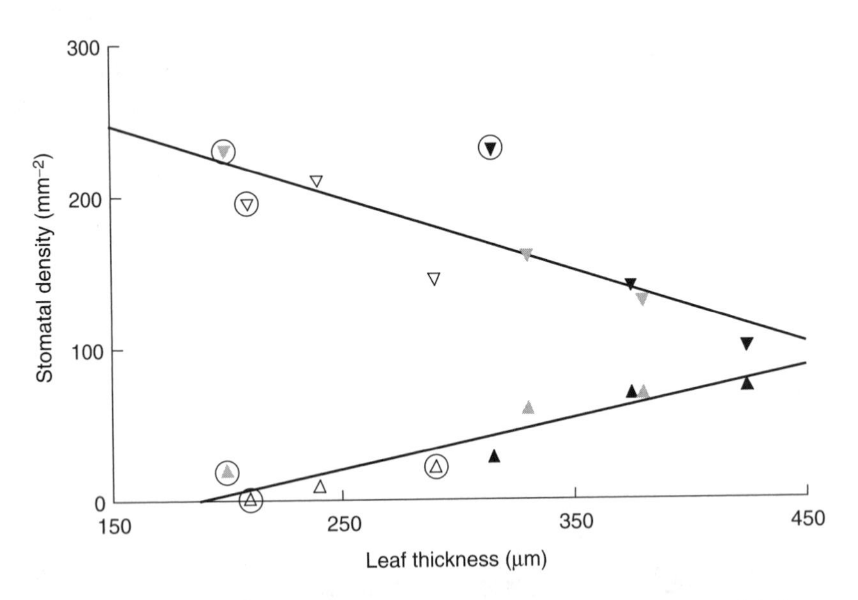

FIGURE 6.7. Changes in leaf thickness with stomatal density for *Eucalyptus globulus* ssp. *globules*, under three different light treatments: 100% sunlight (black symbols), 50% sunlight (grey symbols) and 10% sunlight (open symbols). Upward-facing triangles indicate adaxial surface, downward-facing triangles indicate abaxial surface. Symbols surrounded by circles indicate juvenile leaves (data from James and Bell 2000).

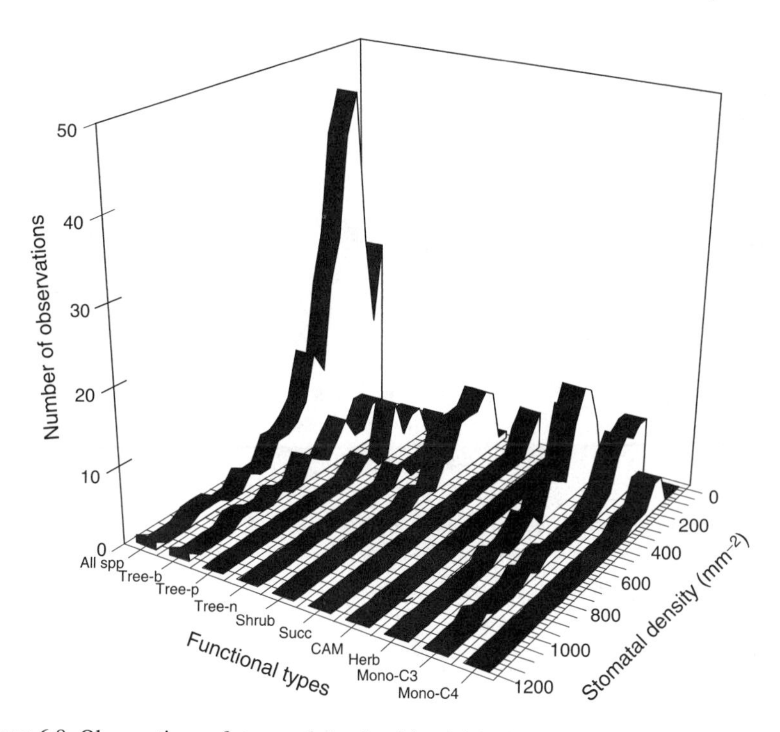

FIGURE 6.8. Observations of stomatal density (abaxial leaf surface) in different functional types of species: *All spp.*, all species; *Tree-b*, broad-leaved trees (not-pioneer); *Tree-p*, pioneer tree species; *Tree-n*, needleaved trees; *Shrub*, shrub species; *Succ*, succulent species, not-CAM; *CAM*, species with CAM photosynthesis; *Herb*, herbs, not including grasses or sedges; *Mono-C3*, monocots with C_3 photosynthesis; Mono-C4, monocots with C_4 photosynthesis (data on 246 different species from references in Fig. 6.5, Woodward and Kelly 1997, Salisbury 1927, and personal observations of F. I. Woodward).

et al. 1982) that pioneer species have lower stomatal densities than species from later in succession. This reflects, to a degree, a higher frequency of amphistomatous pioneer species in this data set. Amphistomatous species tend to have lower stomatal densities on the abaxial surface than hypostomatous species, and consequently larger stomata (see Fig. 6.5). In general, broad-leaved trees species in this sample of 246 species have significantly higher stomatal densities than all

TABLE 6.2. Stomatal densities [means (mm^{-2}) with 95% confidence limits] for nine functional types of plants. All species aggregated into functional types with no analysis by phylogenetic contrasts. Data from references indicated on Figure 6.8.

	Tree-broad	Tree-pioneer	Tree-needle	Shrub	Succulent	CAM	Herb	Mono-C_3	Mono-C_4
Mean	382	241	211	204	41	30	214	184	199
95% C.L.	82	51	135	31	15	13	31	68	50

of the other functional types of species. No other significant differences were observed, beyond that noted for succulent leaves and no attempt has been made to analyze the data by accounting for phylogeny.

Global Scale—Stomata

Global-scale surveys of stomatal density are somewhat incomplete and many biome types appear not to be sampled. However there is a reasonable global coverage of stomatal conductance, using measurements made by porometry. These data (Körner 1994, Schulze et al. 1994, and Woodward and Smith 1994) all demonstrate the rather conservative nature of the maximum stomatal conductance in all biome types. Körner (1994) quotes a global mean leaf conductance for woody vegetation of 218 ± 24 mmol m^{-2} s^{-1}, while the global mean value for all vegetation types in the data set of Woodward and Smith (1994) is 256 ± 15 mmol m^{-2} s^{-1}. The relationship between maximum stomatal conductance and annual precipitation for different vegetation types (Fig. 6.9) also indicates the conservative nature of maximum stomatal conductance, although with a tendency for the highest values to be found in wet tropical vegetation. The measurements of boreal needle-leaved vegetation are calculated on a projected leaf area basis and show little difference from broad-leaved vegetation.

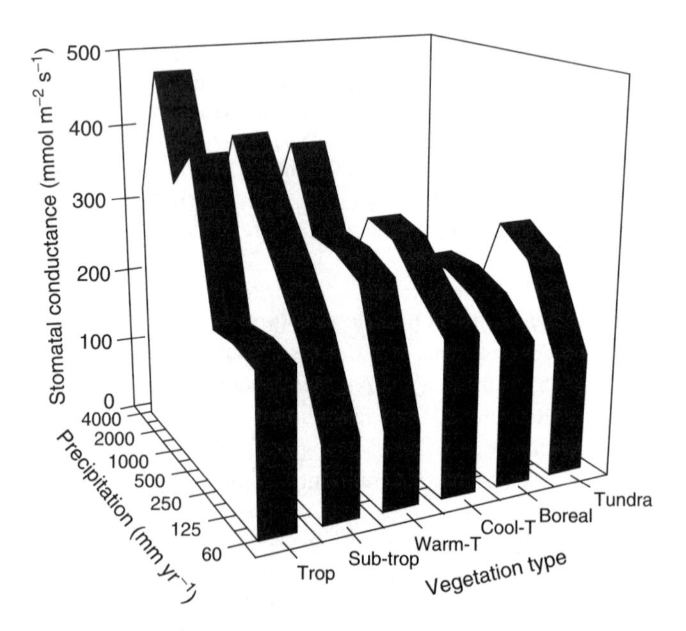

FIGURE 6.9. Maximum stomatal conductance for six major types of vegetation and in relation to annual precipitation: *Trop*, tropical vegetation; *Sub-trop*, subtropical vegetation; *Warm-T*, warm temperate vegetation; *Cool-T*, cool temperate vegetation; *Boreal*, boreal vegetation; *Tundra*, tundra (from Körner 1994, Schulze et al. 1994, and Woodward and Smith 1994).

CO_2 Capture via Diffusion

CO_2 transfer from the surface of the leaf to the leaf interior occurs via diffusion through the stomata. The process of diffusion is driven by the thermal motions of molecules, and, in an inhomogeneous fluid, the resulting random rearrangement of particles results in their mass transfer. Thus, in a still fluid, random diffusive movements will tend to move particles from areas of high to low concentration. This phenomenon is described by Fick's First Law, where the rate of mass transfer (**J**) of substance i through a plane is governed by the concentration gradient across the plane, and a diffusion coefficient (**D**):

$$\mathbf{J}_i = -\mathbf{D}_i \hat{\rho} \frac{\partial m_i}{\partial x} \quad \text{(Eq. 6.4)}$$

The flux is in the direction of decreasing concentration, hence the minus sign. The value of the diffusion coefficient depends on the properties of the diffusing substance and the diffusing medium. For instance, larger molecules diffuse more slowly, so the diffusion coefficient for CO_2 in air is less than that for water. Diffusivity also changes with temperature and atmospheric pressure, because of effects on molecular vibration. Temperature gradients between sources and sinks can thus cause significant errors, and for this reason it is typical to use mole fraction (m), corrected by molar density of air ($\hat{\rho}$), instead of concentration in flux equations (Campbell and Norman 1998).

For the specific case of CO_2 capture for leaves (A), Fick's Law can be rewritten as a function of stomatal conductance (g_s) and the difference between leaf external (c_s) and leaf internal CO_2 mole fraction (c_i):

$$A = g_s(c_s - c_i) \quad \text{(Eq. 6.5)}$$

where the conductance g_s (mol m^{-2} s^{-1}) is defined as a function of diffusion coefficient of CO_2 and the length of the diffusive pathway (l) (see also Eqs. 6.1 to 6.3):

$$g_s = \frac{\mathbf{D}_c \hat{\rho}}{l} \quad \text{(Eq. 6.6)}$$

As described above, by examining stomata microscopically it is possible to determine the ratio of stomatal cross-sectional area to leaf area, and the depth of the stomatal pores to determine l. These data can then be used to construct an estimate of g_s for the entire leaf (Eqs. 6.1 to 6.3). However, the technique is time consuming and not suitable for quantifying the dynamics of stomatal behavior, whereby changes in pore size can have rapid and significant effects on conductance. For this reason, most estimates of g_s are made indirectly with gas exchange systems (Field et al. 1982). Rates of evapotranspiration are recorded for intact leaves in cuvettes, at a known vapor pressure deficit. These data can then be used to calculate stomatal conductance for water vapor, which can be converted (via the ratio in diffusivities for water vapor and CO_2 in air) to an estimate of stomatal conductance for CO_2. Experimental studies have indicated stomatal sensitivity to radiation, temperature, humidity, leaf water status, and atmospheric CO_2

concentration (Jarvis and McNaughton 1986). Many of the responses to these variables are interactive, and individual responses are not always clear because of covariation between many environmental variables under natural conditions (e.g., temperature, vapor pressure deficit, and light). Attempts to describe dynamics of stomatal conductance have thus focused on producing quantitative models that can reproduce and, in some cases, explain observed behaviors.

Models of Stomatal Conductance

Numerous models have been constructed to analyze and synthesize the complex behavior of stomatal opening as a basis for understanding and predicting CO_2 capture. The models range in complexity from the empirical to the highly mechanistic.

Empirical Models

The simplest models are purely empirical or phenomenological, modifying a reference value, specific to a species or functional type, according to changes in environmental parameters. Jarvis's model (1976) is derived from experimentally determined relations between g_s and environmental variables: quantum flux density (I_p), ambient CO_2 concentration (c_a), vapor pressure deficit (D), temperature (T_a), and leaf/soil water potential (Ψ):

$$g_s = f(I) \cdot f(T) \cdot f(D) \cdot f(\Psi) \cdot f(c_a) \qquad \text{(Eq. 6.7)}$$

The form of the functional relations between g_s and each driving variable is best determined in controlled environmental conditions to avoid confusion from multiple interacting constraints. Once the form of the relation has been determined, parameters to best fit the function to field data can be determined by nonlinear least squares. Either a rectangular hyperbola or an asymptotic exponential function can used to represent the light response. Polynomial functions are used for the temperature response, as stomatal conductance is limited by low and high temperatures and exhibits a broad maximum. The response of stomata to vapor pressure deficit [$f(D)$] is complex. Its formulation depends on whether there are feedback or feedforward effects. Changes in D can result in either a linear decrease in g_s, or a decrease to an asymptote:

$$f(D) = 1 - kD \qquad \text{(Eq. 6.8)}$$
$$f(D) = \exp(-kD)$$

When there are feedforward effects, an increase in D causes g_s to decline directly. Evapotranspiration (E) decreases solely in response to the decrease in g_s:

$$g_s \propto \frac{E(g_s)}{D} \qquad \text{(Eq. 6.9)}$$

When there are feedback effects, an increase in D can impose a decrease in g_s, but it also increases the leaf temperature, which can promote D and E:

$$g_s \propto \frac{E(D, g_s)}{D} \qquad \text{(Eq. 6.10)}$$

The algorithm of Jarvis does not consider feedbacks between stomatal conductance and the surface energy balance; g_s is not updated to reflect changes in leaf temperature and humidity. One can also argue, from a modeling standpoint, that it is inappropriate to apply the Jarvis (1976) model in an iterative mode. The Jarvis model is diagnostic and does not include feedback loops among stomatal conductance, internal CO_2, transpiration, humidity deficits, and leaf water potential (Jones and Rawson 1979). The multiplicative, stomatal conductance algorithm requires a considerable amount of tuning and calibration to yield reasonable mass and energy flux densities (Sellers et al. 1989, Baldocchi 1992). Information on maximum conductance and curvature coefficients for the light, temperature, humidity deficit, and soil moisture response functions are needed.

Water-Use Efficiency Models

Cowan and coworkers (Cowan 1977, Cowan and Farquhar 1977) developed theories based on the principle that gas exchange is optimal when the maximal amount of carbon is assimilated (A) for a given amount of water lost (E). Optimal stomatal behavior was defined as the state that maintained $\partial E/\partial A$ constant ($=\lambda$) provided $\partial^2 E/\partial A^2 > 0$. Cowan used economic theory to explain this relationship, with the slope, $\partial E/\partial A$, describing the marginal cost. If this theory is true, then a plant maintains a uniform marginal cost, but the theory says nothing about the value of λ. In practice, the model produces diurnal behavior of A and E that mimics field observations. This theory also led to the insight that stomatal conductance is correlated with photosynthesis, and led to a new generation of stomatal conductance models based on biochemical principles rather than on empirical curve fits. Cowan's theory has not escaped criticism, even from himself. For instance, the theory does not explain how λ varies with soil moisture conditions. The theory is also of limited conceptual value if it does not provide an explanation of why plants are adapted to optimize water use, because the economy of water use seems to be small, and a constant $\partial E/\partial A$ does not account for competition. What value is there in economizing water use if that water can instead be used by a plant's competitors? Also, does short-term regulation of carbon and water exchange relate to long-term adjustments in root and shoot partitioning?

Coupled Photosynthesis-Stomatal Conductance Models

Wong et al. (1979) performed a set of experiments to show that C_3 plants tend to keep the ratio of internal to external CO_2 concentration (c_i/c_a) constant. The metric was near 0.7 for C_3 plants and near 0.4 for C_4 plants. Norman (1982) used this concept to close a set of equations to predict stomatal conductance on the basis of a simple leaf photosynthesis model:

$$g_s = \frac{A}{c_a\left(1 - \frac{c_i}{c_a}\right)} \qquad \text{(Eq. 6.11)}$$

Field experiments (Ellsworth 1999) and elevated CO_2 studies (Sage 1994) have confirmed the occurrence of constant c_i/c_a, although there is no surety that it applies to all species and at all times. Stomatal conductance is controlled in this model largely by assimilation rates. The weaknesses of the constant c_i/c_a ratio approach relate to a lack of detail on soil moisture and plant hydraulic effects. Reduced soil moisture or restrictions in hydraulic conductance are conditions where we might expect a constant c_i/c_a ratio to breakdown.

Ball and coworkers (1987) and Collatz and coworkers (1991) also drew upon the observations of Wong and coworkers (1979) and their own laboratory experiments to publish a model (known as the Ball-Berry model) that linked stomatal conductance to leaf photosynthesis, humidity deficits, and CO_2 concentration at the leaf's surface (c_s).

$$g_s = A \frac{m \cdot rh}{c_s} + g_0 \qquad \text{(Eq. 6.12)}$$

The coefficient m is a dimensionless slope, rh is relative humidity at the leaf surface, g_0 is the zero intercept, and A (μmol m^{-2} s^{-1}) is leaf photosynthesis. Generally, A is derived from a biochemical model of photosynthesis, such as that of Farquhar and von Caemmerer (Farquhar and von Caemmerer 1982). The Ball-Berry model has several appealing attributes. Firstly, this scheme provides us with an algorithm that is able to calculate how stomatal conductance correlates with ecophysiological and biogeochemical factors, such as leaf photosynthetic capacity and nutrition (Schulze et al. 1994, Leuning et al. 1995) and ambient CO_2 concentration. This feature constrains the potential range of values to expect for a given species or plant functional type. Secondly, this model requires fewer tuning parameters than the empirical model above (Eq. 6.7).

The parameter m varies across a range of calculated values (Table 6.3), but it is likely to vary further under conditions of moderate to severe soil moisture deficits (Baldocchi 1997) or in cases involving old trees with decreased hydraulic conductivity (Hubbard et al. 1999). This inability or difficulty in coping with the effects of soil moisture deficits has led to some criticism of the Ball-Berry model. Also, some investigators argue from a mechanistic point of view that stomatal conductance must be dependent on water vapor saturation deficit (Leuning 1995) and transpiration (Monteith 1995a), rather than relative humidity.

TABLE 6.3. Values of the slope parameter m in the Ball-Berry equation as determined in various field studies.

m	Species (location)	Source
9.0	*Glycine max*	Collatz et al. 1991
5.9	*Pinus taeda* (NC, USA)	Katul et al. 2000
9.5	*Quercus/Acer* forest (MA, USA)	Harley and Baldocchi 1995
4.8	*Picea mariana* (Canada)	Dang et al. 1998
7.8	Mixed deciduous forest (TN, USA)	Wilson et al. 2001
5.6	*Pinus banksiana* (Canada)	Dang et al. 1998

Water-Control Feedback Loops

Mott and Parkhurst (1991) provided convincing evidence that the apparent response of stomata to vapor pressure deficit, D, should be interpreted as a response to transpiration rate. Monteith (1995b) has noted that measurements showing a nonlinear relationship between leaf conductance and D can be reinterpreted as a linear relation between g_s and E with the form

$$g_s = a - bE \qquad \text{(Eq. 6.13)}$$

where the constant a is the maximum conductance achieved when E is zero.

Recent experiments have demonstrated a strong coordination between stomatal and hydraulic properties of plants in multiple species (Meinzer 2002). This link is intuitive if we assume that stomata operate to maximize carbon assimilation, while maintaining leaf water balance at a level that ensures maintenance of metabolism. More detailed models of stomata (Williams et al. 1996, Whitehead 1998) incorporate this philosophy by explicitly tracking the water status of leaves, as a function of gas phase losses and liquid phase recharge, and relating water status to stomatal opening.

To link stomatal conductance to leaf water status, the direct connection of g_s to A must be abandoned. Instead, g_s is linked to fluxes of water through the plant (J_p). Liquid fluxes are a function of differences in water potential between soil (Ψ_s) and leaf (Ψ_l), and the resistances of the hydraulic pathway in soil (R_s) and plant (R_p). In tall trees, the gravitational potential on water transport is also significant (h = tree height; g = gravitational acceleration; ρ_w = density of water), and thus;

$$J_p = \frac{\Psi_s - \Psi_l - \rho_w g h}{R_s + R_p} \qquad \text{(Eq. 6.14)}$$

However, the relationship between J_p and the water potential drop is not unique; initially water is drawn from stores within plant tissues, so that liquid flow lags behind the evaporative demand (Schulze et al. 1985). This hysteresis can be modeled by incorporating capacitors into the electrical circuit analog for water flow (Jones 1992). The capacitance (C) of any part of the system is defined as the ratio of the change in tissue water content (W) to the change in water potential:

$$C = \frac{dW}{d\Psi} \qquad \text{(Eq. 6.15)}$$

The rate of change of foliar water content (dW/dt) is given by the difference between the flow of water into the layer and that lost by evaporation:

$$dW_l/dt = J_p - E \qquad \text{(Eq. 6.16)}$$

Thus,

$$dW_l/dt = (\Psi_s - \Psi_l - \rho_w g h)/(R_s + R_p) - E \qquad \text{(Eq. 6.17)}$$

Assuming constant capacitance, the first-order differential equation describing leaf water potential dynamics is

$$\frac{d\Psi_l}{dt} = \frac{\Psi_s - \rho_w g h - E(R_s + R_p) - \Psi_l}{C(R_s + R_p)} \quad \text{(Eq. 6.18)}$$

William et al. (1996) incorporated this equation into a numerical algorithm that varied stomatal conductance so that E (determined by the Penman-Montieth equation) was maintained at the level that kept Ψ_l from falling below a critical threshold value (Ψ_{lmin}). The critical value is a point below which potentially dangerous cavitation of the hydraulic system may occur (Jones and Sutherland 1991). Thus, once $\Psi_l = \Psi_{lmin}$, g_s is adjusted to set E so that $d\Psi_l/dt = 0$. This approach balances atmospheric demand for water with rates of water uptake and supply from soils. In conditions where atmospheric demand exceeds supply (e.g., early morning) stomata open in response to declining intercellular CO_2 concentrations (Mott 1988), but opening is limited to maintain a minimum water use efficiency.

The strength of this approach is that the model integrates the effects on CO_2 capture of both atmospheric demand for water and rates of water supply from soil via the plant hydraulic system. Thus, the model overcomes the problem associated with the more simple approaches that treat water supply as a single tuning factor. Also, the model can simulate stomatal closure as a result of either rate limitation on water supply (the hydraulic pathway has reached its maximum capacity) or an absolute limitation on supply (the soil matrix around the roots has a declining water potential). The difficulty with this approach to simulation is that the complexity of the model raises the parameter requirements. Hydraulic resistance data are becoming more common (Tyree and Ewers 1996, Wullschleger et al. 1998, Becker et al. 2000), but there is considerable uncertainty about the relative importance and the dynamics of the different components in the hydraulic path from soil through roots to stems and leaves (Williams et al. 1998). And assumptions that leaf water potential is maintained above a critical level may not always be valid (Sperry et al. 1998, Williams et al. 2001a).

Leaves

Stomatal conductance describes the limitation on CO_2 transfer from the leaf surface to the leaf interior. However, the CO_2 concentration at the leaf surface will likely differ from that of the bulk atmosphere once a concentration gradient has been generated by CO_2 fixation. The transfer of CO_2 from the canopy airspace to the surface of an individual leaf can occur at the molecular level via diffusion, or via mass movement of air in the process of convection. In the case of diffusion, the movement of air across the leaf surface can speed up diffusive mass transfer considerably by continually replenishing the air. The replenishment maintains the concentration gradient between leaf interior and exterior, and is more likely to occur in heterogeneous environments such as isolated leaves and plants. Air movement via convection may arise in two ways. Firstly, changes in air density resulting from heating or cooling by an adjacent surface cause free convection; heated air expands and rises, chilled air contracts and sinks. Secondly, air movements driven by pressure gradients generate forced convection. Air movements over a surface can generate turbulence as a result of frictional interactions. The size of surface irregularities tends to control the scale of the eddies, and so turbulent motions are generally larger than the movements of molecules in diffusion. In fact, turbulent transfer tends to be orders of magnitude (between 3 and 7) more rapid than diffusion (Jones 1992).

The velocity of the air flowing over a surface is reduced by friction, and the zone of reduced velocity is known as the boundary layer. Even in turbulent conditions, there is still likely to be some distance close to the leaf surface where friction limits turbulence, and diffusion is the dominant mode of mass transfer. The depth of the boundary layer in air tends to be two orders of magnitude smaller than the characteristic dimension of the surface, and so mass transfer across the layer can be regarded as a one-dimensional process at right angles to the surface (Jones 1992). The conductance of the boundary layer is given by

$$g_b = \frac{\mathbf{D}_i \hat{\rho}}{\delta} \qquad \text{(Eq. 6.19)}$$

where δ is the thickness of the boundary layer, itself given by the semi-empirical expression

$$\delta = \sqrt{\frac{vx}{u}} \qquad \text{(Eq. 6.20)}$$

where v is the kinematic viscosity of air, u is the wind speed and x is the distance from the leading edge of the leaf. Typically, boundary-layer conductance under laminar flow conditions is an order of magnitude greater than stomatal conductance. In turbulent conditions, the estimates of conductance given by these equations should be increased by a factor of 1 to 2, and even as much as a factor of 2.5 (Schuepp 1993). The Reynolds number (Re) describes the ratio of inertial to viscous forces, and is a useful indicator of whether the flow over a surface is laminar or turbulent. The thickness of the boundary layer will be considerably reduced by turbulence, with the transition to turbulent flow typically occurring above a Re of 2×10^4 (Grace and Wilson 1976, Monteith and Unsworth 1990), though the actual transfer point will depend on leaf characteristics. For instance, Jones (1992) notes that sparse hairs may increase the likelihood of turbulence, while dense hairs have the opposite effect by increasing the effective depth of the boundary layer. Even within a dense plant canopy, turbulent conditions are highly likely due to intermittent incursions from the planetary boundary layer above the canopy, and from buoyant rising of warmer parcels of air.

Velocity profile measurements above leaf surfaces have confirmed that the thickness of the boundary layer grows with increasing downwind distance from the leading edge of a leaf (Grace and Wilson 1976). The resulting variation in the depth of the boundary layer across the leaf needs to be integrated into any estimation of a single boundary-layer conductance value if this number is used to represent the entire leaf in a model. The usual solution is to employ a leaf characteristic length (**d**) in place of distance from the leading edge (*x*). Theoretical formulations and wind tunnel studies have revealed that **d** is generally between 0.5 and 0.8 of the maximum leaf dimension in the direction of flow (Parkhurst et al. 1968). The derivation of **d** assumes a leaf is infinitely wide perpendicular to the wind stream. In smaller leaves and in leaves with low width-to-length ratios, we can expect increasingly important edge effects.

Leaf shape is an important determinant of boundary-layer characteristics. Boundary-layer thickness is reduced in deeply lobed leaves, while the effects of leaf curvature (deviation from the theoretical planar leaf) indicate an earlier transition to roughness (Schuepp 1993). There is some evidence of a saturation effect on enhancement of boundary-layer thickness. For instance, in turbulent conditions, the addition of leaf curvature or leaf flapping may have an insignificant effect on boundary-layer thickness.

Lobes and serration reduce effective leaf dimension, and so reduce average boundary-layer thickness compared with a non-lobed leaf of similar area. Models have shown that lobed leaves are much more efficient at shedding heat in calm conditions (Vogel 1970). Regular-shaped leaves also tend to display differing boundary-layer thickness depending on their orientation relative to wind velocity. In lobed leaves, these orientation effects are considerably lessened (Schuepp 1993). The ability of leaves to shed heat more effectively has only an indirect effect on CO_2 capture. It may be that leaf shape and size have evolved largely in response to selection for maintaining leaf energy balance within metabolic limits. Baldocchi and coworkers (1985) examined two isolines of *Glycine max* differing only in their leaf width. The canopy with narrower leaves showed enhanced photosynthesis on a leaf area basis, but no significant difference on a ground area basis.

Leaf-level CO_2 capture can be measured in cuvettes with a porometer or a infrared gas analyzer (Jones 1992). Typically, air within the chamber of the cuvette is stirred with a fan to minimize the boundary-layer resistance, so that the measured rate of evaporation can be used to calculate stomatal conductance. Boundary-layer resistance is generally measured using analog; for instance, by measuring water loss from wet surfaces (e.g., blotting paper) of the same size as a leaf, with the same external conditions, or heat loss from heated brass plates (Brenner and Jarvis 1995). These leaf-level data are aggregated summaries of the varied conductance characteristics dependent on patterns of stomatal density and the leaf boundary layer, which vary at scales finer than the leaf itself.

Canopy

Measurements of whole ecosystem exchange of latent energy and CO_2 via eddy covariance are increasingly common (Valentini et al. 2000). Similarly, models of canopy CO_2 and water exchange (Dickinson and Henderson-Sellers 1988, Sellers et al. 1989) are a key component in regional biogeochemical and climate models. These models are generally of two types—"big-leaf" or multilayer. "Big-leaf" models are simpler, ascribing to the entire canopy a single physiological and aerodynamic resistance to water/CO_2 transfer. Multilayer models seek to incorporate at least the vertical variation in canopy structure and microclimate, and simulate the activity of the entire canopy as the summation of activity in multiple canopy layers with individual stomatal and leaf boundary-layer conductances. There has been considerable discussion on the benefits and costs of big-leaf versus multilayer approaches (Raupach and Finnigan 1988, Williams et al. 1996, De Pury and Farquhar 1997, Williams et al. 1997, Baldocchi and Meyers 1998). The big-leaf is computationally simpler, but has difficulties incorporating the nonlinear behavior of leaf-level photosynthesis, for example, coping with the patchy light climate within a canopy. Also, big-leaf model parameters tend to be aggregated and thus more brittle. For example, canopy conductance can be estimated from whole ecosystem measurements, but a model to simulate its dynamics would have to take account of the potentially different response behavior of stomata in the upper and low canopies. Multilayer models have parameters that can be measured and directly related to processes (e.g., stomatal conductance can be recorded by leaf-level gas exchange techniques). But parameter requirements tend to be numerous; if parameters have to be estimated with poorly defined errors, then the resulting model is underdetermined and prone to uncertainty.

To better understand the constraints on CO_2 capture at the canopy scale, we have combined a detailed multilayer model of canopy gas exchange with eddy covariance measurements of net ecosystem CO_2 exchange (NEE) and latent energy (LE) fluxes from a temperate deciduous forest. Our goal is to demonstrate the relationships between stomatal, leaf, and canopy conductance, and their importance in constraining CO_2 capture. We use the Soil-Plant-Atmosphere (SPA) model (Williams et al. 1996), which has a detailed radiative transfer scheme that splits each of four canopy layers into sunlit and shaded fractions (Williams et al. 1998), and uses a previously described stomatal submodel (see Water-Control Feedback Loops above) and leaf boundary layer submodel (see Leaves above). The stomatal submodel is applied separately to the sunlit and shaded fractions. The model attenuates wind speed with depth in the canopy, and we have increased the g_b estimates by a factor of 2.0 to account for turbulence effects.

The eddy covariance data for testing the simulations are from a 60-year-old *Quercus/Acer* stand at the Harvard Forest Long-Term Ecological Research site in central Massachusetts, USA. We used meteorological and flux data from 1995 (data provided by S. C. Wofsy) and the SPA model parametrization previously used at Harvard Forest (Williams et al. 1996) with recent updates to soil heat/water transport routines (Williams et al. 2001b). To compare model predictions of photosynthesis and leaf-level respiration to NEE data, we made a simplification and assumed a background soil/stem respiration of 3 μmol m^{-2} s^{-1}. For full details on the data collection and processing methodology, and associated confidence limits, see Goulden and coworkers (1996a,b).

The model estimates g_s and g_b for each of the four canopy layers each hour of the day, and we looked at 5 days during midsummer for the purposes of this analysis. From both predicted and measured canopy LE flux (λE) we were able to back-calculate the hourly canopy conductance, g_c, using a simplified version of the Penman-Monteith equation:

$$g_c = \frac{\gamma \lambda E}{c_p \rho_w D} \quad \text{(Eq. 6.21)}$$

where γ is the pyschrometer constant, and c_p is the specific heat of air.

The model parametrization was well able to explain the diurnal patterns of CO_2 capture and release recorded by the eddy covariance system (Fig. 6.10).

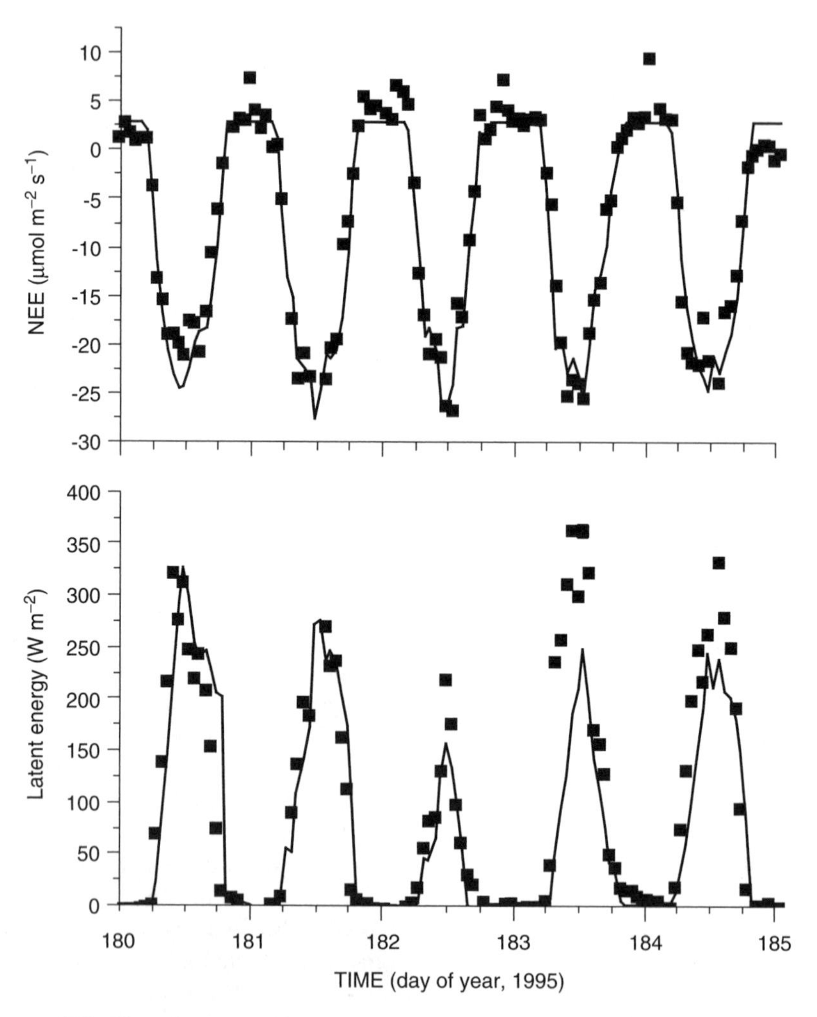

FIGURE 6.10. Diurnal patterns of net ecosystem exchange of CO_2 (top) and latent energy fluxes (bottom), both modeled (lines) and measured (symbols) during five days in the summer of 1995 at Harvard Forest, MA.

There was a tendency for the model to slightly lag the measurements of LE fluxes on some days (e.g., day 180), though not on others (day 181), and the model does not always predict the peak LE fluxes reached on some days (days 183 to 184)—but whether these are errors in model or data are not clear. Most days tend to show peaks in CO_2 capture and water loss during the late morning/noon, and a decline in the early afternoon. The asymmetry in the simulations of gas exchange is governed by hydraulic constraints. Stomata are adjusted to maintain leaf water potential above their critical value, and, in the afternoon, rising vapor pressure deficit (Fig. 6.11) ensures that stomata must close to avoid dangerous

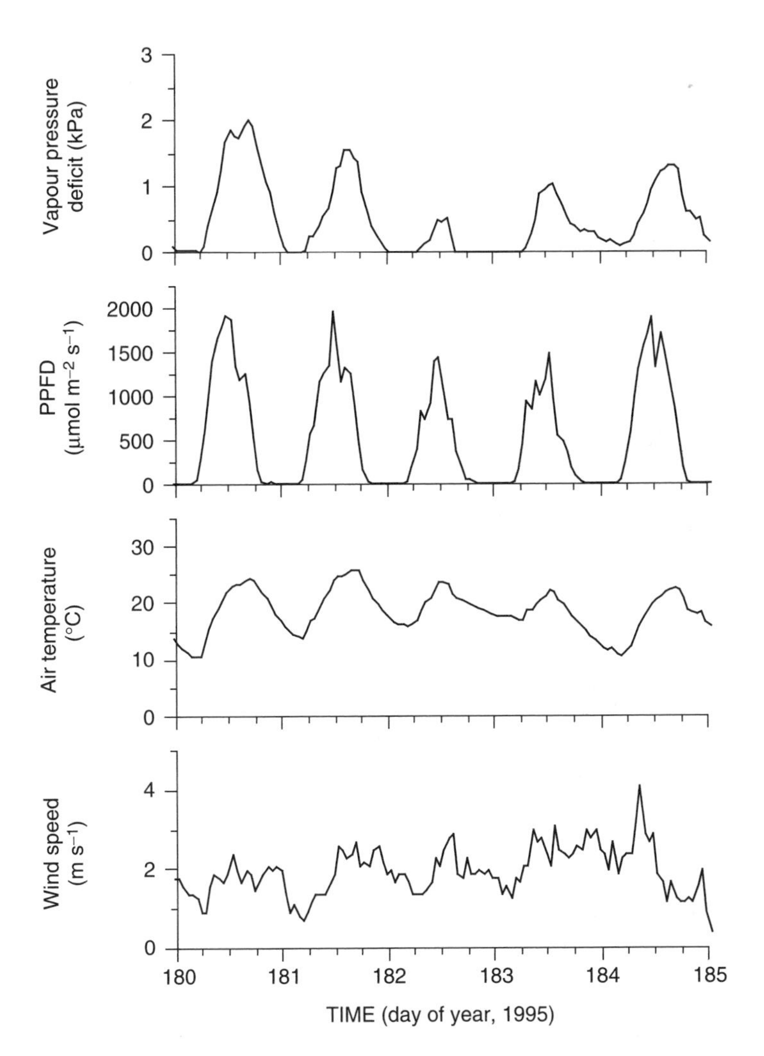

FIGURE 6.11. Hourly meteorological conditions during five days in the summer of 1995 at Harvard Forest, MA. PPFD is photosynthetic photon flux density.

rates of water loss. Peak rates of CO_2 capture are inversely linked to daily maximum of D. Day 180 was brighter than day 182, but maximum D was almost fourfold larger on day 180. The difference in irradiance had insignificant effects on CO_2 capture because of light saturation in the photosynthetic system at relatively low light levels. Instead, greater atmospheric water demand resulted in stomatal closure and this meant that peak CO_2 uptake values were less on day 180 than 182.

Calculations of bulk surface conductance (G_c) from measured LE exchange and canopy conductance (g_c) from simulated evapotranspiration show varying degrees of correspondence (Fig. 6.12). We expect the estimate of canopy conductance to be lower because the estimate of bulk surface conductance includes evaporation from the soil surface (Kelliher et al. 1995). g_c exceeded estimated G_c on some mornings, a result of the tendency for modeled LE fluxes to lag the measurements. On those days where peak LE flux estimates from model and data disagreed, there was also a strong divergence in g_c versus G_c estimates. Because of high LE fluxes and the low D, eddy covariance data suggested G_c exceeded 2000 mmol m^{-2} s^{-1} on days 182 and 183. It is likely that evaporation from the soil or wet canopy surfaces accounts for this discrepancy. Kelliher and cowork-

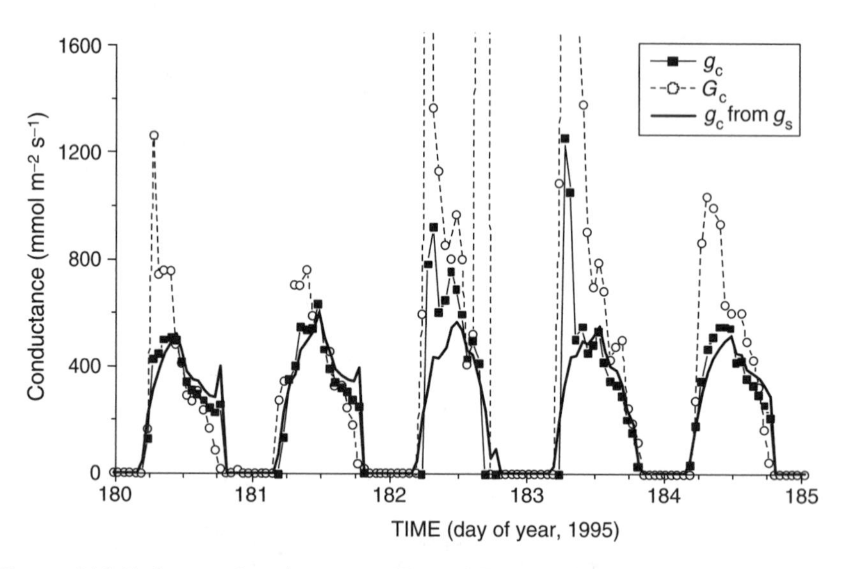

FIGURE 6.12. Estimates of conductance at Harvard Forest during five days in the summer of 1995. Bulk surface conductance, G_c (○), was derived from eddy covariance measurements of latent energy exchange. Canopy conductance, g_c (■), was estimated from simulated evapotranspiration from the vegetation only. Both sets of estimates were constructed using a simplified rearrangement of the Penman-Monteith equation. Also shown (thick line) is the upscaled estimate of total canopy conductance (g_c) derived from total leaf area and modeled stomatal conductance estimates in each canopy layer.

ers (1995), in a global survey, found that bulk surface conductance was conservative against leaf area index, because of the compensating decrease in plant canopy transpiration and increase in soil evaporation as leaf area index diminishes. The modeling results here, and those of Kelliher and coworkers (1995) and Baldocchi and Meyers (1998), confirm that using micrometeorological latent energy data to infer stomatal conductance is complicated by the soil contribution to evaporation.

Model predictions of g_b declined with depth in the canopy, in response to the simulated attenuation of wind speed (Fig. 6.13). Above canopy wind speeds, v varied between 1 and 4 m s^{-1} over the study period (see Fig. 6.11), and variation in g_b was strongly controlled by wind speed. Over just a few hours, v varied by >1 m s^{-1} and there was some tendency for calmer conditions around dawn. However, the effect of wind gusts on g_b was damped, both by the attenuation of speed through the canopy and by the inverse square root relationship of δ with v (see Eq. 19). Thus, the lowest canopy layer had the least variation in g_b. The values of g_b varied between 500 and 2000 mmol m^{-2} s^{-1}, and therefore generally equaled or exceeded estimates of g_c.

Estimates of stomatal conductance followed strong diurnal patterns, but there were clear differences in behavior between days, and between top and bottom of

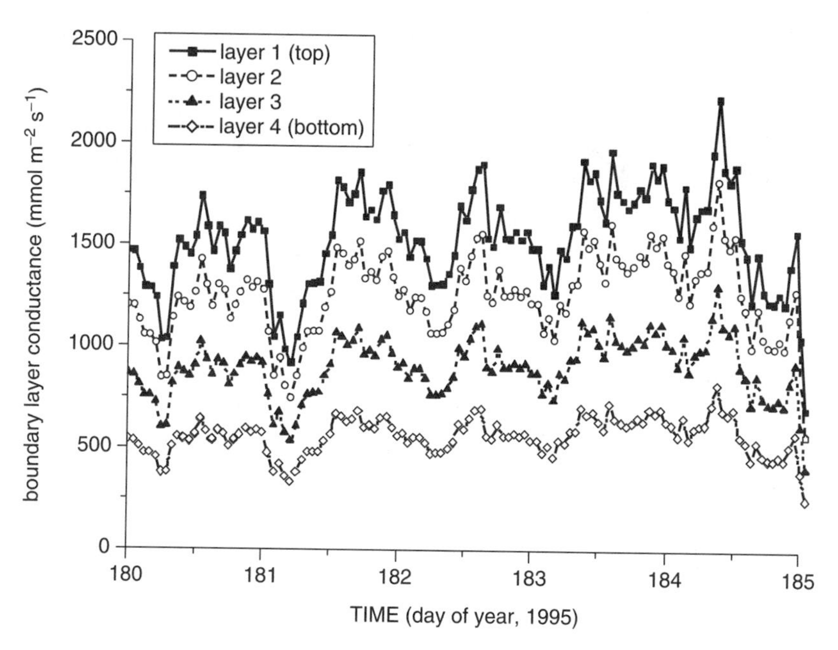

FIGURE 6.13. Leaf boundary-layer conductance as estimated by the SPA model for four canopy layers (layer 1 = top of canopy) during five days in the summer of 1995 at Harvard Forest, MA.

canopy (Fig. 6.14). In the early morning, sunlit leaves at all depths in the canopy tracked increases in solar radiation to reach broadly similar maximum g_s values. D was low each morning and water recharge overnight ensured that hydraulic limitations were at a minimum; light, not water relations, was the limiting factor on CO_2 capture each morning. Once stored water was exhausted, further wa-

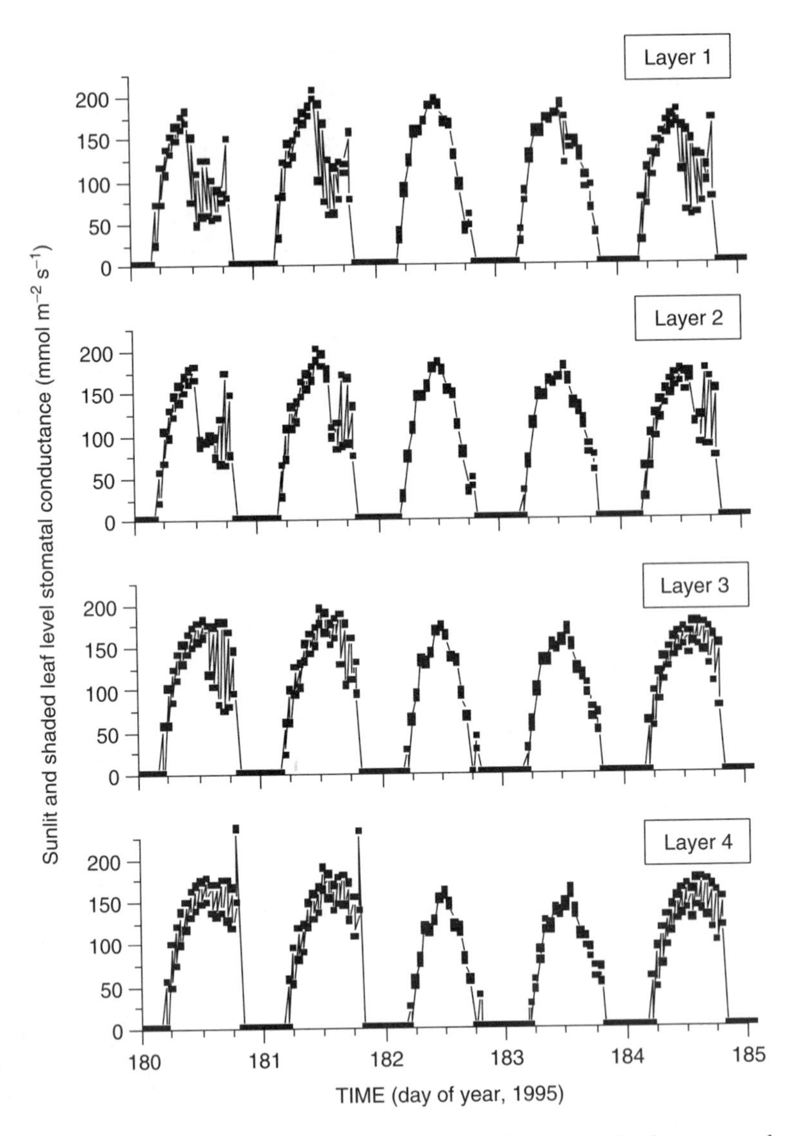

FIGURE 6.14. Stomatal conductance as estimated by the SPA model for four canopy layers (layer 1 = top of canopy) during five days in the summer of 1995 at Harvard Forest, MA. Each time series records stomatal conductance for both sunlit and shaded components of the canopy in each layer, so on sunny days there can be two estimates at each time period.

ter lost from the leaves had to be balanced by transport through the soil-plant system. Atmospheric demand for water (D) rose as temperatures climbed through the day (see Fig. 6.11). On days with high D, water stores were soon exhausted, and the fall in leaf water potential to the critical value caused a decline in g_s values in the afternoon. This decline was most pronounced in the upper canopy, where hydraulic resistance was at its greatest (due to the greater pathlengths required for water transport). Shaded leaf fractions in the upper canopy maintained higher g_s than sunlit fractions, due to differences in energy balance caused by a relatively lower absorption of solar radiation. Energy balance considerations explain why afternoon g_s differs according to solar incidence. Differences between sunlit and shaded fractions were much less noticeable in the early morning when water relations were not limiting, because the photosynthetic metabolism was rapidly light saturated. At the bottom of the canopy (layer 4), the lower hydraulic resistance meant that sun leaves avoided afternoon declines in g_s, while shade leaves at the bottom of the canopy were slightly light limited. During the afternoon in the upper canopy, sun leaves had lower g_s than shade leaves, while the opposite occurred in the lowest canopy layer. On cloudy days (days 182 and 183), leaves in each canopy level were more evenly illuminated. The reduction in D and in incoming radiation meant that water relations were not limiting. It is this sort of complex behavior that challenges the construction of effective big-leaf models.

McNaughton and Jarvis (1983) developed the concept that the degree to which leaves and canopies are coupled to the environment varies significantly among vegetation types. At one extreme, strong coupling occurs when boundary-layer conductance is large and canopy air is well mixed with the overlying atmosphere. Leaf temperature tends to be close to air temperature so that the driving force for transpiration is D. The conditions for Harvard Forest, a tall, rough canopy, are typical of a strongly coupled system. The alternative extreme is where boundary-layer conductance is small. This might be in a low-stature canopy under calm conditions. In this case, poor mixing means that transpiration leads to humidification of the leaf boundary layer. If air temperature within the boundary layer remained unchanged, D would soon drop to 0 and the driving forces for transpiration would disappear. But incoming radiation warms the leaf and its boundary layer, increasing the water-holding capacity of the air and maintaining D in the boundary layer. It is thus the input of radiation that determines transpiration rates in this case rather than the bulk atmospheric D.

If Harvard Forest is strongly coupled to the atmosphere, we would expect that canopy conductance is dominated by stomatal terms. To test this, we estimated g_c from g_s by a simple summation over each of the n canopy layers:

$$g_c = \sum_{i=1,n} g_{si} L_i \qquad \text{(Eq. 6.22)}$$

and we found that g_c estimated from g_s (see Fig. 6.12) closely matched estimates derived from simulation of canopy LE fluxes. So, according to these simulations, stomatal conductance, rather than leaf boundary-layer conductance, is the critical fine-scale control on canopy conductance in this ecosystem. In fact, decoupled canopies are only likely to occur in artificial situations. Given a typical g_b of 400 mmol m^{-2} s^{-1} in a smooth canopy, like a grassland (comparable with the forest understorey g_b in the model simulations above), then g_s must exceed these values to ensure decoupling. This is only likely to happen in well-fertilized and irrigated croplands.

Impacts of Rising CO_2

A major concern for the future is the response of vegetation to continued increases in atmospheric CO_2 concentration and any associated changes in climate. It is established that increasing CO_2 concentration tends to reduce stomatal density, and over temporal scales from days to millions of years (Woodward and Kelly 1995, Beerling and Woodward 1997, Royer 2001). However the impact of these structural changes on actual conductance is a key future concern, with potential for changes in water runoff, and therefore water supplies, on the water status, and productivity of vegetation (Cramer et al. 2001). Curtis and Wang (1998) reported only a small reduction (11%) in stomatal conductance with short-term CO_2 enrichment in a meta-analysis of 48 species of woody plants. However, a subsequent analysis (Medlyn et al. 2001) of longer-term experiments (greater than 1 y) with woody species indicated a much larger decrease in conductance (21%) with CO_2 enrichment. A similar response was also observed for grassland species (Lee et al. 2001).

In vegetation, responses in stomatal conductance may lead to changes in canopy leaf area index such that vegetation water use is unchanged (Field et al. 1995). Therefore, global scale approaches to investigating vegetation responses to changes in CO_2 and climate must consider the whole suite of vegetation responses from gas exchange, to growth and survival. This can be achieved with a dynamic global vegetation model (Cramer et al. 2001), which is designed to incorporate, mechanistically, these features of vegetation responses.

Table 6.4 shows simulations of vegetation transpiration and gross primary production (GPP) averaged for the years 2000 to 2005 and for 2050 to 2055, to provide a snapshot of one possible scenario for future changes in climate and vegetation activity. It is notable that transpiration changes little over this period, unlike temperature, CO_2 concentration, and GPP, which all increase. Figures 6.15

TABLE 6.4. Simulated climates and global exchanges by vegetation of water vapor by transpiration (excluding evaporation) and gross primary production (GPP). Climate data from the HadCM2 simulation modified by Cramer et al. (2001) and vegetation data simulated by the Sheffield Dynamic Global Vegetation Model (SDGVM, Cramer et al. 2001).

Years	Temperature (°C)	Precipitation (mm yr^{-1})	CO_2 (ppm)	Transpiration (10^{18} g)	GPP (10^{15} g)
2000–2005	13.3	777	385	45	194
2050–2055	14.9	801	562	46	249

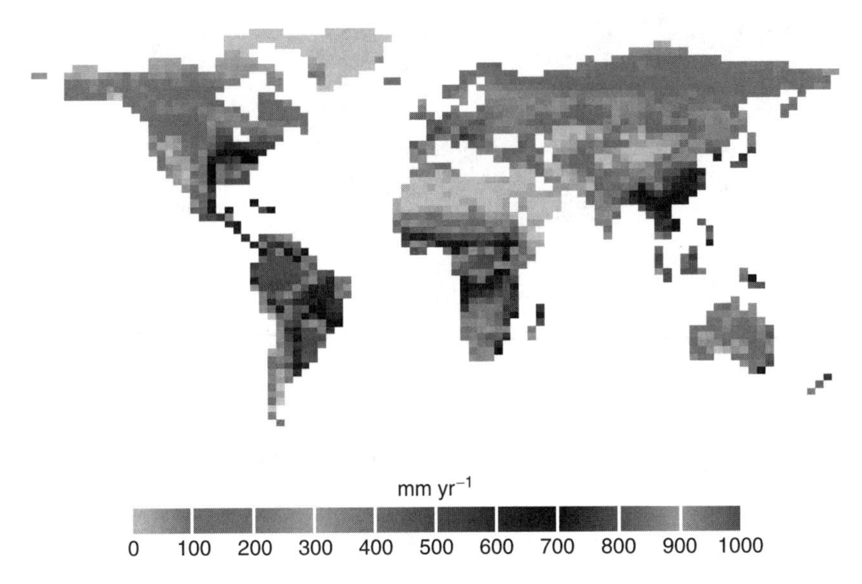

FIGURE 6.15. Simulations of global transpiration (mm y^{-1}) by terrestrial vegetation, using the Sheffield Dynamic Global Vegetation Model, averaged for the years 2000 to 2005.

to 6.18 indicate these simulations at the global scale. Although total transpiration changes little, there is a notable reduction in transpiration at the margins of the tropical, equatorial, and warm temperate forests (the simulations assume no human land-use changes) and increases in transpiration by cool temperate and boreal forests (see Figs. 6.15, 6.16).

In contrast to transpiration, there is a general increase of GPP over much of the world's terrestrial vegetation. Values of GPP for Europe in 2000 to 2005 range between 8 and 16 tC ha^{-1} y^{-1}, which are in the same range as observations made by eddy covariance in forests (Valentini et al. 2000).

Summary and Conclusions

The understanding of stomatal operation has a long history; however, there are still new and surprising observations on the control of stomatal development (Woodward and Kelly 1995). The impacts of CO_2 enrichment on stomatal development and opening are not uniform with species, and these differences could have the capacity to change vegetation activity and composition in the future. However, the effective uniformity of the observed maximum stomatal conductance amongst different vegetation types suggests that species differences may not be very important—but we can't yet be sure.

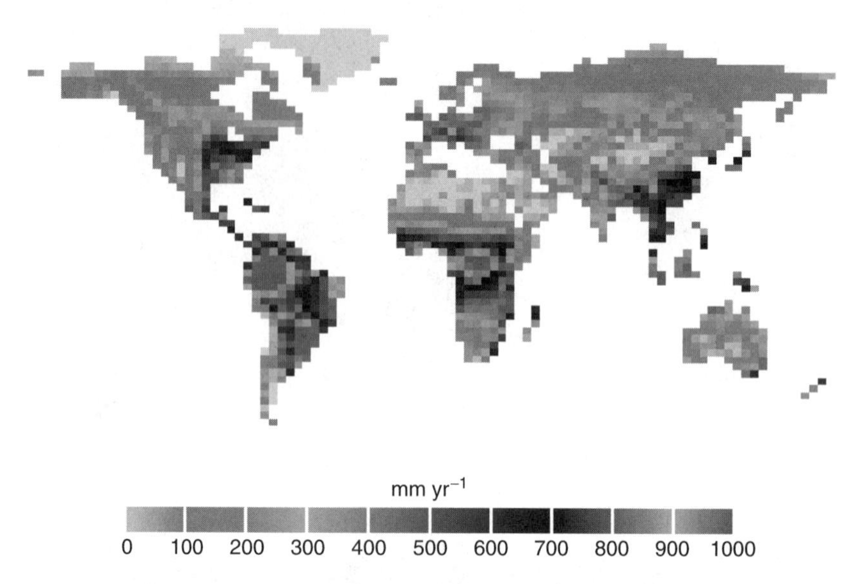

FIGURE 6.16. Simulations of global transpiration (mm y^{-1}) by terrestrial vegetation, using the Sheffield Dynamic Global Vegetation Model, averaged for the years 2050 to 2055.

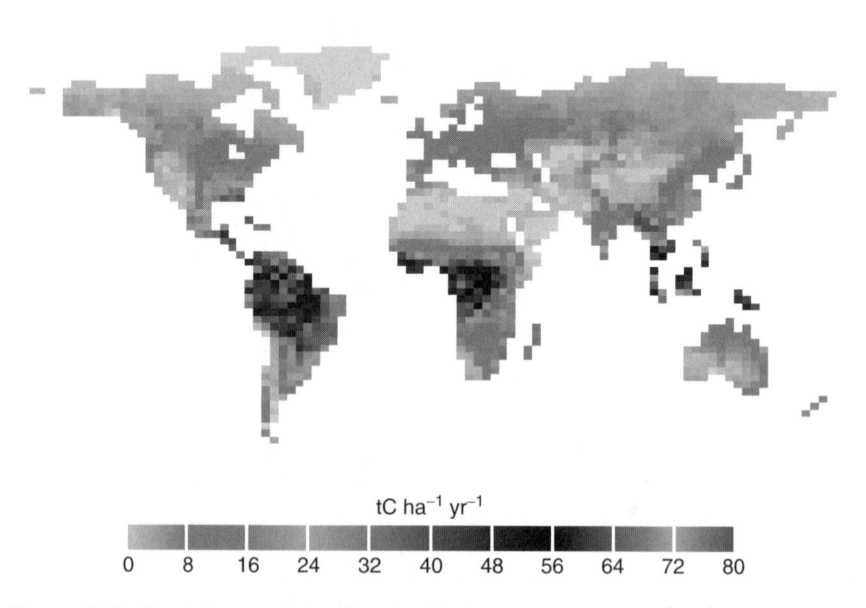

FIGURE 6.17. Simulations of Gross Primary Production (GPP) (tC ha^{-1} y^{-1}) by terrestrial vegetation, using the Sheffield Dynamic Global Vegetation Model, averaged for the years 2000 to 2005.

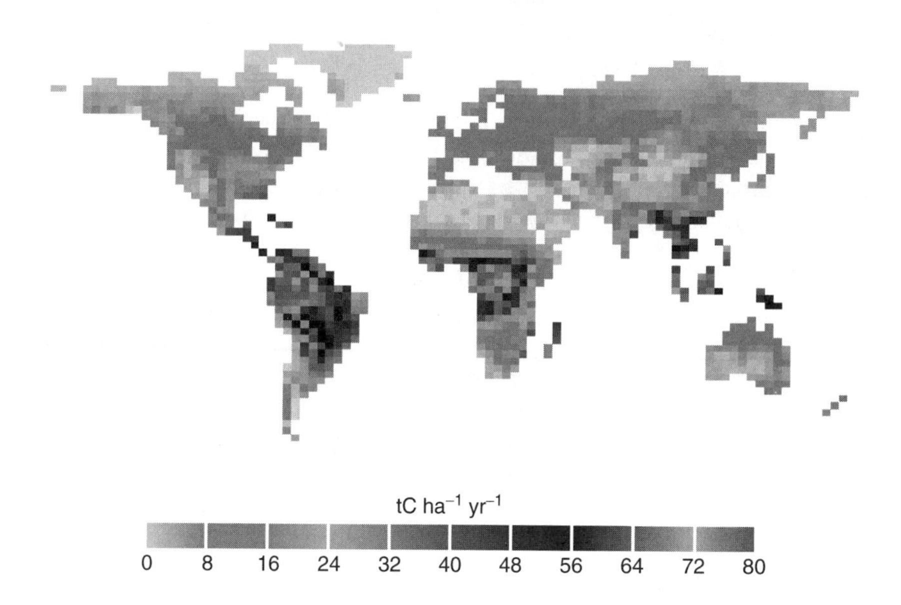

FIGURE 6.18. Simulations of Gross Primary Production (GPP) (tC ha^{-1} y^{-1}) by terrestrial vegetation, using the Sheffield Dynamic Global Vegetation Model, averaged for the years 2050 to 2055.

In the past decade there has been considerable progress in understanding stomatal dynamics and their environmental controls. Initial attempts to scale these patterns to the broader landscape have been broadly successful, but there remain considerable areas of ignorance and uncertainty that should be the focus of future research. Theses areas include:

- *The coupling between CO_2 capture and atmospheric processes.* Atmospheric and climate modelers have recently recognized the need for more detailed, mechanistic representations of biological activity in simulations of land surface processes. The BOREAS experiment in Canada was particularly revealing in this respect (Sellers et al. 1997). Climate models up until the experiment had regarded the boreal forests as analogous to a permanently wet evaporating surface, but the experiment showed that stomatal and canopy conductance were relatively low (a "green desert"), and ratios of sensible to latent heat were high. Incorporating this new information into climate models led to important shifts in global dynamics. As other biomes are studied (e.g., South American rainforests in Large-scale Biosphere Atmosphere (LBA) experiment now underway), we can expect further surprises.
- *The coupling between CO_2 capture and biogeochemical processes.* The focus of this chapter has largely been on the proximate controls on CO_2

capture—energy balance and water relations. Links with biogeochemical cycles occur on decadal and longer timescales, and are thus more difficult both to examine experimentally and to interpret. One explanation of the boreal "green desert" is that cold and anaerobic soils limit rates of nutrient cycling, and thus constrain foliar nitrogen concentrations. With production limited by nutrients, plants invest relatively little in hydraulic structures, because high stomatal conductance will be of limited benefit. This hypothesis suggests powerful links between biogeochemistry, productivity, and land-surface processes that need further study.

- *The coupling between CO_2 capture and hydrological processes.* Forest production tends to decline with stand age, and one explanation is that older, taller stands are hydraulically limited, because longer vascular networks have a larger resistance (Ryan and Yoder 1997, Hubbard et al. 1999). There is also evidence that this age-related hydraulic effect can influence catchment hydrology. Effective management of water resources will rely on better understanding of interactions with plant processes.
- *The relationships between competitive success and CO_2 capture.* Competitive success among plants is not strongly related to the magnitude of photosynthetic rates, particularly in longer-lived individuals (Kuppers 1984). Rather, success is related to the efficiency of resource capture (Tilman 1986), whether of nutrients, space and light, or water. To be able to make accurate predictions of how vegetation will respond to climate change will require considerably better understanding of these determinants of competitive success. For example, we are unable to explain at a mechanistic level the shifts in dominant vegetation and their structural characteristics along the precipitation gradient from tropical rainforest to *cerrado* in South America. Understanding and predicting the transient response of plant communities to global change is a major challenge.

The future of research on CO_2 capture is likely to be focused around these interactions with key global cycles over a range of time scales, and in connecting patterns of CO_2 capture to improved understanding of community dynamics and evolutionary success.

Acknowledgments. We are very grateful to Steve Wofsy for access to the data sets from the Harvard Forest LTER.

References

Abrams, M. D., and Kubiske, M. E. 1990. Leaf structural characteristics of 31 hardwood and conifer species in central Wisconsin: Influence of light regime and shade-tolerance rank. For. Ecol. Manage. 31:245–253.

Baldocchi, D. 1992. A Lagrangian random-walk model for simulating water vapor, CO2 and sensible heat flux densities and scalar profiles over and within a soybean canopy. Bound.-Layer Meteorol. 61:113–144.

Baldocchi, D. D. 1997. Measuring and modelling carbon dioxide and water vapour exchange over a temperate broad-leaved forest during the 1995 summer drought. Plant Cell Environ. 20:1108–1122.

Baldocchi, D., and Meyers, T. 1998. On using eco-physiological, micrometeorological and biogeochemical theory to evaluate carbon dioxide, water vapor and trace gas fluxes over vegetation: A perspective. Agric. For. Meteorol. 90:1–25.

Baldocchi, D. D., Verma, S. B., Rosenberg, N. J., Blad, B. L., and Specht, J. E. 1985. Microclimate plant architectural interactions: Influence of leaf width on the mass and energy exchange of a soybean canopy. Agric. For. Meteorol. 35:1–20.

Ball, J. T., Woodrow, I. E., and Berry, J. A. 1987. A model predicting stomatal conductance and its contribution to the control of photosynthesis under different environmental conditions. In: Progress in Photosynthesis Research, Vol. IV. J. Biggins (ed.), pp. 221–224. Dordrecht, The Netherlands: Martinus Nijhoff.

Becker, P., Meinzer, M. C., and Wullschleger, S. D. 2000. Hydraulic limitation of tree height: A critique. Funct. Ecol. 14:4–11.

Beerling, D. J., and Woodward, F. I. 1997. Changes in land plant function over the Phanerozoic: Reconstructions based on the fossil record. Bot. J. Linnaean Soc. 124:137–153.

Brenner, A. J., and Jarvis, P. G. 1995. A heated leaf replica technique for determination of leaf boundary layer conductance in the field. Agric. For. Meteorol. 72:261–275.

Brown, H. T., and Escombe, F. 1900. Static diffusion of gases and liquids in relation to the assimilation of carbon and translocation in plants. Philos. Trans. R. Soc. Lond. [Biol.] 193:223–291.

Buckley, T. N., Farquhar, G. D., and Mott, K. A. 1999 Carbon-water balance and patchy stomatal conductance. Oecologia 118:132–143.

Campbell, G. S., and Norman, J. M. 1998. Introduction to environmental biophysics. New York: Springer-Verlag.

Collatz, C. G., Ball, J. T., Grivet, C., and Berry, J. A. 1991. Physiological and environmental regulation of stomatal conductance, photosynthesis and transpiration: A model that includes a laminar boundary layer. Agric. For. Meteorol. 54:107–136.

Cowan, I. R. 1977. Stomatal behaviour and environment. Adv. Bot. Res. 4:117–228.

Cowan, I. R., and Farquhar, G. D. 1977. Stomatal function in relation to leaf metabolism and environment: Stomatal function in relation of gas exchange. In: Symposium of the Society for Experimental Botany. D. H. Jennings (ed.), Cambridge: Cambridge University Press.

Cramer, W., Bondeau, A., Woodward, F. I., Prentice, I. C., Betts, R. A., Brovkin, V., Cox, P. M., Fisher, V., Foley, J. A., Friend, A. D., Kucharik, C., Lomas, M. R., Ramankutty, N., Sitch, S., Smith, B., White, A., and Young-Molling, C. 2001. Global response of terrestrial ecosystem structure and function to CO_2 and climate change: Results from six dynamic global vegetation models. Global Change Biol. 7:357–373.

Curtis, P. S., and Wang, X. 1998. A meta-analysis of elevated CO_2 effects on woody plant mass, form and physiology. Oecologia 113:299–313.

Dang, Q. L., Margolis, H. A., and Collatz, G. J. 1998. Parameterization and testing of a coupled photosynthesis stomatal conductance model for boreal trees. Tree Physiol. 18:141–153.

De Pury, D. G. G., and Farquhar, G. D. 1997. Simple scaling of photosynthesis from leaves to canopies without errors of big-leaf models. Plant Cell Environ. 20:537–557.

Dickinson, R. E., and Henderson-Sellers, A. 1988. Modelling tropical deforestation: A study of GCM land-surface parametrizations. Quart. J. R. Meteorol. Soc. 114:439–462.

Ellsworth, D. S. 1999. CO_2 enrichment in a maturing pine forest: Are CO_2 exchange and water status in the canopy affected? Plant Cell Environ. 22:461–472.

Farquhar, G. D. 1989. Models of integrated photosynthesis of cells and leaves. Philos. Trans. R. Soc. Lond. [Biol.] 323:357–367.

Farquhar, G. D., and von Caemmerer, S. 1982. Modelling of photosynthetic response to the environment. In: Encyclopedia of Plant Physiology, New Series, Vol. 12B. Physiological Plant Ecology II. O. L. Lange, P. S. Nobel, C. B. Osmond, and H. Ziegler (eds.), pp. 549–587. Berlin: Springer-Verlag.

Field, C., Berry, J. A., and Mooney, H. A. 1982. A portable system for measuring carbon dioxide and water vapour exchange of leaves. Plant Cell Environ. 5:179–186.

Field, C. B., Jackson, R. B., and Mooney, H. A. 1995. Stomatal responses to increased CO_2: implications from the plant to the global scale. Plant Cell Environ. 18:1214–1225.

Foster, J. R., and Smith, W. K. 1986. Influence of stomatal distribution on transpiration in low-wind environments. Plant Cell Environ. 9:751–759.

Franks, P. J., and Farquhar, G. D. 2001. The effects of exogenous abscisic acid on stomatal development, stomatal mechanics, and leaf gas exchange in *Tradescantia virginiana*. Plant Physiol. 125:935–942.

Gay, A. P., and Hurd, R. G. 1975. The influence of light on stomatal density in the tomato. New Phytologist 75:37–46.

Gibson, A. C. 1996. Structure-Function Relations of Warm Desert Plants. Berlin: Springer-Verlag.

Goulden, M. L., Munger, J. W., Fan, S.-M., Daube, B. C., and Wofsy, S. C. 1996a. Exchange of carbon dioxide by a deciduous forest: Response to interannual climate variability. Science 271:1576–1579.

Goulden, M. L., Munger, J. W., Fan, S.-M., Daube, B. C. and Wofsy, S. C. 1996b. Measurements of carbon storage by long-term eddy correlation: Methods and a critical evaluation of accuracy. Global Change Biol. 2:169–182.

Grace, J., and Wilson, J. 1976. The boundary layer over a *Populus* leaf. J. Exper. Bot. 27:231–241.

Harley, P. C., and Baldocchi, D. D. 1995. Scaling carbon dioxide and water vapour exchange from leaf to canopy in a deciduous forest. I. Leaf model parametrization. Plant Cell Environ. 18:1146–1156.

Hubbard, R. M., Bond, B. J., and Ryan, M. G. 1999. Evidence that hydraulic conductance limits photosynthesis in old *Pinus ponderosa* trees. Tree Physiol. 19:165–172.

James, S. A., and Bell, D. T. 2000. Influence of light availability on leaf structure and growth of two *Eucalyptus globulus* ssp. globulus provenances. Tree Physiol. 20: 1007–1018.

James, S. A., and Bell, D. T. 2001. Leaf morphological and anatomical characteristics of heteroblastic *Eucalyptus globulus* ssp. globulus (Myrtaceae). Aust. J. Bot. 49:259–269.

Jarvis, P. G. 1976. The interpretation of the variations in leaf water potential and stomatal conductance found in the field. Philos. Trans. R. Soc. Lond. [Biol.] 273:593–610.

Jarvis, P. G., and McNaughton, K. G. 1986. Stomatal control of transpiration:scaling up from leaf to region. Adv. Ecol. Res. 15:1–49.

Jones, H. G. 1992. Plants and microclimate. Cambridge: Cambridge University Press.

Jones, H. G. and Sutherland, R. A. 1991. Stomatal control of xylem embolism. Plant Cell Environ. 14:607–612.

Jones, M. M., and Rawson, H. M. 1979. Influence of rate of development of leaf water deficits upon photosynthesis, leaf conductance, water use efficiency, and osmotic potential in sorghum. Physiol. Plant 45:103–111.

Katul, G. G., Ellsworth, D. S., and Lai, C. T. 2000. Modelling assimilation and intercellular CO_2 from measured conductance: A synthesis of approaches. Plant Cell Environ. 23:1313–1328.

Kelliher, F. M., Leuning, R., Raupach, M. R., and Schulze, E.-D. 1995. Maximum conductances for evaporation from global vegetation types. Agric. For. Meteorol. 73:1–16.

Knapp, A. K. 1993. Gas exchange dynamics on C3 and C4 grasses: Consequences of differences in stomatal conductance. Ecology 74:113–123.

Körner, C. 1994. Leaf diffusive conductances in the major vegetation types of the globe. In: Ecophysiology of Photosynthesis. E. D. Schulze, and M. M. Caldwell (eds.), pp. 463–490. Heidelberg: Springer-Verlag.

Kuppers, M. 1984. Carbon relations and competition between woody species in a Central European hedgerow. Oecologia 64:344–354.

Laisk, A., Oja, V., and Kull, K. 1980. Statistical distribution of stomatal apertures of *Vicia faba* and *Hordeum vulgare* and the Spannungsphase of stomatal opening. J. Exp. Bot. 31:49–58.

Lee, T. D., Tjoelker, M. G., Ellsworth, D. S., and Reich, P. B. 2001. Leaf gas exchange responses of 13 prairie grassland species to elevated CO_2 and increased nitrogen supply. New Phytologist 150:405–418.

Leuning, R. 1995. A critical appraisal of a combined stomatal-photosynthesis model for C3 plants. Plant Cell Environ. 18:339–356.

Leuning, R., Kelliher, F. M., de Pury, D. G. G., and Schulze, E. -D. 1995. Leaf nitrogen, photosynthesis, conductance and transpiration: Scaling from leaves to canopy. Plant Cell Environ. 18:1183–1200.

McNaughton, K. G., and Jarvis, P. G. 1983. Predicting effects of vegetation changes on transpiration and evaporation. In: Water Deficits and Plant Growth, Vol. VII. T. T. Kozlowski (ed.), pp. 1–47. New York: Academic Press.

Medlyn, B. E., Barton, C. V. M., Broadmeadow, M. S. J., Ceulemans, R., De Angelis, P., Forstreuter, M., Freeman, M., Jackson, S. B., Kellomaki, S., Laitat, E., Rey, A., Roberntz, P., Sigurdsson, B. D., Strassemeyer, J., Wang, K., Curtis, P. S., and Jarvis, P. G. 2001. Stomatal conductance of forest species after long-term exposure to elevated CO_2 concentration: A synthesis. New Phytologist 149:247–264.

Meinzer, F. C. 2002. Co-ordination of vapour and liquid phase water transport properties in plants. Plant Cell Environ. 25:265–274.

Miranda, V., Baker, N. R., and Long, S. P. 1981. Anatomical variation along the length of the Zea mays leaf in relation to photosynthesis. New Phytologist 88:595–605.

Monteith, J. L. 1995a. Accommodation between transpiring vegetation and the convective boundary layer. J. Hydrol. 166:251–263.

Monteith, J. L. 1995b. A reinterpretation of stomatal responses to humidity. Plant Cell Environ. 18:357–364.

Monteith, J. L., and Unsworth, M. H. 1990. Principles of Environmental Physics. London: Edward Arnold.

Mott, K. A. 1988. Do stomata respond to CO_2 concentrations other than intercellular? Plant Physiol. 86:200–203.

Mott, K. A., and Parkhurst, D. F. 1991. Stomatal responses to humidity in air and helox. Plant Cell Environ. 14:509–515.

Mott, K. A., Gibson, A. C., and O'Leary, J. W. 1982. The adaptive significance of amphistomatic leaves. Plant Cell Environ. 5:455–460.

Norman, J. M. 1982. Simulation of microclimates. In: Biometereology and Integrated Pest Management. J. L. Hatfield and I. Thompson (eds.), pp. 65–99. New York: Academic Press.

Parkhurst. D. F. 1978. The adaptive significance of stomatal occurrence on one or both surfaces of leaves. J. Ecol. 66:367–383.

Parkhurst, D. F., Duncan, P. R., Gates, D. M., and Kreith, F. 1968. Wind tunnel modeling of convection of heat between air and broad leaves of plants. Agric. Meteorol. 5:33–47.

Peat, H. J., and Fitter, A. H. 1994. A comparative study of the distribution and density of stomata in the British Flora. Biol. J. Linnaean Soc. 52:377–393.

Raupach, M. R., and Finnigan, J. J. 1988. "Single-layer models of evaporation from plant canopies are incorrect but useful, whereas multilayer models are correct but useless": Discuss. Aust. J. Plant Physiol. 15:705–716.

Reich, P. B. 1984. Relationships between leaf age, irradiance, leaf conductance, CO_2 exchange and water use efficiency in hybrid poplar. Photosynthetica 18:445–453.

Royer, D. L. 2001. Stomatal density and stomatal index as indicators of paleoatmospheric CO_2 concentration. Rev. Paleobot. Palynol. 114:1–28.

Ryan, M. G., and Yoder, B. J. 1997. Hydraulic limits to tree height and tree growth. BioScience 47:235–242.

Sage, R. F. 1994. Acclimation of photosynthesis to increasing atmospheric CO2: The gas exchange perspective. Photosynth. Res. 39:351–368.

Salisbury, E. J. 1927. On the causes and ecological significance of stomatal frequency, with special reference to the woodland flora. Philos. Trans. R. Soc. Lond. [Biol.] 216:1–65.

Schuepp, P. H. 1993. Tansley Review No. 59. Leaf boundary layers. New Phytologist 125:477–507.

Schulze, E.-D., Cermak, J., Matyssek, R., Penka, M., Zimmermann, R., Vasicke, F., Gries, W., and Kucera, J. 1985. Canopy transpiration and water fluxes in the xylem of the trunk of *Larix* and *Picea* trees—A comparison of xylem flow, porometer and cuvette measurements. Oecologia 66:475–483.

Schulze, E.-D., Kelliher, F. M., Körner, C., Lloyd, J., and Leuning, R. 1994. Relationships among maximum stomatal conductance, ecosystem surface conductance, carbon assimilation rate, and plant nitrogen nutrition: A global ecology scaling exercise. Annu. Rev. Ecol. Systemat. 25:629–660.

Sellers, P. J., Shuttleworth, J. W., Dorman, J. L., Dalcher, A., and Roberts, J. M. 1989. Calibrating the simple biosphere model (SiB) for Amazonian tropical forest using field and remote sensing data: Part 1: Average calibration with field data. J. Appl. Meteorol. 28:727–759.

Sellers, P. J., Hall, F. G., Kelly, R. D., Black, A., Baldocchi, D., Berry, J., Ryan, M. G., Ranson, K. J., Crill, P. M., Lettenmeier, D. P., Margolis, H., Cihlar, J., Newcomer, J., Fitzjarrald, D., Jarvis, P. G., Gower, S. T., Halliwell, D., Williams, D., Goodison, B., Wickland, D. E., and Guertin, F. E. 1997. BOREAS in 1997: Experimental overview, scientific results and future directions. J. Geophys. Res. 102:28731–28769.

Smith, W. K., Bell, D. T., and Shepherd, K. A. 1998. Associations between leaf structure, orientation, and sunlight exposure in five western Australian communities. Am. J. Botany. 85:56–63.

Sperry, S., Adler, F. R., Campbell, G. S., and Comstock, J. P. 1998. Limitation of plant water use by rhizosphere and xylem conductance: Results from a model. Plant Cell Environ. 21:347–359.

Sugden, A. M. 1985. Leaf anatomy in a Venezuelan montane forest. Bot. J. Linnean Soc. 90:231–241.

Tanner, E. V. J., and Kapos, V. 1982. Leaf structure of Jamaican upper montane rainforest trees. Biotropica 14:16–24.

Ticha, I. 1982. Photosynthetic characteristics during ontogenesis of leaves. 7. Stomata density and sizes. Photosynthetica. 16:375–471.

Tilman, D. 1986. Resources, competition and the dynamics of plant communities. In Plant Ecology, ed. M. J. Crawley, pp. 51–75. Oxford: Blackwell.

Tyree, M. T,. and Ewers, F. W. 1996. Hydraulic architecture of woody tropical plants. In Tropical forest plant ecophysiology, eds. S. S. Mulkey, R. L. Chazdon, and A. P. Smith, pp. 217–243. New York: Chapman and Hall.

Valentini, R., Matteucci, G., Dolman, A. J. et al. 2000. Respiration as the main determinant of carbon balance in European forests. Nature 404:861–865.

Van Gardingen, P. R., Jefree, C. E., and Grace, J. 1989. Variation in stomatal aperture in leaves of *Avena fatua* L. observed by low-temperature scanning electron microscopy. Plant Cell Environ. 12:887–898.

Vogel, S. 1970. Convective cooling at low airspeeds and the shapes of broad leaves. J. Exp. Bot. 21:91–101.

Weyers, J. D. B., and Lawson, T. 1997. Heterogeneity in stomatal characteristics. Adv. Bot. Res. 26:317–352.

Whitehead, D. 1998. Regulation of stomatal conductance and transpiration in forest canopies. Tree Physiol. 18:633–644.

Williams, M., Rastetter, E. B., Fernandes, D. N., Goulden, M. L., Wofsy, S. C., Shaver, G. R., Melillo, J. M., Munger, J. W., Fan, S.-M., and Nadelhoffer, K. J. 1996. Modelling the soil-plant-atmosphere continuum in a *Quercus-Acer* stand at Harvard Forest:the regulation of stomatal conductance by light, nitrogen and soil/plant hydraulic properties. Plant Cell Environ. 19:911–927.

Williams, M., Rastetter, E. B., Fernandes, D. N., Goulden, M. L., Shaver, G. R., and Johnson, L. C. 1997. Predicting gross primary productivity in terrestrial ecosystems. Ecol. Applic. 7:882–894.

Williams, M., Malhi, Y., Nobre, A., Rastetter, E. B., Grace, J., and Pereira, M. G. P. 1998. Seasonal variation in net carbon exchange and evapotranspiration in a Brazilian rain forest: A modelling analysis. Plant Cell Environ. 21:953–968.

Williams, M., Bond, B. J., and Ryan, M. G. 2001a. Evaluating different soil and plant hydraulic constraints on tree function using a model and sap flow data from ponderosa pine. Plant Cell Environ. 24:679–690.

Williams, M., Law, B. E., Anthoni, P. M., and Unsworth, M. 2001b. Use of a simulation model and ecosystem flux data to examine carbon-water interactions in ponderosa pine. Tree Physiol. 21:287–298.

Willmer, C. M., and Fricker, M. 1996. Stomata, 2nd edition. New York: Chapman and Hall.

Wilson, K. B., Baldocchi, D. D., and Hanson, P. J. 2001. Leaf age affects the seasonal pattern of photosynthetic capacity and net ecosystem exchange of carbon in a deciduous forest. Plant Cell Environ. 24:571–583.

Wong, W. C., Cowan, I. R., and Farquhar, G. D. 1979. Stomatal conductance correlates with photosynthetic capacity. Nature 282:424–426.

Woodward, F. I., and Kelly, C. K. 1995. The influence of CO_2 concentration on stomatal density. New Phytologist 131:311–327.
Woodward, F. I., and Smith, T. M. 1994. Global photosynthesis and stomatal conductance: Modelling the controls by soil and climate. Adv. Bot. Res. 20:1–41.
Woodward, F. I., Smith, T. M., and Emanuel, W. R. 1995. A global land productivity and phytogeography model. Global Biogeochem. Cycles 9:471–490.
Wullschleger, S. D., Meinzer, F. C., and Vertessy, R. A. 1998. A review of whole-plant water use studies in trees. Tree Physiol. 18:499–512.

Part 5

CO_2 Processing

7
Chloroplast to Leaf

THOMAS D. SHARKEY, SEAN E. WEISE, ANDREW J. STANDISH,
AND ICHIRO TERASHIMA

Introduction

The central reactions of carbon processing in photosynthesis in plants occur in chloroplasts. Chloroplasts are one member of the plastid family of organelles, which are responsible for much of the biochemical flexibility of plants. All plant cells have plastids but the plastids of a cell can develop into a variety of organelles including amyloplasts, which store starch, and chromoplasts, which store pigments. Sugars are quantitatively the most significant result of carbon processing in chloroplasts but by no means are they the only essential products. Plastids are the site of production of the amino acids essential in animal diets, all of the fatty acids in the plants, carotenoids and other isoprenoids, and phenylpropanoids. Plastids also produce vitamins A (or its precursors), C, and E, as well as some of the B vitamins. The biochemical flexibility of carbon processing in plastids is a critical part of the success of plants. Some of the major products of carbon processing in the chloroplast and interactions with the rest of the cell are shown in Figure 7.1. In addition to carbon processing, plastids are the site where nitrogen and sulfur are reduced. Plants require large amounts of nitrogen for chlorophyll and proteins needed for carbon processing.

Chloroplasts have properties of bacteria and probably arose by endosymbiosis of a bacterium (Margulis 1981). Photosynthetic processing of carbon by chloroplasts requires a complex interaction between chloroplasts and the rest of the cell in which they reside. The final steps of photosynthetic carbon processing, sucrose and sugar alcohol synthesis, occurs outside the chloroplast, making photosynthesis a joint venture between the chloroplast and the cytosol of the "host" cell (Walker and Herold 1977). Sugars and sugar alcohols move between cells of the leaf through plasmodesmata up to the phloem. After loading into the phloem, sugars move out of the leaf. In this chapter, we describe photosynthetic carbon processing in the chloroplast and the synthesis of sugars and transport of those sugars to the phloem. The central reactions of carbon processing are highly conserved throughout evolution, the variation among plants in CO_2 processing is how the products of the chloroplast are used, stored, and sensed by the plant, so these topics will be emphasized in this chapter.

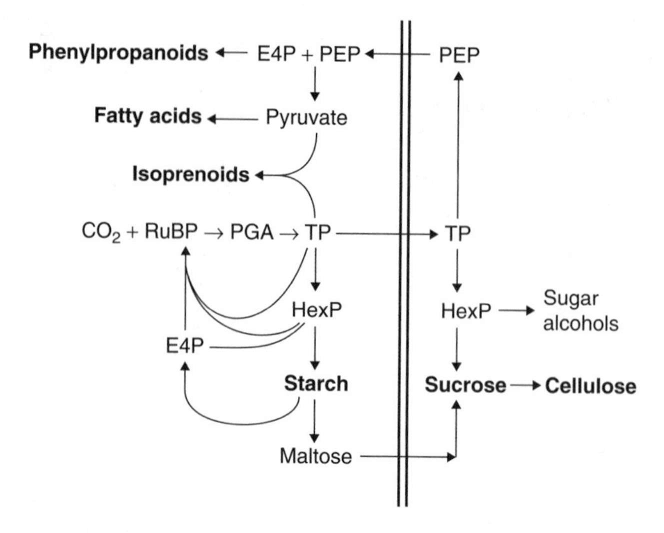

FIGURE 7.1. Some of the major products of carbon processing in chloroplasts Abbreviations: E4P, erythrose 4-phosphate; HexP, hexose phosphate; PEP, phosph*enol*pyruvate; PGA, 3-phosphoglyceric acid; RuBP, ribulose bisphosphate; TP, triose phosphate

Processing by Rubisco

Essentially all CO_2 enters the biosphere through fixation by Rubisco. There are other carbon fixation reactions, notably phospho*enol*pyruvate carboxylase and a reverse Krebs cycle. However, carbon fixed by phospho*enol*pyruvate carboxylase is not subsequently reduced, and only a small number of bacteria use a reverse Krebs cycle for carbon reduction and so the amount of carbon entering the biosphere this way is insignificant.

Rubisco has several problems:

1. It is a very slow enzyme.
2. It has a relatively low affinity for carbon dioxide.
3. It is prone to side reactions, both in terms of substrates and products.

Because Rubisco is slow, plants require very large amounts of Rubisco, leading, in large part, to the high requirement of plants for organic nitrogen. The slow reaction rate and relatively low affinity for CO_2 require that plants that rely on diffusion for CO_2 acquisition must have a substantial gas exchange capacity to allow sufficient CO_2 diffusion (see Chapter 5). This allows large amounts of water vapor to escape the leaf. Finally, in one of the side reactions, Rubisco uses oxygen in place of CO_2, a reaction that begins photorespiration. Thus, the processing of CO_2 by Rubisco is responsible for the high nitrogen requirement, the large water consumption, and photorespiration of plants.

Rubisco protein requires modification by addition of CO_2 (not the CO_2 that will enter the biosphere) and Mg^{2+} (Badger and Lorimer 1976, Roy and Andrews 2000). The CO_2 adds to a lysine residue in the active site to become a carbamate. Many compounds can interfere with carbamylation; for example, the Rubisco substrate ribulose 1,5-bisphosphate (RuBP) binds to uncarbamylated Rubisco to form a dead-end complex. Without additional enzymatic machinery, Rubisco carbamylation, and hence activity, falls to about 20% of the total Rubisco in a leaf (Mate et al. 1993).

In leaves, high levels of carbamylation, and thus Rubisco activity, require another enzyme, Rubisco activase (Portis et al. 1986), that probably works by removing any compound blocking the active site, allowing the enzyme to achieve full activity (Mate et al. 1996). The modulation of Rubisco activity by changes in carbamylation appears to be regulatory (Sharkey 1990). The degree of carbamylation changes through the day so that Rubisco activity does not exceed the ability to regenerate RuBP (Sage 1990). This keeps the steady-state amount of RuBP in a leaf relatively constant. For example, switching to low-light or feedback conditions, which limit the ability of the leaf to regenerate RuBP, causes only a transient decrease in RuBP, and a longer-term decrease in Rubisco decarbamylation (Mott et al. 1984, Sharkey et al. 1986). Therefore, measurements of RuBP levels alone cannot be used to determine what limits photosynthesis. This is illustrated by data of Kobza and Seemann (1988) (Fig. 7.2). At very low light, RuBP levels increased to about two times the concentration of RuBP binding sites on Rubisco. Then RuBP levels were constant as photosynthesis increased with increasing light. Most of the increase in photosynthesis with increasing light was mediated by increasing carbamylation of Rubisco rather than increasing amounts of RuBP. Many enzymes of the Calvin cycle are highly regulated so that the modulation of the Calvin cycle activity results primarily from changes in enzyme activation rather than changes in metabolite concentrations.

Rubisco catalyzes the addition of CO_2 to RuBP and the cleavage of the reaction intermediate into two molecules of phosphoglyceric acid. The affinity of Rubisco for RuBP is very high (K_m is low). In fact, the K_m of Rubisco for RuBP is lower than the concentration of enzyme sites (Farquhar et al. 1980). This un-

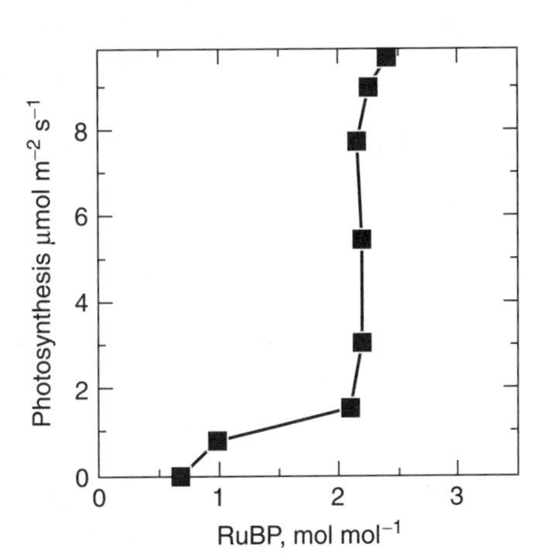

FIGURE 7.2. Relationship between photosynthetic rate and RuBP in a leaf. (data redrawn from Kobza and Seemann 1988).

usual situation causes the rate of carboxylation to increase linearly with increasing total RuBP until all of the RuBP binding sites are filled. Once all of the binding sites are filled, no further increase in rate can be achieved by increasing RuBP. This behavior has often been invoked to explain the sharp transition between Rubisco-limited and RuBP regeneration-limited photosynthesis (Farquhar and von Caemmerer 1982).

Sharp transitions between limiting factors are considered unusual in biological systems but sharp transitions are important in photosynthesis and are more common than sometimes thought (Koshland 1987). It has been found experimentally that leaves can show a profoundly sharp transition between RuBP regeneration-limited (approximately equal to light-limited) photosynthesis and Rubisco-limited (approximately equal to carbon dioxide-limited) photosynthesis. There are two explanations for this sharp transition: the relationship between K_m for RuBP and the concentration of Rubisco described in the preceding paragraph, and the regulation of Rubisco carbamylation. In practice, the concentration of RuBP increases up to saturating levels at low rates of photosynthesis. Then, photosynthesis increases by increasing the carbamylation state (activation state) giving rise to the relationship between photosynthesis and RuBP concentration illustrated in Figure 7.2. Once Rubisco is fully carbamylated, no additional increase in rate is possible and there is a sharp transition to Rubisco-limited photosynthesis. This requires that Rubisco activation state respond to signals that indicate the potential for RuBP regeneration. Rubisco activase is sensitive to adenosine triphosphate (ATP) status (Streusand and Portis 1987, Wang and Portis 1992) and redox status (Zhang and Portis 1999, Zhang et al. 2002) (at least in most plants, which have two forms of Rubisco activase). The sharpness of the transition from the RuBP regeneration-limited to Rubisco-limited state will depend on what has been called the metabolic cost of regulation (Woodrow and Berry 1988). If this cost is low, then the transition is potentially very abrupt, to the point of single limiting factors, as described by Blackman nearly a century ago (Blackman 1905).

As both the temporal and spatial scales under consideration increase, photosynthesis will be limited by both light availability, and CO_2 and it is not possible to blame either limitation for setting the rate of photosynthesis in an ecosystem or in one leaf over a growing season. However, the instantaneous rate of photosynthesis of a chloroplast is normally limited by either Rubisco or RuBP regeneration but not both.

Side Reactions of Rubisco

Once bound to Rubisco, RuBP loses a hydrogen ion in preparation for carboxylation. The resulting ene-diolate of RuBP can be reprotonated and RuBP can be reformed, or CO_2 can be added and, after several more steps on rubisco, two molecules of 3-phosphoglyceric acid (PGA) are formed (Fig. 7.3). However, Rubisco is subject to a number of side reactions. For example, the ene-diolate can be reprotonated in the wrong configuration, leading to the formation of xylulose 1,5-

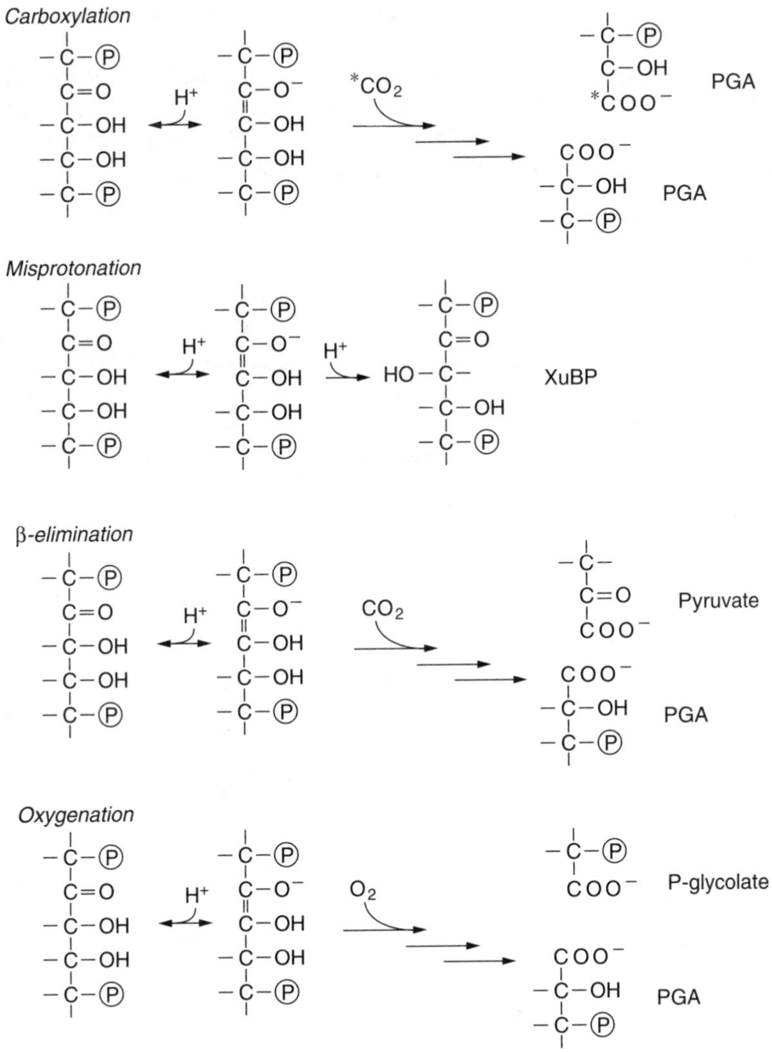

Carboxylation

−C−Ⓟ, C=O, −C−OH, −C−OH, −C−Ⓟ ⇌ (H^+) −C−Ⓟ, C−O⁻, ‖C−OH, −C−OH, −C−Ⓟ → ($^*CO_2$) → → −C−Ⓟ, C−OH, $^*COO^-$ PGA; COO^-, −C−OH, −C−Ⓟ PGA

Misprotonation

−C−Ⓟ, C=O, −C−OH, −C−OH, −C−Ⓟ ⇌ (H^+) −C−Ⓟ, C−O⁻, ‖C−OH, −C−OH, −C−Ⓟ → (H^+) −C−Ⓟ, C=O, HO−C−, −C−OH, −C−Ⓟ XuBP

β-elimination

−C−Ⓟ, C=O, −C−OH, −C−OH, −C−Ⓟ ⇌ (H^+) −C−Ⓟ, C−O⁻, ‖C−OH, −C−OH, −C−Ⓟ → (CO_2) → → −C−, C=O, COO^- Pyruvate; COO^-, −C−OH, −C−Ⓟ PGA

Oxygenation

−C−Ⓟ, C=O, −C−OH, −C−OH, −C−Ⓟ ⇌ (H^+) −C−Ⓟ, C−O⁻, ‖C−OH, −C−OH, −C−Ⓟ → (O_2) → → −C−Ⓟ, COO^- P-glycolate; COO^-, −C−OH, −C−Ⓟ PGA

FIGURE 7.3. Carboxylation and three side reactions of photosynthesis.

bisphosphate (XuBP), a dead-end inhibitor of Rubisco (Edmondson et al. 1990). Xylulose 1,5-bisphosphate is probably removed from Rubisco by Rubisco activase.

Another side reaction of Rubisco is production of pyruvate in place of one of the PGA molecules about 1% of the time (Andrews and Kane 1991). This occurs by β-elimination from a carbocation intermediate just before the last step of the Rubisco reaction. This reaction is not very deleterious and is the only source of pyruvate inside chloroplasts because chloroplasts lack glycolytic enzymes that normally convert PGA to phospho*enol*pyruvate (PEP) and then pyruvate (Stitt and ApRees 1979). Pyruvate is needed for fatty acid synthesis and isoprenoid synthesis.

The most frequent side reaction of Rubisco is oxygenation of RuBP, the reaction that begins photorespiration. When RuBP is oxygenated, the products are one molecule of PGA and one of phosphoglycolate. Phosphoglycolate is toxic to plants and must be metabolized in photorespiration.

Photorespiration

Photorespiration is the oxygenation of RuBP by Rubisco and the subsequent reactions that recover three fourths of the carbon initially lost to phosphoglycolate. It is theorized that Rubisco evolved at a time when there was very little oxygen in the atmosphere and so there was no evolutionary pressure to avoid this reaction (Sage and Pearcy 2000). By the time oxygen accumulated and CO_2 levels in the atmosphere fell, photosynthetic life forms dependent on Rubisco were so widespread that other carboxylation systems have not evolved.

The declining CO_2 and increasing oxygen levels in the atmosphere put substantial pressure on photosynthesis. The evolution of C_4 metabolism was one (recent) result (Sage 2001). Another result was the evolution of new forms of Rubisco that had a higher selectivity for CO_2 over oxygen. Selectivity of Rubisco for CO_2 over oxygen is lowest among bacteria, higher in green algae, and higher still in C_3 plants (Roy and Andrews 2000). However, the cost of the increased selectivity is often the maximum potential rate of carboxylation. In C_4 plants, which actively accumulate CO_2, the selectivity of Rubisco is slightly lower than in most C_3 plants but the maximum rate is higher. Because of the CO_2 pump, C_4 plants are better off trading some selectivity for a higher maximum rate. Except for this slight tradeoff, there is remarkably little variation among plants in the properties of Rubisco and in the Calvin cycle.

An exception to this tradeoff between selectivity and maximum rate occurs in some nongreen algae. These organisms have a more selective Rubisco than in C_3 plants but also a higher catalytic capacity (Badger and Spalding 2000). The relatively high rates of photorespiration are a legacy of land plant evolution from green algae. It may be possible to substantially lower photorespiration rates by engineering a red algal Rubisco into a land plant (Whitney et al. 2001, Whitney and Andrews 2001a,b).

Phosphoglycolate produced by oxygenation of RuBP is metabolized by a series of reactions that occur in three organelles (chloroplast, peroxisome, and mitochondrion) (Husic et al. 1987). Phosphoglycolate is dephosphorylated and then converted to an amino acid (glycine) and is processed in amino acid metabolism, eventually to glycerate but with the release of 0.5 CO_2 per phosphoglycolate. The amino acids can be exported or used directly in protein synthesis and so photorespiration need not be a complete cycle (Madore and Grodzinski 1984, Harley and Sharkey 1991). When glycine and serine are used in protein synthesis rather than completing the photorespiratory pathway, they represent another way that fixed carbon can be withdrawn from the chloroplast. If starch and sucrose synthesis limit photosynthesis, then glycine and serine export can increase the rate of photosynthesis because they are another end product of photosynthesis useful to the plant. This gives rise to reverse sensitivities to CO_2 and oxygen, where switching to low oxygen inhibits photosynthesis because of the loss of this extra pathway for reduced carbon to leave the chloroplast (Harley and Sharkey 1991).

Photorespiration is not the only way plants make glycine and serine. Nor is photorespiration required for other aspects of plant growth. Mutants deficient in any of seven different steps required for photorespiration can survive and reproduce, as long as they are grown in conditions that do not cause photorespiration (Somerville 1984). Therefore, while photorespiration can supply glycine and serine to the plant, the plant does not need photorespiration for this purpose.

Because amino acids are involved in photorespiration, there is intense recycling of reduced nitrogen, especially ammonia. Ammonia enters metabolism through the glutamine synthetase/glutamate synthase (GS/GOGAT) pathway. Because ammonia is a gas it can be taken up by plants from the atmosphere (Farquhar et al. 1980a, Lohaus and Heldt 1997). It is estimated that 90% of the GS/GOGAT activity in leaves is needed to recycle ammonia released in photorespiration and only 10% for building new protein. There appear to be separate enzymes for recovering ammonia released during photorespiration (GS2 in chloroplasts and ferredoxin-dependent GOGAT) versus primary ammonia assimilation (GS1 in the cytosol and NADH-dependent GOGAT) (Tobin and Yamaya 2001).

The reactions of photorespiration (1) require energy, (2) release CO_2, and (3) interfere with the carboxylation reactions of Rubisco. For these three reasons, photorespiration reduces the net rate of photosynthesis (Sharkey 1988). Because of the substantial losses of photosynthetic capacity to photorespiration, and because increasing CO_2 reduces photorespiration, increasing atmospheric CO_2 enhances plant growth both by its effect as a substrate for photosynthesis and by its effect on suppressing photorespiration. Nevertheless, photorespiration will continue to be a substantial problem for C_3 plants for the foreseeable future.

C_4 Photosynthesis

The balance sheet of the Calvin-Benson cycle is:

$$\text{RuBP} + CO_2 + H_2O + 1/3\text{Pi} + 3\ \text{ATP} + 2\ (\text{NADPH} + H^+) \rightarrow$$
$$\text{RuBP} + 1/3\ \text{triose-phosphate} + 3\ \text{ADP} + 3\ \text{Pi} + 2\text{NADP}^+ + O_2$$

Photorespiratory pathway can be expressed as:

$$\text{RuBP} + O_2 + 0.5\ O_2 + 3.5\ \text{ATP} + 2\ (\text{NADPH} + H^+) + 1/6\ \text{triose-phosphate} \rightarrow$$
$$\text{RuBP} + 0.5CO_2 + 3.5\ \text{ADP} + 3.5\ \text{Pi} + 2\text{NADP}^+ + 1/6\ \text{Pi}$$

In photorespiration, 3.5 ATP and 2 NADPH are used. Moreover, because 0.5 CO_2 is released in the photorespiration, 1/6 triose-phosphate is consumed to regenerate RuBP. If the energy and reducing power needed for synthesis of 1/6 triose-phosphate are taken into account, photorespiration pathway uses 5 ATP and 3 NADPH. Thus, C_3 plants lose much energy by photorespiration. The ratio of the rate of RuBP oxygenation to that of carboxylation increases with temperature and when CO_2 inside the leaf is low as happens when stomata close in dry environments. The loss of C and energy is particularly large under the dry and warm conditions.

Plants that can fix inorganic carbon by PEP carboxylase and supply CO_2 to Rubisco at high concentrations evolved in geological periods with low atmospheric CO_2 concentration (such as Pleistocene Era between 0.02 and 2 million years ago when atmospheric CO_2 fell below 200 μmol mol^{-1}). Plants solved the oxygenase problem by (1) localizing Rubisco internal compartment that is separated from the rest of the cell or tissue by a barrier that restricts CO_2 efflux, and then (2) biochemically pumping CO_2 into that compartment at the expense of ATP.

About 8000 C_4 species are known (Sage and Pearcy 2000). These species are not taxonomically close, indicating evolution of C_4 plants occurred independently in various taxa. C_4 plants are divided to three subtypes according to the decarboxylation enzymes: NADP-malic enzyme (NADP-ME), NAD-malic enzyme (NAD-ME), and (PEP caroboxykinase) PCK subtypes (Furbank et al. 2000). Carboxylation of PEP by PEP carboxylase, the first reaction of C_4 photosynthesis, occurs in cytoplasm of mesophyll cells. Unlike Rubisco, PEPCase has high affinity to its substrate, HCO_3^-. The product of PEP carboxylation is OAA, which is then converted to malate or aspartate.

The enzyme that carries out malate formaton is NADP malate dehydrogenase, which catalyses the reaction: OAA + NADPH $\rightarrow$ malate + $NADP^+$. Malate is then transported to bundle sheath cells. In NADP-malic enzyme types, malate undergoes decarboxylation reaction catalyzed by NADP-malic enzyme. Rubisco fixes the CO_2 released. Another product, pyruvate, goes to mesophyll chloroplasts and by means of pyruvate Pi dikinase, PEP is regenerated. Because bundle sheath cells are covered by suberin layer and mesophyll cells and are not directly exposed to intercellular spaces, CO_2 concentration in bundle sheath cells can be high enough to saturate carboxylation reaction of Rubisco and suppress oxygenation. Thus, C_4 plants show little photorespiration. The CO_2 concentration mechanism is carried out at the expense of two ATP, in the pyruvate-P_i dikinase reaction. The conditions in which the loss of energy due to photorespiration is greater than the cost of CO_2 concentrating, C_4 plants are favored. The distribution of C_4 plants is largely consistent of this idea (Sage and Pearcy 2000).

Trafficking of these compounds is through plasmodesmata between the mesophyll cells and bundle sheath cells (Leegood 1997, 2000). In addition to metabolites involved in the CO_2-concentrating mechanism, many other compounds are transported. For example, chloroplasts in the bundle sheath cells in some NADP-ME plants such as maize do not have much photosystem II (PS II) activity. To supply reducing power to the bundle sheath, malate can carry one NADPH equivalent from the mesophyll to bundle sheath cells. In addition, PGA is transported from bundle sheath chloroplasts to mesophyll chloroplasts and PGA is reduced to triose phosphate there and triose phosphate goes back to bundle sheath chloroplasts for the rest of Calvin cycle reactions. Enzymes required for conversion of PGA to DHAP [phosphoglycerate kinase and NADP-glyceraldehyde 3-phosphate dehydrogenase (NADP-GDH)] are located in chloroplasts of both cell types (Furbank et al. 2000).

To maintain efficient transport traffic, it would be better to have no barrier between mesophyll cells and bundle sheath cells, while to maintain high CO_2 concentration in the bundle sheath cells, the cell would be better to be sealed. The transport through plasmodesmata may be a compromise of these two requirements. There is some leak of CO_2 from bundle sheath cells. This can be estimated by means of carbon isotope discrimination and the leak decreases the quantum yield of photosynthesis (for a review see Brugnoli and Farquhar 2000, Evans and Loreto 2000).

Processing in the Chloroplast

Once captured by the reactions of Rubisco, carbon is processed to a wide variety of compounds inside chloroplasts. The Rubisco reaction produces two molecules of phosphoglyceric acid that are reduced by glycolysis reactions similar to those found in nearly all organisms, though in this case going in the opposite (gluconeogenic) direction. The result of this reduction is the production of triose phosphates (glyceraldehyde 3-phosphate and subsequently dihydroxyacetone phosphate). The triose phosphates can be used either for regeneration of RuBP or for other plant metabolism. The gluconeogenic reactions, combined with pentose phosphate pathway reactions to regenerate pentose phosphates, and three unique reactions, sedoheptulose-1,7-bisphosphatase, ribulose 5-phosphate kinase, and Rubisco, constitute the Calvin cycle, also called the photosynthetic carbon reduction cycle (Robinson and Walker 1981). The reactions of the Calvin cycle are highly conserved across a wide range of photosynthetic organisms, even if evolutionary source of the genes coding for the enzymes may not be conserved (Martin et al. 2000). The entire Calvin cycle is most easily described following three carboxylation events. Three RuBP molecules are carboxylated to give six triose phosphates. Five of the triose phosphates must be used to regenerate the three RuBP, leaving one triose phosphate molecule as the net production. Triose phosphate is the starting point for examining the products of the chloroplast.

Export of Triose Phosphate Molecules During the Day

Of the net increase in triose phosphate inside the chloroplast, typically one half is immediately exported to the cytosol. This export is catalyzed by an antiporter that requires a strict counter-exchange of phosphate for triose phosphate (Flügge and Heldt 1991). This carrier is called the triose phosphate/phosphate transporter (TPT) and accounts for a substantial fraction of the protein in the chloroplast envelope (Flügge 1999). It has been sequenced and its activity demonstrated in artificial vesicles. This carrier, like most carriers in the chloroplast envelope, exerts little control on the rate or direction of exchange. Therefore, the control of the export of carbon from chloroplasts rests with the metabolic reactions both inside and outside of the chloroplast. In particular, the fructose bisphosphatases (FBPases)(one inside and one outside the chloroplast) are very highly controlled and determine how much carbon stays in the chloroplast and how much is exported through the TPT (Stitt 1990b).

The chloroplastic FBPase is activated by thioredoxin (Leegood et al. 1982) providing a link between electron transport and carbon metabolism of photosynthesis. This activation depends on the presence of substrate and so the more substrate present, the more activated the enzyme becomes. The cytosolic version of the enzyme is regulated by the presence or absence of the inhibitor fructose 2,6-bisphosphate (F2,6-BP). A specific fructose-6-phosphate-2-kinase makes F2,6-BP when PGA levels are low and phosphate is high, and a phosphatase cleaves the 2 phosphate when PGA is high or phosphate is low (Kruger and Scott 1995, Stitt 1990a). In this way, the faster photosynthesis goes, the less F2,6BP is present (Stitt et al. 1984, Sicher et al. 1986), and the faster cytosolic FBPase goes. This ensures that exactly the right amount of triose phosphate stays in the chloroplast. If the cytosolic FBPase goes too fast, too much triose phosphate is withdrawn from the cycle and not enough triose phosphate is available for regeneration of RuBP. If not enough triose phosphate is withdrawn, not enough phosphate is available for ATP synthesis and photosynthesis becomes limited by lack of ATP.

Starch Synthesis

Triose phosphates not exported to the cytosol are primarily made into transitory starch. Transitory starch is the starch that is synthesized in the chloroplasts of leaves during the day and broken down at night (Trethewey and Smith 2000). Sucrose synthesis is preferred until the photosynthetic rate gets above a certain threshold. At this point, the ratio of starch to sucrose increases rapidly as the photosynthetic rate increases. At some photosynthetic rate, this ratio levels off and remains constant, with approximately half the carbon being partitioned to sucrose synthesis and the other half to starch synthesis (Sharkey et al. 1985). There is substantial variation among plants in the ratio of starch to sucrose during primary partitioning. For example, spinach leaves store significant quantities of sucrose in their vacuoles before starch synthesis begins (Gerhardt et al. 1987, Servaites et al. 1989) while *Phaseolus* makes starch as soon as photosynthesis exceeds a certain threshold rate (Sharkey et al. 1985).

Transitory starch is found in most land plant groups including liverworts (Table 7.1), indicating that the transitory, circadian function of starch during photosynthesis is universal among land plant groups and predates its function as a storage molecule in seeds. Starch is not found in bacteria, animals, or fungi, but is found in some photosynthetic algae, often surrounding the Rubisco-containing

TABLE 7.1. Starch content of various basal land plants at the end of the night or end of the day.

	End of	mg g^{-1} dry wt	SE
Polypodium punctatum	Day	406	± 7
	Night	267	± 7
Drynaria quercifolia	Day	133	± 4
	Night	96	± 6
Marchantia polymorpha	Day	20.5	± 0.8
	Night	6.5	± 1.3
Lycopodium obscurum	Day	8.3	± 0.9
	Night	10.1	± 1.1
Selaginella rupestris	Day	1.4	± 0.2
	Night	0.5	± 0.1
Thuidium delicatulum	Day	1.6	± 0.04
	Night	0.9	± 0.1
Equisetum equinum	Day	17.3	± 0.6
	Night	11.6	± 0.6

pyrenoid. *Chlamydomonas* may have two types of starch, and the function of transitory starch may originate in the algae (Libessart et al. 1995). In angiosperms, transitory starch is not significantly different from storage starch (Zeeman et al. 2002). The amylose part of starch is almost completely α-1,4-linked polyglucan. However, starch is mostly amylopectin, which has α-1,6 branch points. The balance between amylose and amylopectin is altered in mutants of starch synthase (Patron et al. 2002). In glycogen (the reserve polyglucan in animals) the branches occur randomly, but in amylopectin they occur at regular intervals and give amylopectin a highly ordered structure (Fig. 7.4). It is presumed that plants evolved the ability to make starch from the machinery that makes glycogen (Ball and Morrell 2003), and an isoamylase (starch debranching enzyme) may be involved in converting glycogen metabolism to starch metabolism in angiosperms (James et al. 1995, Zeeman et al. 1998b, Burton et al. 2002) while in algae there is evidence for involvement of a glucanotransferase in the shift from glycogen to starch (Colleoni et al. 1999a,b).

Transitory starch serves two functions (Trethewey and Smith 2000). It can act as a carbon overflow allowing photosynthesis to proceed faster than would be allowed by the capacity for triose phosphate export from the chloroplast and sucrose synthesis. In other words the synthesis of starch lessens the frequency of feedback limitations on photosynthesis. Slightly lower photosynthetic rates in starch-less mutants have been observed in previous studies (Caspar et al. 1991, Huber and Hanson 1992). The difference in photosynthetic rate between starch-deficient and wild-type (WT) plants becomes quite significant when the photosynthetic rate is increased by elevated CO_2 (Ludewig et al. 1998) and especially if nitrogen nutrition is sufficient (Sun et al. 2002).

The second function of transitory starch is as an energy reserve providing the plant with carbohydrate during the night when sugars cannot be made by pho-

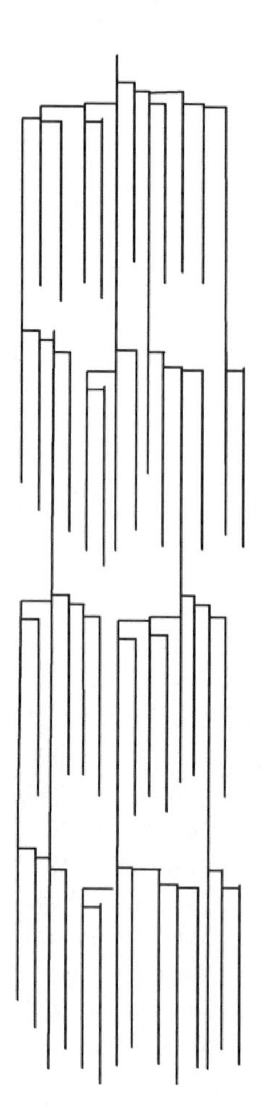

FIGURE 7.4. The branching of amylopectin results in a highly ordered structure in starch.

tosynthesis. Plants made starch-less by antisense or knockout technologies are often unaffected when grown in continuous light but grow less under short photoperiods (Caspar et al. 1991, Huber and Hanson 1992).

Regulation of starch synthesis appears to rest primarily with ADPglucose pyrophosphorylase (AGPase). Low phosphate and high PGA levels stimulate this enzyme so that as photosynthesis increases and PGA accumulates at the expense of inorganic phosphate, starch synthesis is stimulated. AGPase converts glucose 1-phosphate plus one ATP into ADPglucose, which is used to elongate starch molecules by starch synthase. Regulation by chloroplast phosphoglucoisomerase and phosphoglucomutase is also evident (Dietz 1985, Schleucher et al. 1999, Lytovchenko et al. 2002). Both of these enzymes are needed for starch synthesis but not for Calvin cycle reactions.

The biochemistry of starch synthesis is less certain. Recent data indicate that the α-1,4 dextrins of starch may grow in the direction of the reducing end of the molecule (Mukerjea et al. 2002). There are two major forms of starch synthase in plants. Soluble starch synthase is thought to make amylopectin while granule-bound starch synthase is thought to make amylose. The role of branching enzymes and starch synthases is an active area of research at least in part because of the industrial uses of starch (Kossmann and Lloyd 2000).

Starch Breakdown and Export at Night

The intact starch grain in leaves is relatively difficult for degrading enzymes to attack. The question of how starch degradation begins may be answered soon as a result of the discovery of a starch-water dikinase that donates the β-phosphate of ATP to starch. The enzyme, known as R1, can increase the phosphate content of starch, perhaps making it easier to break down (Lorberth et al. 1998). Mutants lacking this enzyme accumulate large amounts of starch. The importance of this finding in explaining the early events of starch breakdown is only now being explored (Reimann et al. 2002).

During the day, carbon is exported from chloroplasts on the TPT. However, this is not the route by which carbon is exported from chloroplasts at night. In a *Flaveria linearis* mutant with no cytosolic FBPase, an enzyme required for the synthesis of sucrose from triose phosphate, no carbon was exported from leaves during the day and large amounts of starch accumulated; at night the starch was broken down (Sharkey et al. 1992). These authors concluded that there exists a pathway for exporting carbon at night that does not involve triose phosphates. In potato plants with antisense genes for cytosolic FBPase, carbon export was shifted from day to night, with plant growth and tuber yield remaining unchanged from the WT (Zrenner et al. 1996). Bypassing cytosolic FBPase makes sense given that leaves have high nighttime concentrations of the FBPase inhibitor F2,6BP (Stitt 1985, Usuda et al. 1987, Servaites et al. 1989, Stitt 1990b, Scott and Kruger 1995, Trevanion 2002).

Further evidence of the limited role of triose phosphate export at night came from a series of experiments using potato and tobacco plants with antisense genes for the TPT. Starch accumulation was greater than WT during the day followed by increased starch breakdown during the night (Heineke et al. 1994, Häusler et al. 1998). The TPT antisense tobacco began starch breakdown approximately 5 h earlier than did the WT, while the time of synthesis remained unchanged (Häusler et al. 1998). Similar results have now been reported for a TPT knock-out *Arabidopsis* (Schneider et al. 2002). However, when ADPglucose pyrophosphorylase activity was reduced using antisense technology, less starch was formed and more sucrose was produced during the day (Leidreiter et al. 1995). Plants with antisense genes to both ADPglucose pyrophosphorylase and TPT had severe growth effects as would be expected if both major carbon sinks in photosynthesis were disrupted (Hattenbach et al. 1997).

From these experiments it is clear that there are two alternate pathways by which carbohydrates can leave the chloroplast: (1) triose phosphate export during the day, and (2) export of carbon resulting from starch breakdown. These pathways can compensate for each other to provide the plant with sugars. This was clearly seen in antisense potato plants with reduced TPT. The antisense TPT plants showed only a slight reduction in height, and tuber yield was unchanged compared to WT (Riesmeier et al. 1993). However, CO_2 assimilation rates in transgenic tobacco plants with antisense repression of TPT were reduced by about 40% under non-photorespiratory conditions in comparison with WT (Häusler et al. 2000a,b). This shows that maximal photosynthetic rates can only be achieved when both pathways for the export of carbon from chloroplasts are fully operational.

The question still remained, in plants with normal levels of FBPase and TPT, do the products of starch breakdown leave the chloroplast by the alternate pathway? To answer this question nuclear magnetic resonance (NMR) spectroscopy was used to examine the incorporation of deuterium from deuterium-enriched water into the glucose moiety of sucrose at night. The experiments demonstrated that in tomato and bean, at least 75%, if not all, of the carbon converted from starch to sucrose at night is never broken down to level of a triose phosphate (Schleucher et al. 1998). This demonstrated that during darkness, carbon is exported from the chloroplast as a hexose, a hexose phosphate, or higher maltodextrin. There are hexose phosphate/phosphate transporters in plants but they are only expressed in heterotrophic plastids, and measured rates of hexose phosphate uptake by isolated chloroplasts are low (Flügge and Heldt 1991, Kammerer et al. 1998).

Maltose is now known to be the predominant molecule exported from chloroplasts at night (Weise et al. 2003). Chloroplasts isolated from leaves in the early part of the dark period export 60 to 80% maltose and 20 to 40% glucose with low levels of PGA as well. Chloroplasts are permeable to maltose (Rost et al. 1996, Servaites and Geiger 2002) and a specific maltose transporter has been identified and shown to be essential for normal starch metabolism in leaves (Nittylä et al. 2004). Maltose has been observed as an export product of photosynthesis in the past (Levi and Gibbs 1976, Stitt and Heldt 1981, Neuhaus and Schulte 1996, Servaites and Geiger 2002) but its importance has only recently been recognized.

With this information, the following model for starch breakdown and conversion to sucrose can be proposed (Fig. 7.5). It is known that α-amylase, debranching enzyme, and perhaps α-glucosidase will attack the intact starch granule (Stitt and Steup 1985, Li, Geiger and Shieh 1992, Sun et al. 1995, Witt and Sauter 1995) producing maltodextrins, linear chains of up to 20 glucose molecules. A deficiency in a chloroplastic α-amylase was found to impair starch breakdown at night and caused the starch excess phenotype *sex4* (Zeeman et al. 1998a, Zeeman and ApRees 1999). It is not clear which enzyme begins the process but the role of the R1 protein (Ritte, Lorberth and Steup 2000, Yu et al. 2001, Reimann et al. 2002, Ritte et al. 2002) and phosphorylation of starch have not yet been explored.

The maltodextrins can be acted upon by β-amylase to produce maltose. Lao and coworkers (1999) demonstrated that there is a gene in *Arabidopsis* encoding a chloroplast-targeted β-amylase. A gene very similar to this has been found in

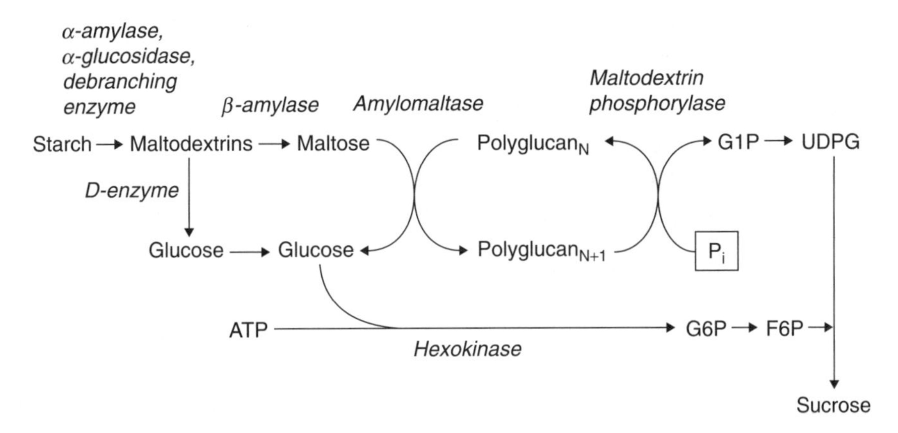

FIGURE 7.5. Pathway for transitory starch conversion to sucrose in photosynthetic leaves Maltose formation exceeds glucose formation. Blocking maltose export (Nittylä et al. 2004) or metabolism (Lu and Sharkey 2003, Chia et al. 2004) results in starch accumulation. Abbreviations: F6P, fructose 6-phosphate; G1P, glucose 1-phosphate; G6P, glucose 6-phosphate; UDPG, uridinediphosphate glucose

potato, and the protein product was imported and processed into pea chloroplasts. When this β-amylase activity id reduced by antisense technology, potato leaves accumulate starch (Scheidig et al. 2002).

The glucose exported from the chloroplast could be produced by the plastidic disproportionating enzyme (D enzyme). D enzyme rearranges maltodextrins resulting in the production of some glucose and maltodextrins with a higher degree of polymerization that can once again serve as substrates for β-amylase or starch phosphorylase (Kakefuda and Duke 1989, Critchley et al. 2001). The glucose is then phosphorylated by hexokinase, which is located on the outside of the chloroplast envelope (Wiese et al. 1999). This hexokinase is one of two major sugar-sensing mechanisms known in plants (Xiao, Sheen and Jang 2000, Koch et al. 2000, Tiessen et al. 2003).

In the cytosol, maltose is apparently converted to higher maltodextrins by amylomaltase (similar to the plastidic D enzyme)(Lu and Sharkey 2003, Chia et al. 2004). These are then acted on by a maltodextrin phosphorylase in a system similar to that used by *E. coli* to process maltose (Boos and Shuman 1998, Lu and Sharkey 2003).

While it appears that the pathway for carbon export is hydrolytic rather than phosphorolytic, a phosphorolytic pathway exists as well. There is a plastidic form of starch phosphorylase which catalyzes the phosphorolytic cleavage of a glucose residue from the nonreducing end of a maltodextrin to form glucose 1-phosphate (Steup and Latzko 1979, Steup and Schächtele 1981). Stitt has found a substantial activity of the oxidative pentose pathway during starch breakdown (Stitt and ApRees 1979, Stitt and ApRees 1980). Starch phosphorylase would provide substrate for this pathway. This could allow metabolism such as the shikimic acid pathway, needed for some essential amino acids and phenylpropanoids, to proceed at night. Starch phosphorylase activity could also provide substrate for the Calvin cycle (i.e., RuBP) when there is a net loss of carbon caused by excessive photorespiration or emission of hydrocarbons. Starch phosphorylase is the only way for starch breakdown products to be used to supplement the Calvin cycle because plastid-associated hexokinase activity is attached to the outside chloroplasts (Wiese et al. 1999) and phosphorylates hexoses in the cytosol, not inside the chloroplast.

Starch metabolism exhibits much mare variation among plants than does the Calvin cycle but it is also one of the least understood aspects of carbon processing in the leaf. Thanks to improved carbohydrate detection assays and the use of molecular tools such as knockout and antisense plants, tremendous progress has been made in the last 10 years.

Other Products of the Chloroplast

Fatty Acids

Fatty acid synthesis in plants takes place in plastids (Harwood 1996). Fatty acids are made from pyruvate, which is converted to acetyl CoA. Carbon dioxide is added to acetyl CoA to make malonyl CoA. However, that carbon is lost in the next step and so is best considered to have a catalytic role.

Even though so much carbon is being reduced during photosynthesis, chloroplasts must rely on import of pyruvate from the cytosol to support fatty acid synthesis. A small amount of pyruvate is made in a side reaction of Rubisco (Andrews and Kane 1991), but this may not be sufficient to support all of the needs of the chloroplast for pyruvate. In most tissues pyruvate is easily made by glycolysis, but in chloroplasts some of the glycolytic enzymes are missing, including phosphoglycerate 2-mutase (Stitt and ApRees 1979). It is suspected that if the PGA 2-mutase were present in chloroplasts it could lead to a futile cycle, with PGA being converted to pyruvate instead of being reduced to triose phosphates.

Pyruvate gets into chloroplasts on another transporter, the phospho*enol*pyruvate/phosphate transporter (PPT) (Fischer et al. 1997). This transporter can transport pyruvate in addition to PEP and PEP can be converted to pyruvate. The pyruvate is converted to acetyl CoA by the pyruvate dehydrogenase complex known to occur in plastids (Williams and Randall 1979, Camp and Randall 1985, Treede and Heise 1986). Plastids can unsaturate the fatty acids but a variable proportion are exported to the endoplasmic reticulum for unsaturation (Somerville and Browse 1991). Some of the important vegetable oils are triglycerides (three fatty acids linked to glycerol) with a high proportion of unsaturated fatty acids. For example, olive oil is over 50% oleic acid, an 18-carbon fatty acid with one unsaturation. Attempts are under way to control the chain length and degree of unsaturation of plant fatty acids to increase their economic value and possibly replacing some uses of fossil fuel.

Amino Acids

Plastids are the site of synthesis of the aromatic amino acids, tyrosine, tryptophan, and phenylalanine. It is also the site of synthesis of the sulfur-containing amino acids (methionine and cysteine), the branched-chain amino acids (leucine, isoleucine, and valine), and lysine. Typically about 10% of photosynthesis is converted to amino acids that are exported to the rest of the plant. However, the flux to amino acids can be quite high because photorespiration makes use of some amino acid metabolism to recover the carbon lost to glycolate when Rubisco uses oxygen instead of carbon dioxide.

Phenylpropanoids

Phenylpropanoids are a large class of compounds that include many defensive chemicals and are the source for lignin. The shikimic acid pathway makes phenylpropanoids from PEP and erythrose 4-phosphate. The PEP comes into the chloroplast on the PPT transporter and if that transporter is knocked out, phenylpropanoid production declines (Fischer et al. 1997). The phosphorolytic breakdown of starch can provide substrate for the production of erythrose 4-phosphate at night while during the day erythrose 4-phosphate is available because it is an intermediate in the Calvin cycle.

Many of the phenylpropanoids can contribute to the pool of carbon that turns over very slowly in the environment. So, while the phenylpropanoids may be a minor sink for carbon normally, the carbon that does end up in phenylpropanoids can be sequestered for many years.

Terpenoids

The plastid is the site of synthesis of many terpenoids, especially carotenoids and monoterpenes (Kleinig 1989, McGarvey and Croteau 1995). Carotenoids serve critical roles as protectants of chlorophyll from destruction by radicals and reactive oxygen species. When carotenoid synthesis is blocked, leaves exposed to light are bleached white because all other pigments are destroyed. Carotenoids are also responsible for many of the colors of flowers.

While all plants have carotenoids, the occurrence of other terpenoid compounds varies greatly from one species to the next. Monoterpenes are responsible for many of the smells associated with plants, such as lemon and pine scent. The appealing smell of orange that can fill a room when one orange is peeled results from monoterpene release. Some monoterpenes are made and stored in specialized glands but in other cases, monoterpenes are made in photosynthesizing chloroplasts from recent photosynthate (Loreto et al. 1996).

Isoprene, the simplest terpenoid, is made in chloroplasts of some species while other species lack the ability to make isoprene. Isoprene may help chloroplasts cope with short high-temperature episodes (Sharkey, Chen and Yeh 2001). It is hypothesized that the constant loss of isoprene from leaves allows for rapid changes in the amount of isoprene dissolved in membranes. In some cases, isoprene synthesis can be the fate for several percent of the carbon fixed in photosynthesis. The monoterpenes that are emitted immediately upon synthesis probably serve the same function as isoprene (Loreto et al. 1998).

The terpenoids in plastids are derived from dimethylallyl pyrophosphate and isopentenyl pyrophosphate. In animals and the cytosol of plants these compounds are made by the mevalonic acid pathway, but in bacteria and plastids an alternative pathway that includes the metabolite methyl erythritol 4-phosphate (MEP) is used (Lichtenthaler 1999). The MEP pathway is more efficient than the mevalonic acid pathway when it can draw on the products of photosynthesis (Sharkey and Yeh 2001).

Processing in the Cytosol

Sucrose Synthesis

Among the best-known products of photosynthesis, and quantitatively often the most important, is sucrose. Sucrose is made from triose phosphates exported from chloroplasts during the day (Winter and Huber 2000). In addition to the cytosolic FBPase, the enzyme sucrose-phosphate synthase (SPS) exerts the most control over the rate of sucrose synthesis. Sucrose-phosphate synthase is regulated by, among other things, phosphorylation (Huber and Huber 1992). The phosphorylation status depends on the balance between a specific SPS kinase and a protein phosphatase 2a (PP2a) (Foyer, Ferrario-Mery and Huber 2000). Nitrate reductase is also regulated by phosphorylation, providing a mechanism for coordinating nitrogen metabolism with carbon metabolism.

The formation of sucrose is the primary interface between photosynthetic metabolism and the physiology of the rest of the plant. For this reason, SPS has been a target for transformation with the hope of manipulating yield. In some cases, increasing SPS activity has had little effect on yield, which has been ascribed to deactivation of the engineered protein or to lack of expression of the protein in roots and or fruits (Foyer, Galtier and Quick 1994). However, Laporte and coworkers (2001) found that doubling the amount of SPS in tomato plants increased yield (up to 50%) but at higher levels of SPS the enhancement was lost. They interpreted this to mean that there is an optimum activity of SPS and that normally, plants are operating below that optimum, either to be conservative allowing for environmental stress or because plants are adapted to lower CO_2 and have not yet adjusted to the increased CO_2 of today's Earth (Sage and Cowling 1999).

Other Products

Many plants produce sugar alcohols in addition to sucrose (Loescher and Everard 2000). The precise role of sugar alcohol metabolism is not known but many sugar alcohols, especially mannitol, quench reactive oxygen species. Sugar alcohol synthesis can be correlated with stress. Mannitol is made from the corresponding phosphomonosaccaride in the cytosol of photosynthetic leaves (Everard et al. 1993).

Some plants, especially in cold climates, make fructans from sucrose (Chatterton et al. 1987). While fructan synthesis can be substantial in stems, some fructans can be made in photosynthesizing cells (Pollock et al. 2003). Another class of componds made in photosynthesizing leaves, the raffinose series of sugars, are also made from sucrose (Haritatos et al. 2000).

Plasmodesmata

Once sucrose is made in the cytosol, it can move from cell to cell through plasmodesmata on its way to the phloem (Schobert et al. 2000). The role of plasmodesmata in sugar transport was demonstrated by the naturally occurring mutant of corn called sucrose export deficient (*sxd*) (Russin et al. 1996). The *sxd1* corn plants lacked plasmodesmata at one specific interface and could not export sucrose. The *sxd1* gene has been cloned and appears to be involved in signaling from plastids to the nucleus (Provencher et al. 2001, Rodermel 2001).

Plasmodesmata play critical roles not only in sugar transport, but also in movement of proteins and RNA from cell to cell (Lucas 1995). The macromolecular trafficking was first discovered as the way that some viruses spread through plants but soon it was recognized that plant proteins and RNA also move through plasmadesmata (Lucas et al. 1995). The proteins that control movement through plasmadesmata can affect carbohydrate partitioning (Balachandran et al. 1995, Lucas et al. 1993, Herbers et al. 1997). Precisely how the plasmodesmatal-movement proteins affect carbohydrate partitioning is one of the important areas of current research into carbon processing in leaves.

Phloem Loading

Sugars move from the mesophyll cells, the sites of photosynthesis, to the phloem. Most of the movement is through plasmodesmata but the last steps of transport are variable. In the simplest case, willow (and possibly other plants) use the concentration of sucrose built up in the mesophyll cells to establish a gradient in sucrose concentration from sources to sinks (Turgeon and Medville 1998).

The more common mechanism includes an active transport step and at least one step during which sucrose leaves the symplast and is then taken up from the apoplast by sucrose-proton cotransport. In plants called apoplastic loaders, sieve tubes that transport sucrose to other parts of the plant lie adjacent to companion cells. Sucrose-proton cotransport in the companion cells actively accumulate sucrose to a much higher concentration that found in the surrounding mesophyll (Lemoine et al. 1996, Turgeon and Medville 1998). This allows a much higher concentration gradient in sucrose from the source regions to the sink regions. A pressure-flow mechanism originally proposed by Mönch then helps move sucrose through the phloem from source to sink regions.

A third system for phloem loading used by some plants, especially in the cucurbits, is called the polymer trap system (Turgeon 1996). In this model, sucrose diffuses into the companion cell of the phloem, and there galactinol reacts with the sucrose to make raffinose, stachyose, and verbascose (sucrose plus one, two, or three galactose units). These large sugars (polymers) cannot return through the plasmodesmata and are thus trapped in the companion cell-sieve tube complex (Oparka and Turgeon 1999). This system requires galactinol synthase, found only in minor veins of leaves that are photosynthetically competent (Haritatos et al. 2000).

Feedback Effects

A long-debated question is whether photosynthesis is limited by growth or growth is limited by photosynthesis. In other words, are there feedback mechanisms that regulate photosynthesis so that it matches growth capacity? There is evidence for feedback at some levels but not at other levels. The mere fact that plants grow faster in elevated carbon dioxide is proof that plant growth does not exert absolute control over the rate of photosynthesis (Farquhar and Sharkey 1994). On the other hand, it has recently been argued that the limiting altitude on mountains for tree growth results when growth limits photosynthesis (Hoch et al. 2002). More generally, growth is highly temperature sensitive while photosynthesis is not (Grace et al. 2002). Simplistically, growth may limit photosynthesis at low temperature while photosynthesis limits growth at high temperature. In truth, the situation is more complex. Most of the mass of a plant is photosynthetic organs or support, both physical (stem) and nutritional (roots) of the photosynthetic organs. Thus, growth is required for photosynthesis and photosynthesis is required for growth. In contrast to the processing of carbon in the chloroplast, mechanisms for matching growth capacity to photosynthetic capacity, and vice versa, vary among plants. This causes much of the rich variety of plant forms that exist. One of the easiest ways to match photosynthesis to growth is feedback of the products of photosynthesis on the capacity for photosynthesis.

A simple mode of feedback would be sucrose repression of some of the reactions of photosynthesis by mass action or allosteric effect. This appears not to occur. There is little evidence for sucrose accumulation exerting any short-term effect on photosynthetic rate (Neales and Incoll 1968, Sharkey 1985). However, there are three levels of feedback within carbon-processing reactions in leaves that have been studied. They are (1) limitation in the rate of use of triose phosphates relative to the capacity to make triose phosphate, (2) effects on gene expression resulting from sugar-sensing mechanisms, and (3) effects on carbohydrate partitioning resulting from plasmodesmatal movement proteins (described above).

The first level of feedback occurs when the chloroplast can make triose phosphates faster than they can be converted to end products. When this happens, phosphorylated products accumulate and the chloroplast soon is starved for phosphate. While this is called a triose phosphate use limitation, it is not triose phosphate that accumulates but primarily phosphoglyceric acid (Sharkey et al. 1986). This is not a phosphate nutrition problem. Both starch and sucrose synthesis are regulated in part by inorganic phosphate. As the rate of photosynthesis increases, inorganic phosphate is converted to organic phosphate, and the lowered concentration of phosphate increases the rates of starch and sucrose synthesis. At the same time, inorganic phosphate is needed for ATP synthesis. So, as the rate of photosynthesis increases, the level of inorganic phosphate required for the increasing rate of ATP synthesis increases. Feedback occurs when the level of inorganic phosphate cannot be both high enough for ATP synthesis and low enough to sufficiently stimulate starch and sucrose synthesis. This is an imbalance between the ability of the chloroplast to make triose phosphate and the ability to process those triose phosphates.

When leaves are artificially put into this kind of feedback condition, for example by switching to low oxygen to increase the efficiency of triose phosphate production, a series of events occur that illustrate the feedback chain. These include a drop in phosphate level (Sharkey and Vanderveer 1989) followed by a decline in ATP, a decline in RuBP, and finally, deactivation of Rubisco, which allows the RuBP to recover (Sharkey 1989,1990). The deactivation of Rubisco ensures that triose phosphates are made only at the rate they can be processed to unphosphorylated products such as starch, sucrose, sugar alcohols, and amino acids.

The triose phosphate use limitation is the mechanism for restricting photosynthesis when it exceeds the ability of the rest of the plant to immediately use the products of photosynthesis. However, it represents lost opportunity, the sunlight not used because of this regulation is lost to the plant forever. This limitation rarely occurs in natural settings. This indicates that there are significant feedback mechanisms that divide resources between production (photosynthesis) and consumption (maintenance, growth, and development) so that consumption rarely limits production. These are longer-term feedback mechanisms based on gene expression and plant hormone status.

The longer-term feedback mechanisms act through sugar sensing. One of the earliest proven mechanisms is sensing by hexokinase. In some way, when hexokinase 1 of *Arabidopsis thaliana* undergoes a catalytic cycle, a signal is generated that can interact with many other signaling pathways in the plant (Veramendi et al. 1999, Dai et al. 1999, Xiao, Sheen and Jang 2000, Rolland et al. 2002, Moore et al. 2003). This serves to balance the ability to make sugar with the ability to process it. A large number of mutant phenotypes involved in sugar sensing have been identified (Pego et al. 2000, Rolland et al. 2002). These phenotypes have different names, including glucose insensitive (*gin*), carbohydrate insensitive (*cai*), impaired sugar induction (*isi*), sucrose insensitive (*sun*) and so forth. (Rolland et al. 2002). Some of the genes responsible for these phenotypes have been isolated because of their involvement in plant hormone signaling. Thus, some of the *gin1* alleles have been described as abscisic acid insensitive (*abi*).

In the case of *gin2*, the gene is now known to be AtHXK1, the gene that codes for the sugar-sensing hexokinase. Hexokinase-based sugar sensing interacts directly with both abscisic acid regulation of plant growth and development (Arenas-Huertero et al. 2000, Xiao et al. 2000) and with ethylene-responsive regulatory pathways (Sheen et al. 2003). The convergence of sugar-sensing and plant hormone studies is very recent and will likely be an area of important new insights in carbon processing in the coming years.

Photosynthesis during the day does not require hexokinase activity. However, the recently proposed pathway of starch conversion to sucrose requires at least one turnover of hexokinase per sucrose molecule made (see Fig. 7.5). Thus, sucrose synthesis at night is directly linked to sugar sensing and both abscisic acid and ethylene responses, but carbon processing during the day is not. As CO_2 levels rise, more carbon is processed into starch during photosynthesis during the day and this carbon is then sensed the following night. If that extra carbon were instead made into sucrose during the day, it would not affect plant development. This appears to happen when plants are transformed with SPS. By stimulating the capacity for sucrose synthesis, partitioning is altered away from starch, toward daytime sucrose synthesis (Laporte et al. 2001). This resulted in substantial yield increase because the normal senescence occurring during fruit development was delayed and we hypothesize this was because of less signal to abscisic acid and ethylene regulatory pathways.

Another sugar-sensing mechanism in plants occurs by proteins similar to a protein whose mutation makes yeast unable to ferment sucrose (sucrose nonfermenting, *snf*). In plants there are kinases related to *snf* and so are called SNF-related kinases (SnRKs). SnRKs phosphorylate critical regulatory proteins such as SPS and nitrate reductase. Once phosphorylated, these proteins bind yet another class of proteins called 14-3-3 proteins, discovered in brain tissue and named for the fractions they are found in from columns used to separate them from other proteins (Sehnke et al. 2002). But, while hexokinase is known to be inside the cytosol, attached to the outer membrane of the chloroplast (Wiese et al. 1999), the SnRKs appear to be on the outside of the cell membrane. In this location they sense sugar levels in the cell wall. This sensing was demonstrated by expressing a gene that coded for an extracellular invertase. These plants had a severe yellow phenotype (Von Schaewen et al. 1990, Heineke et al. 1992) even though invertase expressed in the cytosol had little effect (Hajirezaei et al. 2000).

The large number of proteins and sensing pathways involved in sugar sensing are not surprising given the critical importance of matching photosynthesis and growth of the plant. It can reasonably be anticipated that over the next few years, substantial progress in this area will improve our understanding of how carbon processing in photosynthesis is coordinated with the physiology of the rest of the plant. This is perhaps the most important and exciting frontier of cell-level photosynthesis physiology today.

The interaction between growth and photosynthesis determines the resources allocated to three component processes of photosynthesis as described by the Far-

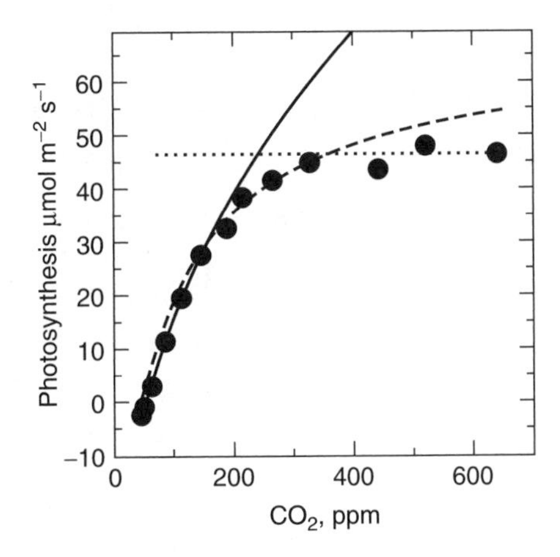

FIGURE 7.6. Response of photosynthesis to CO_2 inside the stroma. Data (●) are for *Gossypium barbadense* growing under agronomic conditions in Arizona with a leaf temperature of 29° C (Wise et al. 2004); —, the minimum Rubisco-limited photosynthetic capacity needed; dashed line is the minimum RuBP regeneration-limited photosynthetic capacity needed (reflecting mostly electron transport), and dotted line is the minimum triose phosphate use required. Photosynthesis is most efficient when the three capacities intersect at the stromal CO_2 concentration of the leaf.

quhar model of photosynthesis (as extended by Sharkey) (Farquhar et al. 1980b, Sharkey 1985). This model accounts for the activity of Rubisco, of electron transport reactions, and triose phosphate use. These three components respond to CO_2 differently and so a CO_2 response curve can be interpreted in terms of a mix of these three capacities (Fig. 7.6). Recent publications of the values needed to accurately describe the CO_2 and temperature responses of these component processes (Bernacchi et al. 2001,2002,2003) have made it easy to model how carbon processing reactions change if the plant has more or less Rubisco, more or less electron transport activity, or more or less capacity to use triose phosphates.

Nitrogen Metabolism

Nitrogen assimilation is essential to make the enzymes needed for carbon processing but at the same time nitrogen metabolism competes with carbon metabolism for the products of photosynthetic electron transport. Most plants take up nitrate (NO_3^-) or ammonium (NH_3 or NH_4^+) as their nitrogen source. When nitrate is absorbed, it is first reduced to ammonium and then assimilated into amino acids such as glutamine and asparagine. In woody plants, these processes mainly occur in roots, while in herbaceous plants, nitrate is transported to leaves via xylem transpiration stream and thereby nitrate reduction, ammonium assimilation, and various nitrogen metabolisms occur in the leaves.

Nitrate Assimilation in the Leaf

Nitrate is reduced to nitrite by cytosolic nitrate reductase (NR):

$$\text{nitrate} + \text{NAD(P)H} \rightarrow \text{nitrite} + \text{NAD(P)}^+$$

Nitrite is further reduced to ammonium by chloroplast nitrite reductase (NiR):

$$\text{nitrite} + 6\text{Fd}_{\text{red}} + 8\text{H}^+ \rightarrow \text{ammonium} + 6\text{Fd}_{\text{ox}} + 2\text{H}_2\text{O}.$$

Ammonium is then assimilated by GS/GOGAT system (Lea and Miflin 1974) in chloroplasts. Chloroplast GS catalyzes:

$$\text{glutamate} + \text{ammonium} + \text{ATP} \rightarrow \text{glutamine} + \text{ADP} + \text{Pi}$$

Chloroplast GOGAT catalyzes:

$$\text{glutamine} + \text{2-oxoglutarate} + 2\text{Fd}_{\text{red}} \rightarrow 2\ \text{glutamate} + 2\text{Fd}_{\text{ox}}$$

The sum of these reactions is:

$$\text{2-oxoglutarate} + \text{ammonium} + \text{ATP} + 2\text{Fd}_{\text{red}} \rightarrow \text{glutamate} + \text{ADP} + \text{Pi} + 2\text{Fd}_{\text{ox}}$$

Besides these chloroplastic enzymes, there are cytosolic isozymes of GS and GOGAT. The cytosolic GOGAT uses NADH instead of Fd_{red} for reducing power. This cytosolic GS/GOGAT system is mainly found in the vascular tissue. The vascular GS/GOGAT system would be important for translocation of nitrogenous compound to the developing leaves and from the senescing leaves (Yamaya et al. 2002).

Gross and Fine Control of Nitrate Assimilation

Expression of genes involved in nitrate assimilation, NT, NR, and NiR, is enhanced by nitrate and suppressed by glutamine or its metabolites. For example, NR mRNA increases 5- to 100-fold within minutes after the plants that have been starved for nitrate are treated with nitrate. This initial transcriptional response is insensitive to protein synthesis inhibitors (Redinbaugh and Campbell 1993). If plastids were not able to reduce nitrite, enhanced NR activity would lead to accumulation of nitrite, which is very toxic to the cells. Sucrose and glucose also enhance NR synthesis (Cheng et al. 2002). On the other hand, glutamine inhibits expression of NR but, this effect is only seen when the concentration of 2-oxoglutarate is low (Deng et al. 1991).

The activity of NR protein is finely modulated (Kaiser and Huber 2001). The most important feature is light/dark modulation. NR is activated by dephosphorylation in the light and inactivated by phosphorylation in the dark. Inactivation occurs in two steps. Firstly, NR kinase phosphorylates a specific Ser-residue. Secondly, 14-3-3 inhibitor protein binds to phosphorylated NR. The binding occurs in the presence of Mg^{2+}. NR kinase is activated by Ca^{2+} and inhibited by triose phosphate. Thus, NR can be active when triose phosphate is available in the light.

How Are Carbon and Nitrogen Metabolism Balanced in the Mesophyll Cells?

There are several coordination mechanisms for carbon and nitrogen metabolism (Foyer et al. 2000, Stitt et al. 2002). First, nitrate, the nitrogen source, acts as a signal. NR, NiR, ammonium assimilation system, and probably NTs in leaf cells are increased in response to nitrate supplied by the xylem sap. On the other hand, nitrate deficiency is a signal for increased expression of ADP-Glc PPase and thereby increases the capacity for starch synthesis (Scheible et al. 1997, Stitt and Krapp 1999, Stitt et al. 2002).

Glutamine or its metabolite suppresses nitrate assimilation. Glutamine has two amino-groups and its abundance can indicate the status that nitrogen is more than sufficient. On the other hand, triose phosphate and 2-oxoglutarate stimulate nitrogen assimilation. Thus, active photosynthesis and abundance of C-skeleton for glutamine formation by the GS/GOGAT system stimulate nitrogen assimilation. Light/dark modulation of NR, discussed above, is one step of such trends.

In maize, a C_4 species, the condition of high glutamine/glutamate ratio induces PEPCase synthesis. When maize seedlings are grown without exogenous nutrients, the leaves are metabolically C_3-like and their main carboxylating enzyme is rubisco. Addition of nitrate or ammonium to the intact nitrogen-deficient maize plants leads to a rapid and marked increase of the Ppc transcript and of PEPCase activity. This is not due to a nitrate signal. When nitrogen-deficient maize shoots without roots were fed with nitrate, PEPCase synthesis was not enhanced. Recent studies showed that cytokinin synthesized in roots in response to nitrate or ammonium application is responsible for the enhanced PEPCase synthesis (Takei et al. 2002). The concentration of cytokinin increases on application of nitrate or ammonium. This connects photosynthesis and chloroplast metabolism to yet a third major plant hormone.

Summary and Conclusions

Once captured by cells, CO_2 is processed to a wide variety of compounds. The single biggest product of processing in photosynthetic cells is sucrose. Much of the sucrose is transported to other parts of the plant and further processed. Many of the compounds needed in the human diet are necessary precisely because their synthesis is unique to plastids among eukaryotes. In terms of amount and variety, carbon processing by plastids is the central hub of biochemical activity in the eukaryotic world. The basic reactions of carbon metabolism in photosynthesis are now well known and there is remarkably little variation in the central carbon processing reactions (i.e., Calvin cycle) among plants. Inefficiencies of the central CO_2-fixing enzyme Rubisco causes plants to require substantial amounts of water and organic nitrogen and the ways that plants deal with these requirements account for much of the variation we see among plants. There are Rubiscos in nongreen algae that appear to be better adapted for the current and future atmosphere. Moving these to land plants is a daunting challenge but the potential payoff is large. Starch metabolism and C_4 metabolism show substantially more variation among plants than do the Calvin cycle reactions. Recent work has connected carbon processing with plant growth and development by finding direct links between sugar sensing and ethylene and abscisic acid responses and between nitrogen status and cytokinins. Understanding the interactions between photosynthetic carbon accumulation and the signals that regulate plant growth and development will be an important advance in understanding how carbon processing in the chloroplast exerts effects at other levels of organization within the plant and biosphere.

References

Andrews, T. J., and Kane, H. J. 1991. Pyruvate is a by-product of catalysis by ribulose-bisphosphate carboxylase/oxygenase. J. Biol. Chem. 266:9447–9452.

Arenas-Huertero, F., Arroyo, A., Zhou, L., Sheen, J., and León, P. 2000. Analysis of *Arabidopsis* glucose insensitive mutants, *gin5* and *gin6*, reveals a central role of the plant hormone ABA in the regulation of plant vegetative development by sugar. Genes Dev. 14:2085–2096.

Badger, M. R., and Lorimer, G. H. 1976. Activation of ribulose-1,5-bisphosphate oxygenase. The role of Mg^{2+}, CO_2 and pH. Arch. Biochem. Biophys. 175:723–729.

Badger, M. R., and Spalding, M. H. 2000. CO_2 acquisition, concentration and fixation in cyanobacteria and algae. In: Advances in Photosynthesis, Vol. 9. Photosynthesis: Physiology and Metabolism. R. C. Leegood, T. D. Sharkey, and S. von Caemmerer (eds.), pp. 369–397. Dordrecht, The Netherlands: Kluwer Academic Publishing.

Balachandran, S., Hull, R. J., Vaadia, Y., Wolf. S., and Lucas, W. J. 1995. Alteration in carbon partitioning induced by the movement protein of tobacco mosaic virus originates in the mesophyll and is independent of change in the plasmodesmal size exclusion limit. Plant Cell Environ. 18:1301–1310.

Ball, S. G., and Morrell, M. K. 2003. From bacterial glycogen to starch: Understanding the Biogenesis of the plant starch granule. Annu. Rev. Plant Biol. 54:207–233.

Bernacchi, C. J., Singsaas, E. L., Pimentel, C., Portis, A. R., Jr., and Long, S. P. 2001. Improved temperature response functions for models of Rubisco-limited photosynthesis. Plant Cell Environ. 24:253–259.

Bernacchi, C. J., Portis, A. R., Nakano, H., von Caemmerer, S., and Long, S. P. 2002. Temperature response of mesophyll conductance. Implications for the determination of Rubisco enzyme kinetics and for limitations to photosynthesis in vivo. Plant Physiol. 130:1992–1998.

Bernacchi, C. J., Pimentel, C., and Long, S. P. 2003. In vivo temperature response functions of parameters required to model RuBP-limited photosynthesis. Plant Cell Environ. 26:1419–1430.

Blackman, F. F. 1905. Optima and limiting factors. Ann. Bot. 19:281–295.

Boos, W., and Shuman, H. 1998. Maltose/maltodextrin system of *Escherichia coli*: Transport, metabolism, and regulation. Microbiol. Molec. Biol. Rev. 62:204–229.

Brugnoli, E., and Farquhar, G. D. 2000. Photosynthetic fractionation of carbon isotopes. In: Photosynthesis: Physiology and Metabolism. R. C. Leegood, T. D. Sharkey, and S. von Caemmerer (eds.), pp. 399–434. Dordrecht, The Netherlands: Kluwer Academic Publishers.

Burton, R. A., Jenner, H., Carrangis, L., Fahy, B., Fincher, G. B., Hylton, C., Laurie, D. A., Parker, M., Waite, D., Van Wegen, S., Verhoeven, T., and Denyer. K. 2002. Starch granule initiation and growth are altered in barley mutants that lack isoamylase activity. Plant J. 31:97–112.

Camp, P. J., and Randall, D. D. 1985. Purification and characterization of the pea chloroplast pyruvate dehydrogenase complex. Plant Physiol. 77:571–577.

Caspar, T., Lin, T.-P., Kakefuda, G., Benbow, L., Preiss, J., and Somerville, C. 1991. Mutants of *Arabidopsis* with altered regulation of starch degradation. Plant Physiol. 95:1181–1188.

Chatterton, N. J., Harrison, P. A., Bennet, J. H., and Thornley, W. R. 1987. Fructan, starch and sucrose concentrations in crested wheatgrass and redtop as affected by temperature. Plant Physiol. Biochem. 25:617–623.

Cheng, S. H., Willmann, M. R., Chen, H. C., and Sheen, J. 2002. Calcium signaling through protein kinases. The Arabidopsis calcium-dependent protein kinase gene family. Plant Physiol. 129:469–485.

Chia, T., Thorneycraft, D., Chapple, A., Messerli, G., Chen, J., Zeeman, S., Smith, S. M., and Smith, A. M. 2004. A cytosolic glycosyltransferase is required for conversion of starch to sucrose in *Arabidopsis* leaves at night. Plant Biol. 37:853–863.

Colleoni, C., Dauvillee, D., Mouille, G., Buleon, A., Gallant, D., Bouchet, B., Morell, M., Samuel, M., Delrue, B., D'Hulst, C., Bliard, C., Nuzillard, J. M., and Ball, S. 1999a. Genetic and biochemical evidence for the involvement of alpha -1,4 glucanotransferases in amylopectin synthesis. Plant Physiol. 120:993–1004.

Colleoni, C., Dauvillee, D., Mouille, G., Morell, M., Samuel, M., Slomiany, M. C., Lienard, L., Wattebled, F., D'Hulst, C., and Ball, S. 1999b. Biochemical characterization of the chlamydomonas reinhardtii alpha-1,4 glucanotransferase supports a direct function in amylopectin biosynthesis. Plant Physiol. 120:1005–1014.

Critchley, J. H., Zeeman, S. C., Takaha, T., Smith, A. M., and Smith, S. M. 2001. A critical role for disproportionating enzyme in starch breakdown is revealed by a knockout mutation in *Arabidopsis*. Plant J. 26:89–100.

Dai, N., Schaffer, A., Petreikov, M., Shahak, Y., Giller, Y., Ratner, K., Levine, A., and Granot, D. 1999. Overexpression of Arabidopsis hexokinase in tomato plants inhibits growth, reduces photosynthesis, and induces rapid senescence. Plant Cell 11:1253–1266.

Deng, M. D., Moureaux, T., Cherel, I., Boutin, J. P., and Caboche, M. 1991. Effects of nitrogen metabolites on the regulation and circadian expression of tobacco nitrate reductase. Plant Physiol. Biochem. 29:239–247.

Dietz, K. J. 1985. A possible rate limiting function of chloroplast hexosemonophosphate isomerase in starch synthesis of leaves. Biochim. Biophys. Acta 839:240–248.

Edmondson, D. L., Kane, H. J., and Andrews, T. J. 1990. Substrate isomerization inhibits ribulosebisphosphate carboxylase-oxygenase during catalysis. FEBS Lett. 260:62–66.

Evans, J. R., and Loreto, F. 2000. Acquisition and diffusion of CO_2 in higher plant leaves. In: Photosynthesis: Physiology and Metabolism. R. C. Leegood, T. D. Sharkey, and S. von Caemmerer (eds.), pp. 321–351. Dordrecht, The Netherlands: Kluwer Academic Publishers.

Everard, J. D., Franceschi, V. R., and Loescher, W. H. 1993. Mannose-6-phosphate reductase, a key enzyme in photoassimilate partitioning, is abundant and located in the cytosol of photosynthetically active cells of celery (*Apium graveolens* L.) source leaves. Plant Physiol. 102:345–356.

Farquhar, G. D. and Sharkey, T. D. 1994. Photosynthesis and carbon assimilation. In:

Physiology and Determination of Crop Yield. K. J. Boote et al. (eds.), pp. 187–210. Madison: ASA, CSSA, SSSA.

Farquhar, G. D. and von Caemmerer, S. 1982. Modelling of photosynthetic response to environmental conditions. In: Encyclopedia of Plant Physiology, New Series, Vol. 12B. Physiological Plant Ecology II. Water Relations and Carbon Assimilation. O. L. Lange et al. (eds.), pp. 549–587. Berlin: Springer-Verlag.

Farquhar, G. D., Firth, P. M., Wetselaar, R., and Weir, B. 1980a. On the gaseous exchange of ammonia between leaves and the environment: Determination of the ammoinia compensation point. Plant Physiol. 66:710–714.

Farquhar, G. D., von Caemmerer, S., and Berry, J. A. 1980b. A biochemical model of photosynthetic CO_2 assimilation in leaves of C_3 species. Planta 149:78–90.

Fischer, K., Kammerer, B., Gutensohn, M., Arbinger, B., Weber, A., Häusler, R. E., and Flügge, U.-I. 1997. A new class of plastidic phosphate translocators: A putative link between primary and secondary metabolism by the phosphoenolpyruvate/phosphate antiporter. Plant Cell 9:453–462.

Flügge, U.-I. 1999. Phosphate translocators in plastids. Annu. Rev. Plant Physiol. Plant Mol. Biol. 50:27–45.

Flügge, U.-I., and Heldt, H. W. 1991. Metabolite translocators of the chloroplast envelope. Annu. Rev. Plant Physiol. Plant Mol. Biol. 42:129–144.

Foyer, C., Galtier, N., and Quick, P. 1994. Modifications in carbon assimilation, carbon partitioning and total biomass as a result of over-expression of sucrose phosphate synthase in transgenic potato plants. Plant Physiol. 105(Suppl.):23.

Foyer, C. H., Ferrario-Mery, S., and Huber, S. C. 2000. Regulation of carbon fluxes in the cytosol: Coordination of sucrose synthesis, nitrate reduction and organic and amino acid biosynthesis. In: Photosynthesis: Physiology and Metabolism. R. C. Leegood, T. D. Sharkey, and S. von Caemmerer (eds.), pp. 177–203. Dordrecht, The Netherlands: Kluwer Academic Publishers.

Furbank, R. T., Hatch, M. D., and Jenkens, C. L. D. 2000. C_4 photosynthesis: Mechanism and regulation. In: Photosynthesis: Physiology and Metabolism. R. C. Leegood, T. D. Sharkey, and S. von Caemmerer (eds.), pp. 435–457. Dordrecht, The Netherlands: Kluwer Academic Publishers.

Gerhardt, R., Stitt, M., and Heldt, H. W. 1987. Subcellular metabolite levels in spinach leaves. Regulation of sucrose synthesis during diurnal alterations in photosynthetic partitioning. Plant Physiol. 83:399–407.

Grace, J., Berninger, F., and Nagy, L. 2002. Impacts of climate change on the tree line. Ann. Bot. 90:537–544.

Hajirezaei, M. R., Takahata, Y., Trethewey, R. N., Willmitzer, L., and Sonnewald, U. 2000. Impact of elevated cytosolic and apoplastic invertase activity on carbon metabolism during potato tuber development. J. Exp. Bot. 51:439–445.

Haritatos, E., Ayre, B. G., and Turgeon, R. 2000. Identification of phloem involved in assimilate loading in leaves by the activity of the galactinol synthase promoter. Plant Physiol. 123:929–938.

Harley, P. C. and Sharkey, T. D. 1991. An improved model of C_3 photosynthesis at high CO_2: Reversed O_2 sensitivity explained by lack of glycerate reentry into the chloroplast. Photosynth. Res. 27:169–178.

Harwood, J. L. 1996. Recent advances in the biosynthesis of plant fatty acids. Biochim. Biophys. Acta (BBA) - Lipids and Lipid Metabolism 1301:7–56.

Hattenbach, A., Müller-Röber, B., Nast, G., and Heineke, D. 1997. Antisense repression of both ADP-glucose pyrophosphorylase and triose phosphate translocator modifies carbohydrate partitioning in leaves. Plant Physiol. 115:471–475.

Häusler, R. E., Schlieben, N. H., Schulz, B., and Flügge, U.-I. 1998. Compensation of decreased triose phosphate-phosphate translocator activity by accelerated starch turnover and glucose transport in transgenic tobacco. Planta 204:366–376.

Häusler, R. E., Schlieben, N. H., and Flügge, U.-I. 2000a. Control of carbon partitioning and photosynthesis by the triose phosphate/phosphate translocator in transgenic tobacco plants (*Nicotiana tabacum*). II. Assessment of control coefficients of the triose phosphate/phosphate translocator. Planta 210:383–390.

Häusler, R. E., Schlieben, N. H., Nicolay, P., Fischer, K., Fischer, K. L., and Flügge, U.-I. 2000b. Control of carbon partitioning and photosynthesis by the triose phosphate/phosphate translocator in transgenic tobacco plants (*Nicotiana tabacum* L.). I. Comparative physiological analysis of tobacco plants with antisense repression and overexpression of the triose phosphate/phosphate translocator. Planta 210:371–382.

Heineke, D., Sonnewald, U., Büssis, D., Günter, G., Leidreiter, K., Wilke, I., Raschke, K., Willmitzer, L., and Heldt, H. W. 1992. Apoplastic expression of yeast-derived invertase in potato. Effects on photosynthesis, leaf solute composition, water relations, and tuber composition. Plant Physiol. 100:301–308.

Heineke, D., Kruse, A., Flügge, U.-I., Frommer, W. B., Riesmeier, J. W., Willmitzer, L., and Heldt, H. W. 1994. Effect of antisense repression of the chloroplast triose phosphate translocator on photosynthetic metabolism in transgenic potato plants. Planta 193:174–180.

Herbers, K., Tacke, E., Hazirezaei, M., Krause, K. P., Melzer, M., Rohde, W., and Sonnewald, U. 1997. Expression of a luteoviral movement protein in transgenic plants leads to carbohydrate accumulation and reduced photosynthetic capacity in source leaves. Plant J. 12:1045–1056.

Hoch, G., Popp, M., and Korner, C. 2002. Altitudinal increase of mobile carbon pools in *Pinus cembra* suggests sink limitation of growth at the Swiss treeline. Oikos 98:361–374.

Huber, J. L. A., and Huber, S. C. 1992. Site-specific serine phosphorylation of spinach leaf sucrose-phosphate synthase. Biochem. J. 283:877–882.

Huber, S. C., and Hanson, K. R. 1992. Carbon partitioning and growth of a starchless mutant of *Nicotiana sylvestris*. Plant Physiol. 99:1449–1454.

Husic, D. W., Husic, H. D., and Tolbert, N. E. 1987. The oxidative photosynthetic carbon cycle or C2 cycle. CRC Crit. Rev. Plant Sci. 5:45–100.

James, M. G., Robertson, D. S., and Myers, A. M. 1995. Characterization of the maize gene sugary1, a determinant of starch composition in kernels. Plant Cell 7:417–429.

Kaiser, W. M., and Huber, S. C. 2001. Post-translational regulation of nitrate reductase: Mechanism, physiological relevance and environmental triggers. J. Exp. Bot. 52:1981–1989.

Kakefuda, G., and Duke, S. H. 1989. Characterization of pea chloroplast D-enzyme (4-a-D-glucanotransferase). Plant Physiol. 91:136–143.

Kammerer, B., Fischer, K., Hilpert, B., Schubert, S., Gutensohn, M., Weber, A., and Flügge, U.-I. 1998. Molecular characterization of a carbon transporter in plastids from heterotrophic tissues: The glucose 6-phosphate/phosphate antiporter. Plant Cell 10:105–117.

Kleinig, H. 1989. The role of plastids in isoprenoid synthesis. Annu. Rev. Plant Physiol. Plant Mol. Biol. 40:39–59.

Kobza, J., and Seemann, J. R. 1988. Mechanisms for the light-dependent regulation of ribulose-1,5-bisphosphate carboxylase activity and photosynthesis in leaves. Proc. Natl. Acad. Sci. USA 85:3815–3819.

Koch, K. E., Ying, Z., Wu, Y., and Avigne, W. T. 2000. Multiple paths of sugar-sensing and a sugar/oxygen overlap for genes of sucrose and ethanol metabolism. J. Exp. Bot. 51:417–427.

Koshland, D. E. 1987. Switches, thresholds and ultrasensitivity. Trends Biochem. Sci. 12:225–229.

Kossmann, J., and Lloyd, J. 2000. Understanding and influencing starch biochemistry. Crit. Rev. Plant Sci. 19:171–226.

Kruger, N. J., and Scott, P. 1995. Integration of cytosolic and plastidic carbon metabolism by fructose 2,6-bisphosphate. J. Exp. Bot. 46:1325–1333.

Lao, N. T., Schoneveld, O., Mould, R. M., Hibberd, J. M., Gray, J. C., and Kavanaugh, T. A. 1999. An *Arabidopsis* gene encoding a chloroplast-targeted b-amylase. Plant J 20:519–527.

Laporte, M. M., Galagan, J. A., Prasch, A. L., Vanderveer, P. J., Hanson, D. T., Shewmaker, C. K., and Sharkey, T. D. 2001. Promoter strength and tissue specificity effects on growth of tomato plants transformed with maize sucrose-phosphate synthase. Planta 212:817–822.

Lea, P. J., and Miflin, B. J. 1974. An alternative route for nitrogen assimilation in plants. Nature 251:614–616.

Leegood, R. C. 1997. The regulation of C_4 photosynthesis. Adv. Bot. Res. 26:251–316.

Leegood, R. C. 2000. Transport during C_4 photosynthesis. In: Photosynthesis: Physiology and Metabolism. R. C. Leegood, T. D. Sharkey, and S. von Caemmerer (eds.), pp. 459–469. Dordrecht, The Netherlands: Kluwer Academic Publishers.

Leegood, R. C., Kobayashi, Y., Neimanis, S., Walker, D. A., and Heber, U. 1982. Cooperative activation of chloroplast fructose-1,6-bisphosphatase by reductant, pH, and substrate. Biochim. Biophys. Acta 682:168–178.

Leidreiter, K., Heineke, D., Heldt, H. W., Müller-Röber, B., Sonnewald, U., and Willmitzer, L. 1995. Leaf-specific antisense inhibition of starch biosynthesis in transgenic potato plants leads to an increase in photoassimilate export from source leaves during the light period. Plant Cell Physiol. 36:615–624.

Lemoine, R., Kühn, C., Thiele, N., Delrot, S., and Frommer, W. B. 1996. Antisense inhibition of the sucrose transporter in potato: Effects on amount and activity. Plant Cell Environ. 19:1124–1131.

Levi, C., and Gibbs, M. 1976. Starch degradation in isolated chloroplasts. Plant Physiol. 57:933–935.

Li, B., Geiger, D. R., and Shieh, W.-J. 1992. Evidence for circadian regulation of starch and sucrose synthesis in sugar beet leaves. Plant Physiol. 99:1393–1399.

Libessart, N., Maddelein, M. L., Van den Koornhuyse, N., Decq, A., Delrue, B., Mouille, G., D'Hulst, C., and Ball, S. 1995. Storage, photosynthesis, and growth: The conditional nature of mutations affecting starch synthesis and structure in *Chlamydomonas*. Plant Cell 7:1117–1127.

Lichtenthaler, H. K. 1999. The 1-deoxy-D-xylulose-5-phosphate pathway of isoprenoid biosynthesis in plants. Annu. Rev. Plant Physiol. Plant Mol. Biol. 50:47–65.

Loescher, W. H., and Everard, J. D. 2000. Regulation of sugar alcohol biosynthesis. In: Photosynthesis: Physiology and Metabolism. R. C. Leegood, T. D. Sharkey, and S. von Caemmerer (eds.), pp. 275–299. Dordrecht, The Netherlands: Kluwer Academic Publishers.

Lohaus, G., and Heldt, H. W. 1997. Assimilation of gaseous ammonia and the transport of its products in barley and spinach leaves. J. Exp. Bot. 48:1779–1786.

Lorberth, R., Ritte, G., Willmitzer, L., and Kossmann, J. 1998. Inhibition of a starch-granule-bound protein leads to modified starch and repression of cold sweetening. Nat. Biotechnol. 16:473–477.

Loreto, F., Ciccioli, P., Cecinato, A., Brancaleoni, E., Frattoni, M., Fabozzi, C., and Tricoli, D. 1996. Evidence of the photosynthetic origin of monoterpenes emitted by *Quercus ilex* L leaves by ^{13}C labeling. Plant Physiol. 110:1317–1322.

Loreto, F., Förster, A., Dürr, M., Csiky, O., and Seufert, G. 1998. On the monoterpene emission under heat stress and on the increased thermotolerance of leaves of *Quercus ilex* L. fumigated with selected monoterpenes. Plant Cell Environ. 21:101–107.

Lu, Y., and Sharkey, T. D. 2003. The role of amylomaltase in maltose metabolism in the cytosol of photosynthetic cells. Planta 218:466–473.

Lucas, W. J. 1995. Plasmodesmata: Intercellular channels for macromolecular transport in plants. Curr. Opin. Cell Biol. 7:673–680.

Lucas, W. J., Olesinski, A., Hull, R. J., Haudenshield, J. S., Deom, C. M., Beachy, R. N., and Wolf, S. 1993. Influence of the tobacco mosaic virus 30-kDa movement protein on carbon metabolism and photosynthate partitioning in transgenic tobacco plants. Planta 190:88–96.

Lucas, W. J., Bouché-Pillon, S., Jackson, D. P., Nguyen, L., Baker, L., Ding, B., and Hake, S. 1995. Selective trafficking of KNOTTED1 homeodomain protein and its mRNA through plasmodesmata. Science 270:1980–1983.

Ludewig, F., Sonnewald, U., Kauder, F., Heineke, D., Geiger, M., Stitt, M., Müller-Röber, B. T., Gillissen, B., Kühn, C., and Frommer, W. B. 1998. The role of transient starch in acclimation to elevated atmospheric CO_2. FEBS Lett. 429:147–151.

Lytovchenko, A., Bieberich, K., Willmitzer, L., and Fernie, A. R. 2002. Carbon assimilation and metabolism in potato leaves deficient in plastidial phosphoglucomutase. Planta 215:802–811.

Madore, M., and Grodzinski, B. 1984. Effect of oxygen concentration on ^{14}C-photoassimilate transport from leaves of *Salvia splendens* L. Plant Physiol. 76:782–786.

Margulis, L. (1981) Symbiosis in Cell Evolution, pp. 1–419. San Francisco: W. H. Freeman and Company.

Martin, W., Scheibe, R., and Schnarrenberger, C. 2000. The Calvin cycle and its regulation. In: Photosynthesis: Physiology and Metabolism. R. C. Leegood, T. D. Sharkey, and S. von Caemmerer (eds.), pp. 16–31. Dordrecht, The Netherlands: Kluwer Academic Publishers.

Mate, C. J., Hudson, G. S., Von Caemmerer, S., Evans, J. R., and Andrews, T. J. 1993. Reduction of ribulose bisphosphate carboxylase activase levels in tobacco (*Nicotiana tabacum*) by antisense RNA reduces ribulose bisphosphate carboxylase carbamylation and impairs photosynthesis. Plant Physiol. 102:1119–1128.

Mate, C. J., von Caemmerer, S., Evans, J. R., Hudson, G. S., and Andrews, T. J. 1996. The relationship between CO_2-assimilation rate, Rubisco carbamylation and Rubisco activase content in activase-deficient transgenic tobacco suggests a simple model of activase action. Planta 198:604–613.

McGarvey, D. J., and Croteau, R. 1995. Terpenoid metabolism. Plant Cell 7:1015–1026.

Moore, B., Zhou, L., Rolland, F., Hall, Q., Cheng, W. H., Liu, Y. X., Hwang, I., Jones, T., and Sheen, J. 2003. Role of the *Arabidopsis* glucose sensor HXK1 in nutrient, light, and hormonal signaling. Science 300:332–336.

Mott, K. A., Jensen, R. G., O'Leary, J. W., and Berry, J. A. 1984. Photosynthesis and ribulose 1,5-bisphosphate concentrations in intact leaves of *Xanthium strumarium* L. Plant Physiol. 76:968–971.

Mukerjea, R., Yu, L. L., and Robyt, J. F. 2002. Starch biosynthesis: Mechanism for the elongation of starch chains. Carbohydr. Res. 337:1015–1022.

Neales, T. F., and Incoll, L. D. 1968. The control of leaf photosynthesis rate by the level of assimilate concentration in the leaf: a review of the hypothesis. Bot. Rev. 34:107–125.

Neuhaus, H. E., and Schulte, N. 1996. Starch degradation in chloroplasts isolated from C_3 or CAM (crassulacean acid metabolism)-induced *Mesembryanthemum crystallinum* L. Biochem. J. 318:945–953.

Nittylä, T., Messerli, G., Trevisan, M., Chen, J., Smith, A. M., and Zeeman, S. C. 2004. "A Previously Unknown Maltose Transporter Essential for Starch Degradation in Leaves." Science 303:87–89.

Oparka, K. J., and Turgeon, R. 1999. Sieve elements and companion cells: Traffic control centers of the phloem. Plant Cell 11:739–750.

Patron, N. J., Smith, A. M., Fahy, B. F., Hylton, C. M., Naldrett, M. J., Rossnagel, B. G., and Denyer, K. 2002. The altered pattern of amylose accumulation in the endosperm of low-amylose barley cultivars is attributable to a single mutant allele of granule-bound starch synthase I with a deletion in the 5′-non-coding region. Plant Physiol. 130:190–198.

Pego, J. V., Kortstee, A. J., Huijser, G., and Smeekens, S. G. M. 2000. Photosynthesis, sugars and the regulation of gene expression. J. Exp. Bot. 51:407–416.

Pollock, C., Farrar, J., Tomos, D., Gallagher, J., Lu, C. G., and Koroleva, O. 2003. Balancing supply and demand: the spatial regulation of carbon metabolism in grass and cereal leaves. J. Exp. Bot. 54:489–494.

Portis, A. R., Jr., Salvucci, M. E., and Ogren, W. L. 1986. Activation of ribulose bisphophate carboxylase/oxygenase at physiological CO_2 and ribulose bisphosphate concentrations by rubisco activase. Plant Physiol. 82:967–971.

Provencher, L. M., Miao, L., Sinha, N., and Lucas, W. J. 2001. *Sucrose export defective 1* encodes a novel protein implicated in chloroplast-to-nucleus signaling. Plant Cell 13:1127–1141.

Redinbaugh, M. G., and Campbell, W. H. 1993. Glutamine synthetase and ferredoxin-dependent glutamate synthase expression in the maize (*Zea mays*) root primary response to nitrate (evidence for an organ-specific response). Plant Physiol. 101:1249–1255.

Reimann, R., Ritte, G., Steup, M., and Appenroth, K. J. 2002. Association of a-amylase and the R1 protein with starch granules precedes the initiation of net starch degradation in turions of *Spirodela polyrhiza*. Physiol. Plant. 114:2–12.

Riesmeier, J. W., Flügge, U.-I., Schulz, B., Heineke, D., Heldt, H. W., Willmitzer, L., and Frommer, W. B. 1993. Antisense repression of the chloroplast triose phosphate translocator affects carbon partitioning in transgenic potato plants. Proc. Natl. Acad. Sci. USA 90:6160–6164.

Ritte, G., Lorberth, R., and Steup, M. 2000. Reversible binding of the starch-related R1 protein to the surface of transitory starch granules. Plant J. 21:387–391.

Ritte, G., Lloyd, J. R., Eckermann, N., Rottmann, A., Kossmann, J., and Steup, M. 2002. The starch-related R1 protein is an a-glucan, water dikinase. Proc. Natl. Acad. Sci. USA 99:7166–7171.

Robinson, S. P., and Walker, D. A. 1981. Photosynthetic carbon reduction cycle. In: The Biochemistry of Plants. A Comprehensive Treatise. M. D. Hatch and N. K. Boardman (eds.), pp. 193–236. New York: Academic Press.

Rodermel, S. 2001. Pathways of plastid-to-nucleus signaling. Trends Plant Sci. 6:471–478.

Rolland, F., Moore, B., and Sheen, J. 2002. Sugar sensing and signaling in plants. Plant Cell 14(Suppl.):185–205.

Rost, S., Frank, C., and Beck, E. 1996. The chloroplast envelope is permeable for maltose but not for maltodextrins. Biochim. Biophys. Acta 1291:221–227.

Roy, H., and Andrews, T. J. 2000. Rubisco: Assembly and mechanism. In: Photosynthesis:Physiology and Metabolism. R. C. Leegood, T. D. Sharkey, and S. von Caemmerer (eds.), pp. 53–83. Dordrecht, The Netherlands: Kluwer Academic Publishers.

Russin, W. A., Evert, R. F., Vanderveer, P. J., Sharkey, T. D., and Briggs, S. P. 1996. Modification of a specific class of plasmodesmata and loss of sucrose export ability in the *sucrose export defective 1* maize mutant. Plant Cell 8:645–658.

Sage, R. F. 1990. A model describing the regulation of ribulose-1,5-bisphosphate carboxylase, electron transport, and triose phosphate use in response to light intensity and CO_2 in C_3 plants. Plant Physiol. 94:1728–1734.

Sage, R. F. 2001. Environmental and evolutionary preconditions for the origin and diversification of the C_4 photosynthetic syndrome. Plant Biol. 3:202–213.

Sage, R. F., and Cowling, S. A. 1999. Implications of stress in low CO_2 atmospheres of the past: Are today's plants too conservative for a high CO_2 world? In: Carbon Dioxide and Environmental Stress, pp. 289–308. San Diego: Academic Press.

Sage, R. F., and Pearcy, R. W. 2000. The physiological ecology of C_4 photosynthesis. In: Photosynthesis: Physiology and Metabolism. R. C. Leegood, T. D. Sharkey, and S. von Caemmerer (eds.), pp. 497–532. Dordrecht, The Netherlands: Kluwer Academic Publishers.

Scheible, W. R., Lauerer, M., Schulze, E. D., Caboche, M., and Stitt, M. 1997. Accumulation of nitrate in the shoot acts as a signal to regulate shoot-root allocation in tobacco. Plant J. 11:671–691.

Scheidig, A., Fröhlich, A., Schulze, S., Lloyd, J. R., and Kossmann, J. 2002. Downregulation of a chloroplast-targeted b-amylase leads to a starch-excess phenotype in leaves. Plant J. 30:581–591.

Schleucher, J., Vanderveer, P., Markley, J. L., and Sharkey, T. D. 1999. Intramolecular deuterium distributions reveal disequilibrium of chloroplast phosphoglucose isomerase. Plant Cell Environ. 22:525–533.

Schleucher, J., Vanderveer, P. J., and Sharkey, T. D. 1998. Export of carbon from chloroplasts at night. Plant Physiol. 118:1439–1445.

Schneider, A., Häusler, R. E., Kolukisaoglu, Ü., Kunze, R., van der Graaf, E., Schwacke, R., Catoni, E., Desimone, M., and Flügge, U.-I. 2002. An *Arabidopsis thaliana* knockout mutant of the chloroplast triosephosphate/phosphate translocator is severely compromised only when starch synthesis, but not starch mobilisation is abolished. Plant J 32:1–15.

Schobert, C., Lucas, W. J., Franceschi, V. R., and Frommer, W. B. 2000. Intercellular transport and phloem loading of sucrose, oligosaccharides and amino acids. In: Photosynthesis: Physiology and Metabolism. R. C. Leegood, T. D. Sharkey, and S. von Caemmerer (eds.), pp. 249–274. Dordrecht, The Netherlands: Kluwer Academic Publishers.

Scott, P., and Kruger, N. J. 1995. Influence of elevated fructose-2,6-bisphosphate levels on starch mobilization in transgenic tobacco leaves in the dark. Plant Physiol. 108:1569–1577.

Sehnke, P. C., DeLille, J. M., and Ferl, R. J. 2002. Consummating signal transduction: The role of 14-3-3 protyeins in the signal-induced transitions in protein activity. Plant Cell 14(Suppl.):339–354.

Servaites, J. C., and Geiger, D. R. 2002. Kinetic characteristics of chloroplast glucose transport. J. Exp. Bot. 53:1581–1591.

Servaites, J. C., Fondy, B. R., Li, B., and Geiger, D. R. 1989. Sources of carbon for export from spinach leaves throughout the day. Plant Physiol. 90:1168–1174.

Sharkey, T. D. 1985. Photosynthesis in intact leaves of C_3 plants: Physics, physiology and rate limitations. Bot. Rev. 51:53–105.

Sharkey, T. D. 1988. Estimating the rate of photorespiration in leaves. Physiol. Plant. 73:147–152.

Sharkey, T. D. 1989. Evaluating the role of rubisco regulation in C_3 photosynthesis. Philos. Trans. R. Soc. Lond. [Biol.] 323:435–448.

Sharkey, T. D. 1990. Feedback limitation of photosynthesis and the physiological role of ribulose bisphosphate carboxylase carbamylation. Bot. Mag. Tokyo (special issue) 2:87–105.

Sharkey, T. D., and Vanderveer, P. J. 1989. Stromal phosphate concentration is low during feedback limited photosynthesis. Plant Physiol. 91:679–684.

Sharkey, T. D., and Yeh, S. S. 2001. Isoprene emission from plants. Annu. Rev. Plant Physiol. Plant Mol. Biol. 52:407–436.

Sharkey, T. D., Berry, J. A., and Raschke, K. 1985. Starch and sucrose synthesis in Phaseolus vulgaris as affected by light, CO_2, and abscisic acid. Plant Physiol. 77:617–620.

Sharkey, T. D., Stitt, M., Heineke, D., Gerhardt, R., Raschke, K., and Heldt, H. W. 1986. Limitation of photosynthesis by carbon metabolism. II O_2 insensitive CO_2 uptake results from limitation of triose phosphate utilization. Plant Physiol. 81:1123–1129.

Sharkey, T. D., Savitch, L. V., Vanderveer, P. J., and Micallef, B. J. 1992. Carbon partitioning in a *Flaveria linearis* mutant with reduced cytosolic fructose bisphosphatase. Plant Physiol. 100:210–215.

Sharkey, T. D., Chen, X. Y., and Yeh, S. 2001. Isoprene increases thermotolerance of fosmidomycin-fed leaves. Plant Physiol. 125:2001–2006.

Sheen, J., Yanagisawa, S., Yoo, S.-D., and Sheen. 2003. Differential regulation of EIN3 stability by glucose and ethylene signalling in plants. Nature 425:521–525.

Sicher, R.C., Kremer, D. F., and Harris, W. G. 1986. Control of photosynthetic sucrose synthesis in barley primary leaves role of fructose 2,6-bisphosphate. Plant Physiol. 82:15–18.

Somerville, C. R. 1984. The analysis of photosynthetic carbon dioxide fixation and photorespiration by mutant selection. Ox. Sur. Plant Mol. Cell Biol. 1:103–131.

Somerville, C., and Browse, J. 1991. Plant lipids: metabolism, mutants, and membranes. Science 252:80–87.

Steup, M., and Latzko, E. 1979. Intracellular location of phosphorylases in spinach and pea leaves. Planta 145:69–75.

Steup, M., and Schächtele, C. 1981. Mode of glucan degradation by purified phosphorylase forms from spinach leaves. Planta 153:351–361.

Stitt, M. 1985. Fine control of sucrose synthesis by fructose-2,6-bisphosphate. In: Regulation of Carbon Partitioning in Photosynthetic Tissue. R. L. Heath and J. Preiss (eds.), pp. 109–126. Rockville: American Society of Plant Physiologists.

Stitt, M. 1990a. Fructose-2,6-bisphosphate as a regulatory molecule in plants. Annu. Rev. Plant Physiol. Plant Mol. Biol. 41:153–185.

Stitt, M. 1990b. The flux of carbon betweeen the chloroplast and cytosol. In: Plant Physiology, Biochemistry and Moleculr Biology. D. T. Dennis and D. H. Turpin (eds.), pp. 309–326. Essex: Longman Scientific and Technical.

Stitt, M., and ApRees, T. 1979. Capacities of pea chloroplasts to catalyse the oxidative pentose phosphate pathway and glycolysis. Phytochemistry 18:1905–1911.

Stitt, M., and ApRees, T. 1980. Estimation of the activity of the oxidative pentose phosphate pathway in pea chloroplasts. Phytochemistry 19:1583–1585.

Stitt, M., and Heldt, H. W. 1981. Physiological rates of starch breakdown in isolated intact spinach chloroplasts. Plant Physiol. 68:755–761.

Stitt, M., Herzog, B., and Heldt, H. W. 1984. Control of photosynthetic sucrose synthesis by fructose-2,6-bisphosphate. I. Coordination of CO_2 fixation and sucrose synthesis. Plant Physiol. 75:548–553.

Stitt, M., and Steup, M. 1985. Starch and sucrose degradation. In: Encyclopedia of Plant Physiology, Vol. 18. R. Douce and D. A. Day (eds.), pp. 347–390. Berlin: Springer-Verlag.

Stitt, M., and Krapp, A. 1999. The interaction between elevated carbon dioxide and nitrogen nutrition: The physiological and molecular background. Plant Cell Environ. 22:583–621.

Stitt, M., Müller, C., Matt, P., Gibon, Y., Carillo, P., Morcuende, R., Scheible, W. R., and Krapp, A. 2002. Steps towards an integrated view of nitrogen metabolism. J. Exp. Bot. 53:959–970.

Streusand, V. J., and Portis, A.R.J.r. 1987. Rubisco activase mediates ATP-dependent RuBPCase activation. Plant Physiol. 85:152–154.
Sun, Z., Duke, S. H., and Henson, C. A. 1995. The role of pea chloroplast a-glucosidase in transitory starch degradation. Plant Physiol. 108:211–217.
Sun, J. D., Gibson, K. M., Kiirats, O., Okita, T. W., and Edwards, G. E. 2002. Interactions of nitrate and CO_2 enrichment on growth, carbohydrates, and rubisco in arabidopsis starch mutants. Significance of starch and hexose. Plant Physiol. 130:1573–1583.
Takei, K., Takahashi, T., Sugiyama, T., Yamaya, T., and Sakakibara, H. 2002. Multiple routes communicating nitrogen availability from roots to shoots: A signal transduction pathway mediated by cytokinin. J. Exp. Bot. 53:971–977.
Tiessen, A., Prescha, K., Branscheid, A., Palacios, N., McKibbin, R., Halford, N. G., and Geigenberger, P. 2003. Evidence that SNF1-related kinase and hexokinase are involved in separate sugar-signalling pathways modulating post-translational redox activation of ADP-glucose pyrophosphorylase in potato tubers. Plant J. 35:490–500.
Tobin, A. K., and Yamaya, T. 2001. Cellular compartmentation of ammonium assimilation in rice and barley. J. Exp. Bot. 52:591–604.
Treede, H.-J., and Heise, K.-P. 1986. Purification of the chloroplast pyruvate dehydrogenase complex from spinach and maize mesophyll. Z. Naturforsch. 41c:1011–1017.
Trethewey, R. N., and Smith, A. M. 2000. Starch metabolism in leaves. In: Photosynthesis: Physiology and Metabolism. R. C. Leegood, T. D. Sharkey, and S. von Caemmerer (eds.), pp. 205–231. Doedrecht, The Netherlands: Kluwer Academic Publishers.
Trevanion, S. J. 2002. Regulation of sucrose and starch synthesis in wheat (*Triticum aestivum* L.) leaves: Role of fructose 2,6-bisphosphate. Planta 215:653–665.
Turgeon, R. 1996. Phloem loading and plasmodesmata. Trends Plant Sci. 1:403–411.
Turgeon, R., and Medville, R. 1998. The absence of phloem loading in willow leaves. Proc. Natl. Acad. Sci. USA 95:12055–12060.
Usuda, H., Kalt-Torres, W., Kerr, P. S., and Huber, S. C. 1987. Diurnal changes in maize leaf photosynthesis. 2. Levels of metabolic intermediates of sucrose synthesis and the regulatory metabolite fructose 2,6-bisphosphate. Plant Physiol. 83:289–293.
Veramendi, J., Roessner, U., Renz, A., Willmitzer, L., and Trethewey, R. N. 1999. Antisense repression of hexokinase 1 leads to an overaccumulation of starch in leaves of transgenic potato plants but not to significant changes in tuber carbohydrate metabolism. Plant Physiol. 121:123–133.
Von Schaewen, A., Stitt, M., Schmidt, R., Sonnewald, U., and Willmitzer, L. 1990. Expression of a yeast-derived invertase in the cell wall of tobacco and *Arabidopsis* plants leads to accumulation of carbohydrate and inhibition of photosynthesis and strongly influences growth and phenotype of transgenic tobacco plants. EMBO J. 9:3033–3044.
Walker, D. A., and Herold, A. 1977. Can the chloroplast support photosynthesis unaided? Plant Cell Physiol. 51:295–310.
Wang, Z. Y., and Portis, A. R., Jr. 1992. Dissociation of ribulose-1,5-bisphosphate bound to ribulose-1,5-bisphosphate carboxylase/oxygenase and its enhancement by ribulose-1,5-bisphosphate carboxylase/oxygenase activase-mediated hydrolysis of ATP. Plant Physiol. 99:1348–1353.
Weise, S. E., Weber, A., and Sharkey, T. D. 2003. Maltose is the major form of carbon exported from the chloroplast at night. Planta 218:474–482.
Whitney, S. M., and Andrews, T. J. 2001a. Plastome-encoded bacterial ribulose-1,5-bisphosphate carboxylase/oxygenase (RubisCO) supports photosynthesis and growth in tobacco. Proc. Natl. Acad. Sci. USA 98:14738–14743.
Whitney, S. M., and Andrews, T. J. 2001b. The gene for the ribulose-1,5-bisphosphate car-

boxylase/oxygenase (Rubisco) small subunit relocated to the plastid genome of tobacco directs the synthesis of small subunits that assemble into Rubisco. Plant Cell 13:193–205.

Whitney, S. M., Baldett, P., Hudson, G. S., and Andrews, T. J. 2001. Form I Rubiscos from non-green algae are expressed abundantly but not assembled in tobacco chloroplasts. Plant J. 26:535–547.

Wiese, A., Gröner, F., Sonnewald, U., Deppner, H., Lerchl, J., Hebbeker, U., Flügge, U.-I., and Weber, A. 1999. Spinach hexokinase I is located in the outer envelope membrane of plastids. FEBS Lett. 461:13–18.

Williams, M. and Randall, D. D. 1979. Pyruvate dehydrogenase complex from chloroplasts of *Pisum sativum* L. Plant Physiol. 64:1099–1103.

Winter, H., and Huber, S. C. 2000. Regulation of sucrose metabolism in higher plants: Localization and regulation of activity of key enzymes. Crit. Rev. Plant Sci. 19:31–67.

Wise, R. R., Olson, A. J., Schrader, S. M., and Sharkey, T. D. 2004. Electron transport is the functional limitation of photosynthesis in field-grown Pima cotton plants at high temperature. Plant Cell Environ. (in press).

Witt, W., and Sauter, J. J. 1995. In-vitro degradation of starch grains by phosphorylases and amylases from poplar wood. J. Plant Physiol. 146:35–40.

Woodrow, I. E., and Berry, J. A. 1988. Enzymatic regulation of photosynthetic CO_2 fixation in C_3 plants. Annu. Rev. Plant Physiol. 39:533–594.

Xiao, W. Y., Sheen, J., and Jang, J. C. 2000. The role of hexokinase in plant sugar signal transduction and growth and development. Plant Mol. Biol. 44:451–461.

Yamaya, T., Obara, M., Nakajima, H., Sasaki, S., Hayakawa, T., and Sato, T. 2002. Genetic manipulation and quantitative-trait loci mapping for nitrogen recycling in rice. J. Exp. Bot. 53:917–925.

Yu, T. S., Kofler, H., Häusler, R. E., Hille, D., Flügge, U.-I., Zeeman, S. C., Smith, A. M., Kossmann, J., Lloyd, J., Ritte, G., Steup, M., Lue, W. L., Chen, J. C., and Weber, A. 2001. The Arabidopsis *sex1* mutant is defective in the R1 protein, a general regulator of starch degradation in plants, and not in the chloroplast hexose transporter. Plant Cell 13:1907–1918.

Zeeman, S. C., and ApRees, T. 1999. Changes in carbohydrate metabolism and assimilate export in starch-excess mutants of *Arabidopsis*. Plant Cell Environ. 22:1445–1453.

Zeeman, S. C., Northrop, F., Smith, A. M., and ApRees, T. 1998a. A starch-accumulating mutant of *Arabidopsis thaliana* deficient in a chloroplastic starch-hydrolyzing enzyme. Plant J 15:357–385.

Zeeman, S. C., Umemoto, T., Lue, W. L., Au-Yeung, P., Martin, C., Smith, A. M., and Chen, J. 1998b. A mutant of arabidopsis lacking a chloroplastic isoamylase accumulates both starch and phytoglycogen. Plant Cell 10:1699–1712.

Zeeman, S. C., Tiessen, A., Pilling, E., Kato, K. L., Donald, A. M., and Smith, A. M. 2002. Starch synthesis in arabidopsis. Granule synthesis, composition, and structure. Plant Physiol. 129:516–529.

Zhang, N., Kallis, R. P., Ewy, R. G., and Portis, A. R., Jr. 2002. Light modulation of Rubisco in *Arabidopsis* requires a capacity for redox regulation of the larger Rubisco activase isoform. Proc. Natl. Acad. Sci. USA 99:3330–3334.

Zhang, N., and Portis, A. R., Jr. 1999. Mechanism of light regulation of Rubisco: A specific role for the larger Rubisco activase isoform involving reductive activation by thioredoxin-f. Proc. Natl. Acad. Sci. USA 96:9438–9443.

Zrenner, R., Krause, K. P., Apel, P., and Sonnewald, U. 1996. Reduction of the cytosolic fructose-1,6-bisphosphatase in transgenic potato plants limits photosynthetic sucrose biosynthesis with no impact on plant growth and tuber yield. Plant J. 9:671–681.

8
Leaf to Landscape

David S. Ellsworth, Ülo Niinemets, and Peter B. Reich

Introduction

The capacity of terrestrial plant leaves for photosynthetic CO_2 fixation per unit gram of leaf varies over 10-fold (Reich et al. 1997). The results of CO_2 fixation, processing and subsequent accumulation of mass (Fig. 8.1) gives plants the most enormous variation in size of organisms on earth (Niklas and Enquist 2001). The variation in photosynthetic capacity and in leaf form among species of higher plants attests to strong adaptation to different environments, in combination with exaptation and ecological sorting, by selection for traits that enable plants to survive and thrive in even the earth's extreme climates. Thus, diverse adaptations to permit photosynthetic carbon assimilation and utilization at the leaf and canopy level among different environments, as well as the synchronization of downstream metabolic processing of this carbon for growth and other functions, can be considered important drivers of biological diversity from a functional perspective.

While consideration of the many functional adaptations associated with plant CO_2 fixation and the environmental conditions that produced them is not possible here, we focus on one of the major constraints to photosynthetic carbon assimilation and processing in higher plants: natural selection for leaf and canopy form and structure that enhances net carbon assimilation in specific habitats and environments. Following the plant carbon flow diagram in Figure 8.1, we emphasize sources of variation in CO_2 assimilation capacity and carbon processing among C_3 plant leaves and canopies. Given that the products of photosynthesis in leaves (e.g., photoassimilates) are available for storage, metabolism, and new growth in leaves or export to other parts of the plant (see Fig. 8.1), we consider that coordination of the multiple enzymes involved in carbon processing occurs such that feedback regulation of early steps in the biosynthetic pathways for carbon (e.g., CO_2 fixation and proximal carbohydrate metabolism) is typically minimized. There are cases where feedback inhibition of photosynthetic capacity may be regulated by carbon processing, and one such case that we consider occurs when carbohydrate accumulation occurs under high atmospheric CO_2 levels as are expected in future decades. Very little information on carbon processing is

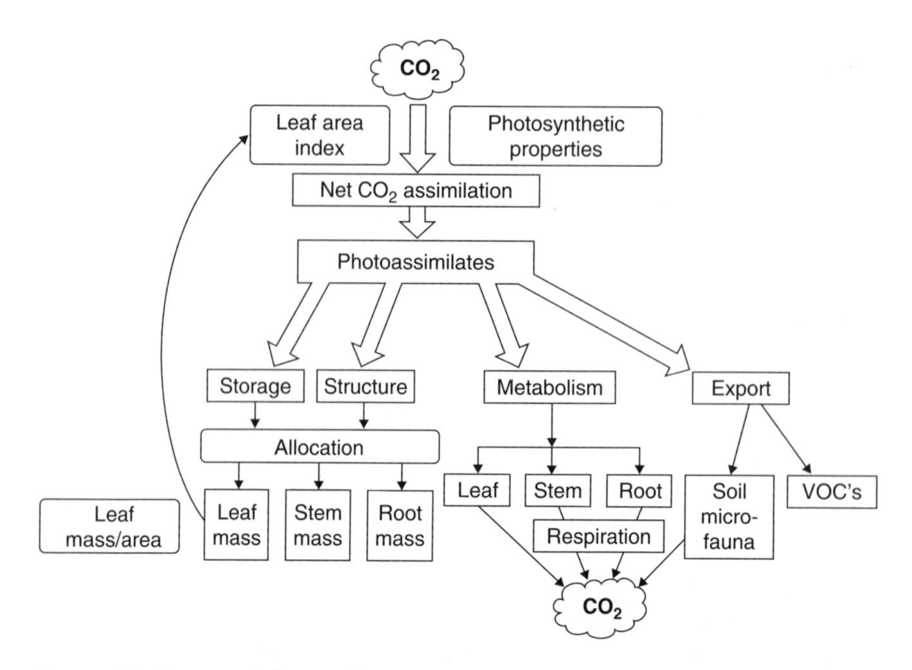

FIGURE 8.1. Conceptual diagram illustrating carbon flows in plants and ecosystems originating from CO_2 assimilation. Photoassimilates are partitioned into stored carbohydrates (total nonstructural carbohydrates), plant structural material (biomass), metabolic functions such as maintenance respiration, and export material such as root exudates and volatile organic carbon molecules (VOC's) (modified from Körner 2003).

available at higher levels of organization than the leaf or branch cluster, though it is recognized that whole-plant integration of carbon storage is central to understanding plant growth (Farrar 1996)

Our treatment here largely focuses on C_3 plants, as aspects of CO_2 fixation of C_4 plants in relation to the ecology and physiology of photosynthesis have been separately reviewed recently (Sage and Monson 1999). The major levels of plant organization to consider are the organ scale and greater; namely, spatial scales of the single leaf to aggregated leaves and branch, branch to plot or ecosystem level, and ecosystem to landscape.

Since the 1980s there has been an increasing emphasis on understanding plant processes such as photosynthesis at higher levels of biological organization than the leaf. For a variety of reasons, including the logistics of human observations and the inconvenience of studying tall plant canopies (Lowman and Nadkarni 1995), there had been a variety of process-level studies focusing on photosynthesis at the leaf level and much less emphasis on leaf clusters on a shoot, branches, tree crowns, and canopies. Active research on whole-plant canopies has been increasing due to the need for (1) basic studies of the productivity of the earth's terrestrial ecosystems to understand why plants vary in photosynthetic capacity and the implications of scale for their ability to process CO_2, (2) understanding the integration of photosynthetic activity of leaves with utilization and rest of plant metabolism and allocation, and (3) practical concern over ability of ecosystems to respond to and absorb excess CO_2 put into the atmosphere through anthropogenic activities (Canadell et al. 2000, Körner 2000). Given the variation among plants for CO_2 fixation, there is considerable potential for manipulation of ecosystems by planting or replacing existing plants with types that either maximize CO_2 fixation per unit ground area or optimize CO_2 fixation for the available resources, as is manifested by natural selection and/or agriculture and forestry. We review the magnitude of variation in leaf net CO_2 assimilation (area-based and mass-based; A_{area} and A_{mass}, respectively) among plants, and then consider how carbon processing rates and consequently stored carbon accumulation varies among leaves and ecosystems.

Determinants of Variation in Leaf CO_2 Assimilation Rates Among Species

The functional relationship between leaf maximum leaf CO_2 assimilation capacity per unit leaf mass (A_{mass}) and bulk nitrogen per unit leaf mass (N_{mass}) provides one means of assessing variation in carbon assimilation and processing among diverse plant species. This relationship has been tested interspecifically for a wide variety of C_3 plant species and growth forms (Field and Mooney 1986, Evans 1989, Reich et al. 1997, Peterson et al. 1999, Canadell et al. 2000, Körner 2000) and also intraspecifically in many studies and is central to current models linking ecosystem carbon and nitrogen cycles. It is important to note that the interspecific correlations, although most frequently cited as evidence for the physiological generality of this relationship, do not necessarily have the same slope as intraspecific relations (Reich et al. 1994). The interspecific relationships include both physiological scaling as well as evolutionary tradeoffs, whereas within a population of leaves of a given species the relationship should be determined much more by physiology per se. Moreover, A_{mass}-N slopes differ among species predictably as a function of their leaf N and specific leaf area (SLA) (Reich and Walters 1994, Reich et al. 1994, 1998).

Increasingly, there is interest in evaluating relationships between A_{mass} and N_{mass} with respect to the framework of the C_3 photosynthesis model of (Farquhar et al. 1980), which provides for the possibility of predicting CO_2 assimilation rates for leaves and whole-canopies at a wide range of conditions given the supply of CO_2 to leaf internal surfaces. The short-term response of net CO_2 assimilation (A_{area}) of leaves to the internal partial pressure of CO_2 in leaves (pC_i) is of interest for several reasons: (1) it describes the fundamental biochemical dependence of A_{area} on the supply of the rate-limiting substrate under physiological temperatures and photosynthetic photon flux density levels (Long 1991) and (2) the CO_2 and temperature-driven instantaneous response of leaves is used to predict CO_2 fluxes in physiologically based plant growth models, stand-level C flux models (Leuning) and larger-scale terrestrial CO_2 flux models using big-leaf types of formulations (Amthor 1994, Woodward et al. 1995, Aber et al. 1996) (see Chapter 6).

A principal physiological characteristic of leaves in the Farquhar et al. (1980) model is the capacity for carboxylation in CO_2 fixation, or V_{cmax}, which depicts the in situ amount and kinetic properties of the primary carboxylating enzyme Rubisco as manifested in whole leaves. Since the A_{mass}-N_{mass} relationship arises from the central role of N in photosynthetic proteins (Evans 1989), V_{cmax} should be strongly correlated with leaf N as has been shown by (Walcroft et al. 1997, Medlyn et al. 1999) and others. We reanalyzed data from the literature for V_{cmax}, leaf N and SLA among different plant types (Table 8.1) when fitting methods for V_{cmax} were similar and where several species were analyzed together, in order to determine if differences in photosynthesis-N relationships among species groups or their leaf properties may be related to differences in biochemical function. We found distinct relationships between V_{cmax} and leaf N that varied among growth forms of plants: herbaceous plants (including graminoids), woody deciduous trees, and evergreen trees (both coniferous trees and Mediterranean trees and shrubs combined) (Fig. 8.2). Despite roughly the same mean V_{cmax} among these groups within the data set (see Table 8.1), different groups of species showed different slopes of the V_{cmax}-N relationship for area-based expressions. These differences match those for A_{area}-N_{area} relations for species groups varying in *SLA* (Reich and Walters 1994, Reich et al. 1998). The differences in slopes can be accounted for by differences in the calculated fraction of N allocated to the Rubisco protein ($f_{N,Rub}$) (see Table 8.1), calculated according to (Niinemets

TABLE 8.1. Photosynthetic characteristics of species groups from compiled literature data[a] including carboxylation capacity (V_{cmax}) and associated leaf properties. V_{cmax} was fitted based on the Farquhar et al. (1980) model with modifications in dePury and Farquhar (1997) and Medlyn et al. (1999), standardized to 25°C. SLA, specific leaf area; $f_{N,Rub}$, estimated fraction of total leaf N allocated to carboxylation.

Species type	A_{area} μmol m^{-2} s^{-1})	V_{cmax} at 25°C (μmol m^{-2} s^{-1})	SLA (cm^2 g^{-1})	$f_{N,Rub}$
Mediterranean trees and shrubs	12.8 ± 2.8	67.5 ± 3.4	47.18 ± 4.4	0.10 ± 0.01
Coniferous trees	9.1 ± 3.3	94.6 ± 7.5	51.26 ± 7.2	0.11 ± 0.01
Deciduous trees and shrubs	17.2 ± 6.3	64.8 ± 10.8	135.22 ± 23.2	0.13 ± 0.01
Annual and perennial herbs	11.8 ± 3.6	74.6 ± 6.3	208.1 ± 12.4	0.16 ± 0.01

[a]Data are for the following species. Deciduous trees: *Betula pendula* (Rey and Jarvis 1998), *Fagus sylvatica* (Matteucci and Valentini, in Medlyn et al. 1999), *Betula papyrifera* and *Populus grandidentata* (Ellsworth, unpublished data); Mediterranean trees and shrubs: *Pistacea, Phillyrea, Quercus* (Scarascia-Mugnozza et al. 1996) and *Hakea* (Ellsworth and Reich, unpubl.); Coniferous trees: *Pinus sylvestris* (Jach and Ceulemans 2000), *Picea abies* (Robernz and Stockfors 1998), *Pinus taeda* (Crous and Ellsworth 2004) and *Pinus strobus* (Ellsworth, unpubl.); Herbs: *Allium, Tiarella*, and *Viola* in Rothstein and Zak (2001), *Podophyllum peltatum* (Ellsworth, unpublished data), and 21 species of herbs and 9 species of grasses (Wohlfahrt et al. 1999).

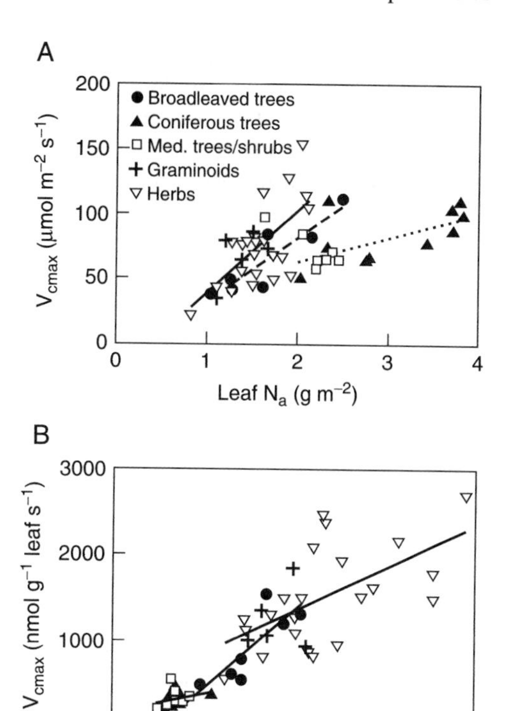

FIGURE 8.2. A. Relationship between carboxylation capacity, V_{cmax}, standardized to 25° C, and leaf nitrogen per unit area among different plant types. V_{cmax} was fitted according to the Farquhar photosynthesis model (Farquhar et al. 1980 and revisions in Medlyn et al. 1999). Fitted V_{cmax} values are largely from Medlyn et al. (1999) or from Wohlfahrt et al. (1999) and other sources (see Table 8.1). The lines show the least-squares regression fit to the data for: herbs and graminoids (—), broadleaved deciduous trees (- - -), and coniferous and mediterranean trees and shrubs (......). B. Mass-based carboxylation capacity (V_{cmax}, nmol g^{-1} s^{-1}) as a function of leaf N per unit mass (N_{mass}) for the data in A. Within-group least-squares regression fits are as in A.

and Tenhunen 1997), assuming standard catalytic properties of Rubisco, full activation of the enzyme, and that this enzyme is comprised of 16.7% N like other proteins (Evans 1989). A greater proportional allocation of leaf N to Rubisco confers greater carboxylation efficiency per unit N, hence the steeper slope of the relationship between V_{cmax} and N in Figure 8.2. Variation in this allocation can be ascribed to relative differences in the content of photosynthetic enzyme complexes relative to Rubisco (Hikosaka and Hirose 1998), differences in the accumulation of N in leaves as free or stored compounds (Bauer et al. 2001), and differences in the content of N-containing herbivore defenses.

While area-based expressions of V_{cmax} are commonly analyzed since CO_2 absorption like light harvesting fundamentally involves surface area, there is strong evidence that leaf structure (e.g., SLA) affects net CO_2 assimilation capacity (Reich et al. 1998, Wright and Westoby 2002), hence analysis of a leaf's carboxylation rate per unit leaf mass (V_{cmax}/mass) is warranted. When carboxylation capacity and leaf N are expressed on a mass basis (see Fig. 8.2B), relative differences in slopes of the V_{cmax}-N relationship in Figure 8.2A among herbs, deciduous trees and evergreen trees are maintained. This is similar to A_{mass}-N relations for species differing in SLA (Reich et al. 1998). This suggests that differences in biochemical function are most important in driving the relationship between A_{area} or A_{mass} and leaf N rather than CO_2 supply to internal leaf surfaces, at least as depicted for the photosynthesis model of (Farquhar et al. 1980).

Variation in leaf structure has been identified as an important factor associated with differences between the area- and mass-based expressions of the A_{mass}-N relationship, and this analysis suggest that there is a biochemical component underlying these differences (complementary to potential gas flux or light gradient components). However, the mass-based V_{cmax} and N relationship among the plant growth forms is compressed relative to the area-basis, such that an overall relationship between mass-based V_{cmax} and N (overall, V_{cmax}/mass $= 49.78 * N_m - 151.96$, $r^2 = 0.76$) is significant across species groups (see Fig. 8.2). These and other results (Reich et al. 1998, Niinemets 1999) suggest that species with low SLA and high-density leaves have lower photosynthetic capacity than those with high SLA in part due to a lower investment in biochemical machinery for CO_2 assimilation. Causes for the difference in N allocation among species that share a common SLA likely relate to the reduced CO_2 assimilation return for these species, and the demands for other organic N components (Bauer et al. 2001) in species growing on poor sites either in terms of water or nutrient availability.

Specific Leaf Area and Foliage Photosynthetic Potential

Whole-leaf structure is most often quantified using SLA or the reciprocal, leaf dry mass per unit area (LMA). Specific leaf area is a composite of leaf thickness and dry mass density, but for different leaf types modifications in SLA are associated with changes in thickness and density to a varying extent (Niinemets 1999, Castro-Díez et al. 2000). Specific leaf area is a component of leaf area ratio in functional growth models and is closely related to variation in plant growth rates among diverse species (Lambers and Poorter 1992). Across different species growing in high light, maximum leaf CO_2 assimilation capacity per unit leaf mass (A_{mass}) is positively and strongly related to SLA (Fig. 8.3). Here we present relationships for A_{mass} rather than the more familiar maximum CO_2 assimilation rate per unit area (A_{area}) because the former shows a stronger empirical relationship to SLA (Field and Mooney 1986, Reich et al. 1991), because it is a useful proxy measure of the instantaneous energetic payback per unit C invested to produce a leaf, and because other components of leaf structure (thickness and air space volume) do not necessarily map directly with the mass investment to produce a unit leaf area (Niinemets 1999). Implicit in the usage of A_{mass} in this analysis is that it is measured in nonstressed conditions, thus providing an estimate of the biochemical potentials of foliage photosynthetic apparatus.

Specific leaf area is associated with leaf carbon processing in that investment in structural materials in a leaf like lignin cause SLA to decline (see Chapter 10). Species with lower SLA generally have thicker and denser leaves than species with high SLA (but see Niinemets 1999, Roderick et al. 2000), with low SLA closely associated with low N concentration (mass-based leaf N; N_{mass}) and low A_{mass} (see Fig. 8.3). There may be large species differences in foliar biomass in-

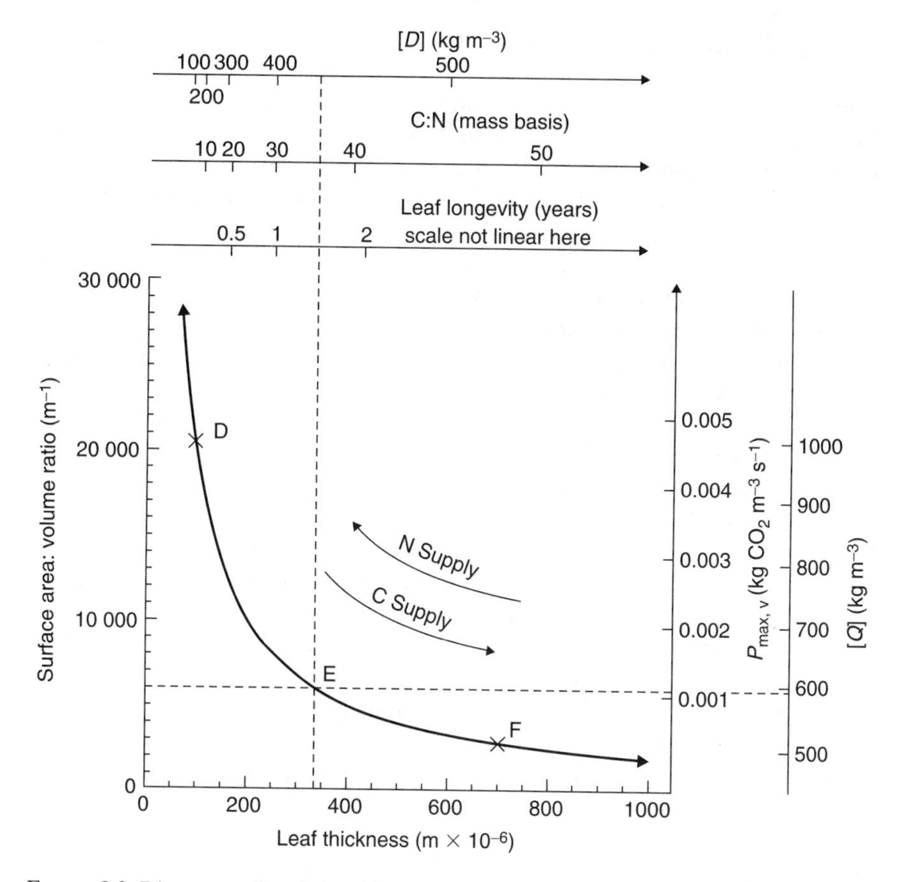

FIGURE 8.3. Diagrammatic relationship between leaf thickness and leaf structural (surface area/volume and liquid content, Q) and functional parameters (photosynthetic capacity per unit volume, $P_{max,v}$, and leaf lifespan) (from Roderick et al. 2000).

vestments in support structures within the leaves. The photosynthetic mesophyll is always embedded within cuticle and epidermal layers that may comprise a significant volume fraction of a leaf. It is particularly relevant that the leaf volume fractions in cuticle and epidermal layers increase with increasing leaf density, thereby leading to greater leaf density and lower A_{mass} (Niinemets 1999). A common observation has been that many species in nutrient-poor or water-stressed environments have low SLA (Grubb 1977, Chapin 1980, Reich et al. 1997, Niinemets 2001, Lamont et al. 2002). Current evidence suggests that while low-SLA leaves incur a physiological disadvantage in being limited in photosynthetic capacity, such leaves are favored when the cost to produce them is amortized over time (Chapin 1980), and when this kind of leaf enhances long-term CO_2 assimilation in the face of physical (wind abrasion), resource availability (low nutrient and water availability), and biotic (herbivory) stresses.

Canopy- and "Big-Leaf"-Level Variation in CO_2 Fixation

With growing information on the magnitude and constraints on ecosystem-scale CO_2 fluxes (Canadell et al. 2000), modeling approaches were developed to extend mechanistic understanding of CO_2 fixation from the leaf to the entire canopy of vegetated ecosystems. The so-called "big-leaf" analogy of the plant canopy was articulated by (Jarvis 1976) and others (Sinclair et al. 1976). The concept of the canopy as a "big-leaf" relies on the substitution of canopy function with a single virtual leaf representing the ensemble behavior of all of the leaves in the canopy (Farquhar 1989, Sellers et al. 1997). Typical problems with the big-leaf convention include the lack of consideration of nonlinearities in environmental responses, and lack of homogeneity in environmental conditions within the canopy. It has been argued that scaling leaf photosynthesis from leaf to canopy involves the same generalizations as scaling chloroplast biochemistry to the leaf level (Farquhar 1989) such that the emergent properties of the big leaf can be analyzed despite knowledge of inhomogeneities within the canopy, since selection will tend to constrain plant canopies in a similar manner. While debate about the usefulness and validity of the big-leaf concept has persisted (Friend 2001), it is possible to examine gross properties of canopies to establish to what extent stacking of photosynthetic units (e.g., leaves) affects the overall magnitude of CO_2 assimilation for canopies.

In an aim to understand canopy CO_2 assimilation and its variability across landscapes and among different ecosystems, tower-based ecosystem CO_2 flux measurements over annual time scales or greater were initiated in the late 1980s and early 1990s (Wofsy et al. 1993, Valentini et al. 2000, Canadell et al. 2000). Most frequently, the focus of these measurements is a comparison of annual integrated fluxes among different vegetation types, and individual hourly flux rates are not compared because of the high variability in measurements due to the stochastic nature of turbulence events affecting CO_2 fluxes. However, a number of analyses have presented maximum summertime flux rates for different ecosystems, most often in the form of a so-called ecosystem "light response" function (Amthor et al. 1994, Goulden 1996, Valentini et al. 1996, Hollinger et al. 1999) in direct analogy with the big-leaf concept. This top-down approach is used to analyze whole-system CO_2 fluxes as a function of photosynthetic photon flux density (PPFD), assuming photon flux is the primary environmental driver for short-term ecosystem CO_2 fluxes via its role as a driver for photosynthetic CO_2 assimilation in leaves and its importance in canopy energy balance. Such an approach is simplistic, and problems associated with nonlinearities in the light response of sun- versus shade-leaf classes, hysteresis with respect to direct versus diffuse light, and other implicit assumptions of the big-leaf approximation have been analyzed (Amthor 1994, dePury and Farquhar 1997, Friend 2001). However, analysis of the maximum rates of ecosystem net CO_2 fixation that arise from this analysis may provide useful insights into the regulation of these fluxes at the ecosystem scale, which has previously been difficult to examine among different species and canopy types.

A compilation of literature data with respect to recently published tower flux data shows considerable site-site variability in maximum CO_2 fluxes in full sun conditions (Table 8.2). While generalization from a rather small data set must be circumspect, there are a number of well-characterized temperate forest sites. Given the site specificity of the small data set, maximum CO_2 fluxes are similar for 8 temperate deciduous and 7 conifer forests on a land area basis (20.6 ± 3.0 μmol CO_2 m^{-2} land area s^{-1} and 17.9 ± 3.5 μmol m^{-2} s^{-1}, respectively) as well as on a per unit leaf area basis (4.4 ± 1.1 μmol m^{-2} leaf s^{-1} and 5.1 ± 2.0 μmol m^{-2} leaf s^{-1}, respectively). This is consistent with similar land-based annual aboveground net primary productivity for deciduous and conifer forests when compared in comparable climate zones (Reich et al. 1997, Reich and Bolstad 2002). There was a significant relationship between leaf-specific maximum CO_2 fluxes and N reported for upper canopy leaves (Fig. 8.4), in direct analogy to the leaf-level CO_2 assimilation-N relationship (see Fig. 8.2). The positive intercept of this relationship can in part be attributed to the fact that between 0.5 to 1 m^2 m^{-2} of the reported vegetation area in the studies represents woody branches and stems that is indirectly sensed along with actual leaf area (Law et al. 2001). Remarkably, the slope of this relationship is similar to the slope for the equivalent leaf-level relationship in (Field and Mooney 1986, Reich et al. 1999, Law et al. 2001) which indirectly suggests that total canopy N_{area} approximately scales with the upper, sunlit canopy N_{area} (Meir et al. 2002).

Ecological scaling of leaf-level measurements to understand ecosystem processes has long been of interest, as a potential means of providing a mechanistic basis (Ehleringer and Field 1993, Baldocchi and Meyers 1998). A number of refinements of analytical and numerical models have been proposed for the big-leaf model in order to permit scaling, with distinction of sun- and shade-leaf classes a major concern (dePury and Farquhar 1997, Friend 2001). Given the importance of leaf overlap in controlling canopy light environment (and the patterns of its variability) and hence photosynthesis within canopies, a poor relationship between ecosystem maximum CO_2 flux and individual leaf A_{area} is not surprising (Fig. 8.5). For instance, even within a site, variation among canopies in leaf overlap [i.e., leaf area index (LAI) or vegetation area index (VAI)] is not independent of variation in species leaf attributes—shade tolerant species (characterized by lower SLA, lower A_{mass} and related traits) have denser canopies that intercept more light (long hypothesized but only recently quantified, see Canham et al. 1994, Reich et al. 2003). Hence the maximum canopy, F_c, would be compensated for by a denser canopy for species with low A_{mass} leaves, leading to this poor relationship. However, among systems where VAI is low (arbitrarily defined here as VAI < 3.0 m^2 m^{-2}), there was a good relationship between maximum F_c and A_{area}. Coincidentally, the data comprising this relationship where VAI is low are for conifers and graminoids, both of which exhibit erectophile leaves and hence can be expected to have less overlap than for broadleaves. As a result, variability among species in both leaf inclination angle (LIA) and species shade tolerance and maximum LAI in stands can diminish the overall relationship between leaf-level maximum A_{area} and the maximum F_c of a vegetation canopy.

TABLE 8.2. Compilation of maximum CO_2 flux (F_c) of different ecosystems from the literature, based on the mean maximum reported F_c at high light (incident photosynthetic photon flux density of 1500 μmol m^{-2} s^{-1}). In cases where several monthly composite F_c-PPFD curves were presented, we selected the time period corresponding to maximum growing season photosynthesis.

Ecosystem type	Site	Latitude	Major species	Max F_c (μmol m^{-2} ground s^{-1})	Data source
Alpine coniferous	Niwot Ridge, CO, USA	40° 01′ 58″ N	*Abies lasiocarpa, Picea engelmannii, Pinus contorta*	7.8	1
Arctic graminoids	U-Pad, Prudhoe Bay, AK, USA	70° 16′ 53″ N	*Eriophorum, Carex*	6	2
Boreal coniferous	BOR-SOJP, Sask, CA	53° 54′ 59″ N	*Pinus banksiana*	9	3
Boreal coniferous	Flakaliden, SWE	64° 06′ 46″ N	*Picea abies*	12	4
Boreal deciduous	Boreas-OA, Sask, CA	53° 37′ 44.0″ N	*Populus trem, Corylus cornuta*	—	5
Boreal deciduous	Yakutsk, Siberia, RUS	60° 51′ N	*Larix gmelinii*	7.5	6
Temperate conifer	Balmoral, Cant, NZ	50° 18′ 32″ S	*Pinus radiata*	19.7	7
Temperate conifer	Duke Forest, NC, USA	35° 58′ 41.40″ N	*Pinus taeda*	16	8
Temperate conifer	Gainesville, FL, USA	29° 48.169′ N	*Pinus elliottii*	18	9
Temperate conifer	Howland, ME, USA	45° 12.25′ N	*Picea rubens, Tsuga canadensis*	24.8	10
Temperate conifer	Le Bray, Bordeaux, FR	44° 43′ 02″ N	*Pinus pinaster*	17.8	11
Temperate conifer	Loobos, NETH	52° 10′ 04″ N	*Pinus sylvestris*	14	12
Temperate conifer	Metolius, OR, USA	44° 29′ 57″ N	*Pinus ponderosa*	15.2	13
Temperate mixed conifer	Vielsalm, BE	50° 18′ 32″ N	*Pseudotsuga, Fagus, Pinus*	16	14
Temperate mixed conifer	Wind River, WA, USA	45° 49′ 14″ N	*Pseudotsuga, Tsuga heterophylla*	8.9	15
Temperate deciduous	Camp Borden, ON, CA	44° 19′ N	*Acer rubrum, Populus trem*	22.7	16
Temperate deciduous	Harvard Forest, MA, USA	42° 32′ 8.6″ N	*Acer rubrum, Quercus rubra*	22.2	17
Temperate deciduous	Hesse, FR	48° 40.45′ N	*Fagus sylvatica*	19	18
Temperate deciduous	Collelongo, IT	41° 52′ N	*Fagus sylvatica*	16.4	19
Temperate deciduous	Lille Bogeskov, DEN	55° 29′ 13″N	*Fagus sylvatica*	17.8	20
Temperate deciduous	Maruia, S. Island, NZ	40° S	*Nothfagus fusca*	—	21
Temperate deciduous	Morgan-Monroe, IN, USA	39° 19′ 17″ N	*Acer sacc, Liriodendron, Quercus*	25	22
Temperate deciduous	Walker Br, TN, USA	35° 57′ 32″ N	*Quercus rubra, Acer rubrum, Carya* spp.	21	23
Tropical rainforest	Jaru, Rhondonia, BR	10° 04′ 59″ S	Rainforest trees	21	24
Tropical rainforest	Manaus, Amazonia, BR	2° 36′ 33″ S	Rainforest trees: *Jacaranda, Inga,* etc.	28	25

[a]List of sources of data from which maximum F_c is compiled is found in Appendix 1.

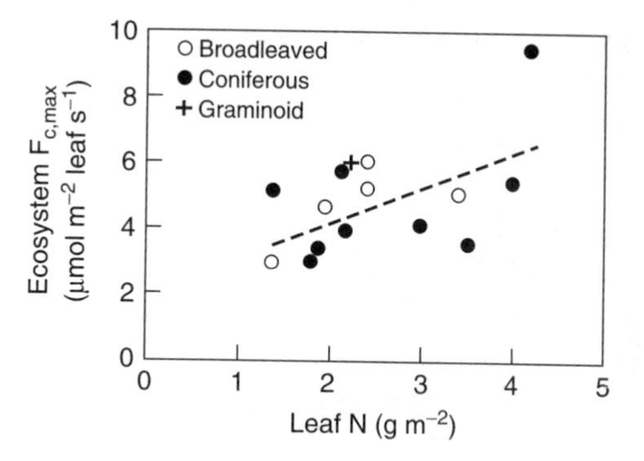

FIGURE 8.4. Variation in whole-ecosystem maximum CO_2 flux (F_c,*max*, expressed on a unit leaf area basis using reported vegetation area index measurements) as a function of leaf *N* at the top of the canopy. The regression line shown is F_c,*max* = 2.059 + 1.058*N, r^2 = 0.30 for C_3-dominated ecosystems. Data sources are listed in Table 8.2.

Carbon Processing by Leaves

Achievement of high carbon assimilation rates in leaves and canopies is beneficial to plants, but the influence on plant growth and competitiveness cannot be viewed apart from the processes involved in downstream carbon utilization for growth and maintenance (see Fig. 8.1). Based on mass-balance principles, a sustained carbon flux into leaves via assimilation requires equivalent utilization through growth and metabolic sinks for this carbon, or accumulation of carbo-

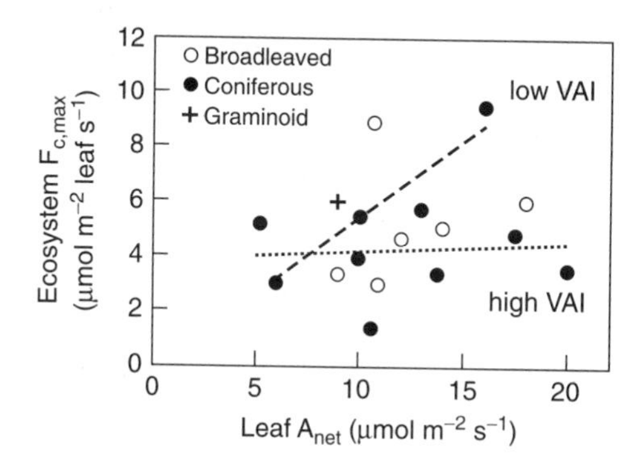

FIGURE 8.5. An examination of simple scaling between single-leaf mean maximum A_{area} reported in the literature, and the whole-ecosystem maximum CO_2 flux in full sun ($F_{c,max}$) from flux tower sites in C_3-dominated ecosystems. The two lines indicate the relationship for fluxes from vegetation with low-VAI (- - -, VAI < 3.0 for 5 sites, r^2 = 0.80, $P < 0.001$) and high-VAI ecosystems (....., VAI > 3.0 for 12 sites, relationship not significant). A_{area} is the maximum reported value for the dominant overstory species reported for the same site as the ecosystem CO_2 fluxes, often in a separate publication. In the case of mixed-species ecosystems, linear averaging of A_{area} was done for the two dominant species for which measurements have been reported. Note that a 1:1 correspondence between F_c,*max* and leaf A_{area} is not expected due to self-shading within canopies, even those with low VAI. Sources of A_{area} data (Appendix 1): Bassow and Bazzaz 1997, Yygodskaya et al. 1997, Walcroft et al. 1997, Robernz and Stockfors 1998, Porté and Lousteau 1998).

hydrates. However, little is know about the necessary coupling between A_{area} and local or whole-plant growth or the signals that achieve this coupling (Farrar et al. 2000). Post-photosynthetic carbohydrate utilization and storage is of particular importance in perennial root crops and trees, which are long-lived and maintain a large biomass pool. Rates of carbon processing by leaves appear to be broadly tuned to the photosynthetic capacity of leaves, since stored carbohydrates in leaves are maintained in a fairly narrow range (~5 to 25% of total leaf mass) despite 10-fold variation in A_{mass} capacity among species. One well-known mechanism for balancing CO_2 assimilation capacity with proximal utilization processes is through biochemical feedback regulation of photosynthesis by carbohydrate signaling (Stitt 1991, Drake et al. 1997) (see Chapter 7). When starch accumulates in leaf chloroplasts, it can inhibit photosynthesis (Azcon-Bieto 1983), which serves as a feedback to regulate source-sink balance in the plant by reducing the translation of photosynthetic proteins (Stitt 1991, Stitt and Krapp 1999). This feedback inhibition effect is most common when photosynthetic productivity greatly exceeds sink strength on low-nutrient or droughty sites, or when photosynthesis is stimulated by long-term exposure to elevated atmospheric CO_2 concentrations (Sage 1994).

While increased growth is associated with sink strength for utilization of soluble sugars, if carbohydrates cannot be utilized for sink growth (for instance, when environmental conditions are unfavorable because of drought or low nutrient availability) then storage reserves such as starch accumulate. The pool of total nonstructural carbohydrates (TNC) in leaves can be used to indicate the balance between supply of carbohydrates by photosynthesis and demand by utilization functions, growth, and maintenance. This represents the mobile storage pool within leaves or other plant organs. As such, high TNC concentrations in leaves may indicate either high photosynthate production rates and low carbon utilization rates, or low production concurrent with low utilization (Körner 2003). Studies analyzing the carbohydrate content of tissues in relation to patterns of growth were formerly more popular than at present, though consideration of plant and ecosystem carbon storage capacity related to environmental issues such as rising atmospheric CO_2 have renewed interest in carbohydrate accumulation as an indicator of plant carbon processing capacity.

Carbon Processing in a Future, Higher-CO_2 Atmosphere

A situation of particular current interest is what happens when plants are grown for long periods under elevated atmospheric CO_2 conditions. Experimental enhancement of the partial pressure of atmospheric CO_2 in experiments is typically done to provide indications of future plant function under global environmental change scenarios (Saxe et al. 1998, Norby et al. 1999). The topic is of applied environmental interest, but can also be viewed in a broader context of identifying the limits to plant productivity, assuming photosynthesis itself was operating at a point near CO_2-saturation. This is generally true at the concentrations of elevated CO_2 concentrations that are used in many recent field environmental change experiments (Saxe et al. 1998, Nowak et al. 2004). A broad question that has been considered in such studies is whether carbon assimilation is regulated synchronously with carbon processing and utilization by growth sinks, such that the accumulation of carbohydrates in leaves remains similar to that at current atmospheric CO_2 (Sage et al. 1989, Arp 1991). Early work on this topic conducted in controlled environments and greenhouse facilities was criticized on the grounds that plants were grown under artificial conditions often including root restrictions that limited the magnitude of a possible sink for processing carbohydrates (Arp 1991). However, countless field studies have now exhibited that increases in soluble carbohydrates occur under long-term elevated atmospheric CO_2 (Drake et al. 1997, Saxe et al. 1998, Körner 2000). Some recent evidence to this effect is shown in Figure 8.6A, where a mean increase in leaf nonstructural carbohydrates of 3.5 ± 0.9% is shown for 10 plant species in long-term elevated CO_2 studies. This suggests that with enhanced carbon assimilation in elevated atmospheric CO_2, carbon processing has not kept pace, resulting in these increases in stored carbohydrate.

A number of recent long-term studies have ascribed changes in the photosynthetic capacity of leaves growing in elevated CO_2, termed photosynthetic down-

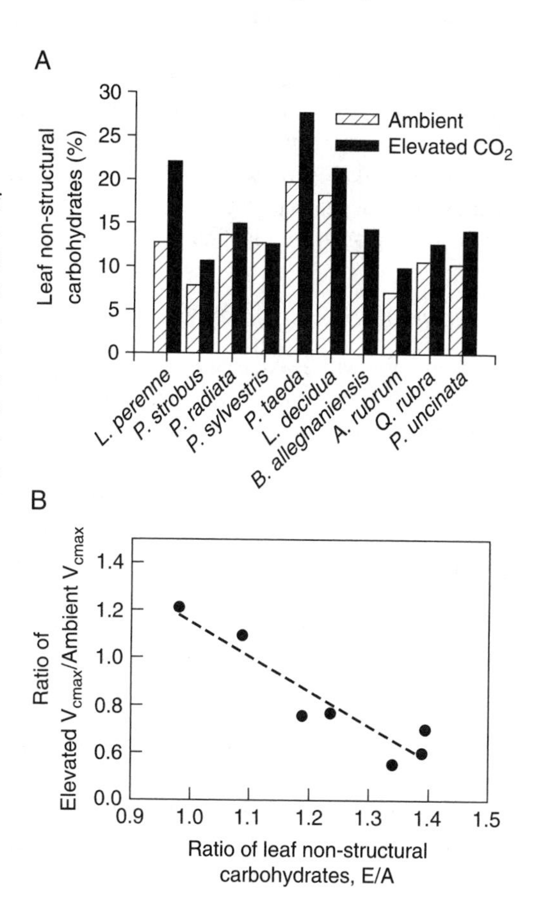

FIGURE 8.6. A. Comparison of leaf nonstructural carbohydrates between ambient CO_2-grown and elevated CO_2-grown plants as an estimate of carbon processing capabilities in elevated atmospheric CO_2 concentrations. Species are arrayed according to their relative photosynthetic capacity in ambient CO_2, from high to low (data from Bauer et al. 2001, Hattenschwiler et al. 2002, Rogers et al. 1998, Rogers and Ellsworth 2002, Tissue et al. 2001). B. Correlation of enhancement of leaf carbohydrates in studies from A and the relative downregulation response of V_{cmax}. Both leaf carbohydrates and V_{cmax} are expressed as ratios of the value for plants grown in elevated CO_2 to the value for plants grown in ambient CO_2. The correlation shown is significant ($p < 0.001$, $r^2 = 0.86$).

regulation, to source-sink relations associated with carbon processing (Tissue et al. 2001, Rogers and Ellsworth 2002, Ainsworth et al. 2003). There are reasonable hypotheses to suggest that the increase in carbohydrates in leaves is mechanistically related to reductions in photosynthetic capacity with long-term elevated CO_2 exposure (Sage 1994, Moore et al. 1999). The downregulation phenomenon is often attributed to source-sink imbalances in the plant, though quantifying plant sink strength is problematic (Farrar 1996) and hence a complete understanding is still lacking. Carbohydrate signaling is an attractive mechanism for CO_2-induced downregulation of the photosynthetic apparatus in part because is explains similar acclimation responses for a wide range of perturbations, including clipping, fertility differences, and so forth. Correlative evidence for recent data from trees (Fig. 8.6B) suggests that the degree of photosynthetic downregulation, indicated by decreases in V_{cmax} between ambient and elevated CO_2-grown plants, is associated with the degree of carbohydrate accumulation in leaves. Thus, reduced local sinks for carbohydrate can promote downregulation (Jach and Ceulemans 2000, Tissue et al. 2001). Because sink limitation occurs when carbohydrate production exceeds carbohydrate utilization, this appears to result in a reduction of Rubisco activity or content, though again the complete causal connection is not elaborated. Although the carbohydrate source-sink hypothesis has a well-defined molecular basis, most studies have not been able to quantitatively substantiate the link between sink capacity and downregulation. Still, the results indicate that plant carbon processing is likely synchronized with carbon assimilation rates through leaf carbohydrates, and that future enhancement of photosynthesis with rising atmospheric CO_2 concentration may be constrained somewhat by plant sink capacity and carbon processing rates.

Carbon Processing by Ecosystems

In an analogy with leaves, carbon processing by ecosystems can be measured via the accumulation of carbon through on-ground techniques of assessing net primary production, though typically the belowground component of this measure is lacking or inadequate (Clark et al. 2001). Another approach relies on the mass balance of CO_2 exchange by ecosystems as is emerging from the FLUXNET initiative (Canadell et al. 2000). From such data, it can be determined whether carbon processing by ecosystems scales as a constant proportion with total carbon income into an ecosystem as has been predicted by theoretical models (Dewar et al. 1998). This has been examined in a recent study using European flux tower data (Janssens et al. 2001), and we have expanded this analysis based on data presented in Falge and coworkers (2002). Because photosynthesis provides the substrate for respiration, these two fluxes are likely to remain in approximate balance in the long term and among different ecosystems. As a result, there is a strong correlation between total ecosystem respiration and gross primary production among grasslands, deciduous forest, and coniferous forests (Fig. 8.7) (see also Janssens et al. 2001). This correlation is related to the observation of constant carbon-use efficiency among ecosystems, where the net primary produc-

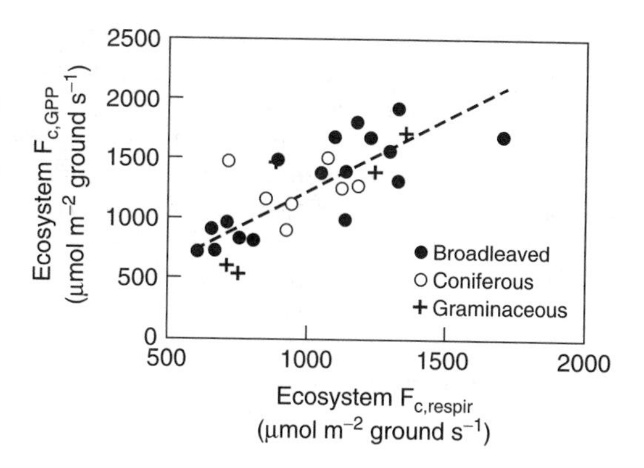

FIGURE 8.7. Relationship between total ecosystem respiration ($F_{c,respir}$) and gross primary production ($F_{c,GPP}$) calculated as the sum of net ecosystem exchange and $F_{c,respir}$ based on flux tower data presented in Falge et al. (2002).

tivity of ecosystems per unit of carbon fixed into the system has been observed to be approximately constant (Waring et al. 1998). These general relationships among ecosystems appear to be the result of an optimal balance between photosynthesis and maintenance respiration that may be a constraint by carbon processing limiting the total amount of carbon stored by ecosystems.

Summary and Conclusions

Carbon assimilation varies severalfold among different plant species and plant types. We found distinct relationships between the carboxylation capacity, V_{cmax}, that underlies photosynthetic capacity in relation to leaf *N* among growth forms of plants. These relationships occur at the leaf scale and at the ecosystem scale, and emerge from the display of *N* via the stacking of photosynthetic proteins in leaves and canopies. Reductions in photosynthetic efficiency due to these stacking considerations represent an important set of constraints on carbon income into ecosystems. However, incoming carbon assimilate appears to be regulated not only by photosynthetic capacity but also by carbon processing and utilization capacity. An important generalization is that carbon processing is synchronized with carbon assimilation among a wide variety of plants and plant types. This relationship is demonstrated at the leaf-level for plants under elevated atmospheric CO_2 that tend to exhibit compensatory decreases in photosynthetic capacity when leaf carbohydrates accumulate. Since photosynthesis provides the substrate for respiration, these two fluxes are likely to remain in approximate balance in the long term, which has important implications for the magnitude of carbon cycling under future, higher atmospheric CO_2 conditions.

References

Aber, J. D., Reich, P. B., and Goulden, M. L. 1996. Extrapolating leaf CO_2 exchange to the canopy: A generalized model of forest photosynthesis validated by eddy correlation. Oecologia 106:267–275.

Ainsworth, E. A., Davey, P. A., Hymus, G. J., Osborne, C. E., Rogers, A., Blum, H., Nosberger, J., and Long, S. E. 2003. Is stimulation of leaf photosynthesis by elevated carbon dioxide concentration maintained in the long term? A test with *Lolium perenne* grown for 10 years at two nitrogen fertilization levels under Free Air CO_2 Enrichment (FACE). Plant Cell Environ. 26:705–714.

Amthor, J. S. 1994. Scaling CO_2-photosynthesis relationships from the leaf to the canopy. Photosynth. Res. 39:321–350.

Arp, W. J. 1991. Effects of source-sink relations on photosynthetic acclimation to elevated CO_2. Plant Cell Environ. 14:869–875.

Azcon-Bieto, J. 1983. Inhibition of photosynthesis by carbohydrates in wheat leaves. Plant Physiol. 73:681–686.

Bauer, G. A., Berntson, G. M., and Bazzaz, F. A. 2001. Regenerating temperate forests under elevated CO_2 and nitrogen deposition: Comparing biochemical and stomatal limitation of photosynthesis. New Phytologist 152:249–266.

Canadell, J. G., Mooney, H. A., Baldocchi, D. D., Berry, J. A., Ehleringer, J. R., Field, C. B., Gower, S. T., Hollinger, D. Y., Hunt, J. E., Jackson, R. B., Running, S. W., Shaver, G. R., Steffen, W., Trumbore, S. E., Valentini, R., and Bond, B. Y. 2000. Carbon metabolism of the terrestrial biosphere: A multitechnique approach for improved understanding. Ecosystems 3:115–130.

Castro-Diaz, P., Puyravaud, J. P., and Cornelissen, J. H. C. 2000. Leaf structure and anatomy as related to leaf mass per area variation in seedlings of a wide range of woody plant species and types. Oecologia 124:476–486.

Clark, D. A., Brown, S., Kicklighter, D. W., Chambers, J. Q., Tomlinson, J. R., and Ni, J. 2001. Measuring net primary production in forests: Concepts and field methods. Ecol. Appl. 11:356–370.

dePury, D. G. G., and Farquhar, G. D. 1997. Simple scaling of photosynthesis from leaves to canopies without the errors of big-leaf models. Plant Cell Environ. 20:537–557.

Dewar, R. C., Medlyn, B. E., and McMurtrie, R. E. 1998. A mechanistic analysis of light and carbon use efficiencies. Plant Cell Environ. 21:573–588.

Drake, B. G., Gonzàlez-Meler, M. A., and Long, S. P. 1997. More efficient plants: A consequence of rising atmospheric CO_2? Annu. Rev. Plant Physiol. Plant Molec. Biol. 48:609–639.

Evans, J. R. 1989. Photosynthesis and nitrogen relationships in leaves of C3 plants. Oecologia 78:9–19.

Falge, E., Baldocchi, D., Tenhunen, J. D., Aubinet, M., Berbigier, P., Bernhofer. C., Burba. G., Clement, R., Davis, K., and et. al. 2002. Seasonality of ecosystem respiration and gross primary production as derived from FLUXNET measurements. Agric. For. Meteorol. 113:53–74.

Farquhar, G. D. 1989. Models of integrated photosynthesis of cells and leaves. Philos. Trans. R. Soc. Lond. [Biol.] 323:357–367.

Farquhar, G. D., von Caemmerer, S., and Berry, J. A. 1980. A biochemical model of photosynthetic CO_2 assimilation in leaves of C_3 species. Planta 149:78–90.

Farrar, J., Pollock, C., and Gallagher, J. 2000. Sucrose and the integration of metabolism in vascular plants. Plant Sci. 154:1–11.

Farrar, J. F. 1996. Sinks—Integral parts of a whole plant. J. Exp. Bot. 47:1273–1279.

Field, C. B., and Mooney, H. A. 1986. The photosynthesis-nitrogen relationship in wild plants. In: On the Economy of Plant Form and Function. T. J. Givnish (ed.). Cambridge: Cambridge University Press.

Friend, A. D. 2001. Modelling canopy CO_2 fluxes: Are "big-leaf" simplifications justified? Global Ecol. Biogeog. 10:603–619.

Hikosaka, K., and Hirose, T. 1998. Leaf and canopy photosynthesis of C_3 plants at elevated CO_2 in relation to optimal partitioning of nitrogen among photosynthetic components: Theoretical prediction. Ecol. Model. 106:247–259.

Jach, M. E., and Ceulemans, R. 2000. Effects of season, needle age and elevated atmospheric CO_2 on photosynthesis in Scots pine (*Pinus sylvestris*). Tree Physiol. 20: 145–157.

Janssens, I. A., Lankreijer, H., Matteucci, G., Kowalski, A. S., and Buchmann, N. 2001. Productivity overshadows temperature in determining soil and ecosystem respiration across European forests. Global Change Biol. 7:269–278.

Jarvis, P. G. 1976. Interpretation of variations in leaf water potential and stomatal conductance found in canopies in field. Philos. Trans. R. Soc. Lond. [Biol.] 273:593–610.

Körner, C. 2000. Biosphere responses to CO_2 enrichment. Ecol. Appl. 10:1590–1619.

Lamhers, H., and Poorter, H. 1992. Inherent variation in growth rate between higher plants: A search for physiological causes and ecological consequences. Advances in Ecological Research 23:188–261.

Law, B. E., Cescatti, A., and Baldocchi, D. D. 2001. Leaf area distribution and radiative transfer in open-canopy forests: Implications for mass and energy exchange. Tree Physiol. 21:777–787.

Long, S. P. 1991. Modification of the response of photosynthetic productivity to rising temperature by atmospheric CO_2 concentration: Has its importance been underestimated? Plant Cell Environ. 14:729–740.

Lowman, M. D., and Nadkarni, N. M. 1995. Forest Canopies. San Diego: Academic Press.

Medlyn, B. E., Badeck, F. W., de Pury, D. G. G., Barton, C. V. M., Broadmeadow, M., Ceulemans, R., De Angelis, P., Forstreuter, M., Jach, M. E., Kellomäki, S., Laitat, E., Marek, M., Philippot, S., Rey, A., Strassemeyer, J., Laitinen, K., Liozon, R., Portier, B., Roberntz, P., Wang, K., and Jarvis, P. G. 1999. Effects of elevated CO_2 on photosynthesis in European forest species: A meta-analysis of model parameters. Plant Cell Environ. 22:1475–1495.

Meir, P., Kruijt, B., Broadmeadow, M., Barbosa, E., Kull, O., Carswell, F., Nobre, A., and Jarvis, P. G. 2002. Acclimation of photosynthetic capacity to irradiance in tree canopies in relation to leaf nitrogen concentration and leaf mass per unit area. Plant Cell Environ. 25:343–357.

Moore, B. D., Cheng, S. H., Sims, D., and Seemann, J. R. 1999. The biochemical and molecular basis for photosynthetic acclimation to elevated atmospheric CO_2. Plant Cell Environ. 22:567–582.

Niinemets, Ü. 1999. Components of leaf dry mass per area—thickness and density—alter leaf photosynthetic capacity in reverse directions in woody plants. New Phytologist 144:35–47.

Niinemets, U., and Tenhunen, J. D. 1997. A model separating leaf structural and physiological effects on carbon gain along light gradients for the shade-tolerant species *Acer saccharum*. Plant Cell Environ. 20:845–866.

Niklas, K. J., and Enquist, B. J. 2001. Invariant scaling relationships for interspecific plant biomass production rates and body size. Proc. Natl. Acad. Sci. USA 98:2922–2927.

Norby, R. J., Wullschleger, S. D., Gunderson, C. A., Johnson, D. A., and Ceulemans, R. 1999. Tree responses to rising CO_2 in field experiments: Implications for the future forest. Plant Cell Environ. 22:683–714.

Nowak, R. S., Ellsworth, D. S., and Smith, S. D. 2004. Plant functional responses to elevated CO_2: Do data from FACE support early predictions? Tansley Review. New Phytologist.

Peterson, A. G., Ball, J. T., Luo, Y., Field, C. B., Curtis, P. S., Griffin, K. L., Gunderson, C. A., Norby, R. J., Tissue, D. T., Forstreuter, M., Rey, A., and Vogel, C. S. 1999.

Quantifying the response of photosynthesis to changes in leaf nitrogen content and leaf mass per area in plants grown under atmospheric CO_2 enrichment. Plant Cell Environ. 22:1109–1119.

Reich, P. B., and Walters, M. B. 1994. Photosynthesis-nitrogen relations in Amazonian tree species. 2. Variation in nitrogen vis-a-vis specific leaf area influences mass-based and area-based expressions. Oecologia 97:73–81.

Reich, P. B., Walters, M. B., Ellsworth, D. S., and Uhl, C. 1994. Photosynthesis-nitrogen relations in Amazonian tree species. 1. Patterns among species and communities. Oecologia 97:62–72.

Reich, P. B., Walters, M. B., and Ellsworth, D. S. 1997. From tropics to tundra: Global convergence in plant functioning. Proc. Natl. Acad. Sci. USA 94:13730–13734.

Reich, P. B., Ellsworth, D. S., and Walters, M. B. 1998. Leaf structure (specific leaf area) modulates photosynthesis-nitrogen relations: Evidence from within and across species and functional groups. Funct. Ecol. 12:948–958.

Reich, P. B., Ellsworth, D. S., Walters, M. B., Vose, J. M., Gresham, C., Volin, J. C., and Bowman, W. D. 1999. Generality of leaf trait relationships: A test across six biomes. Ecology 80:1955–1969.

Roderick, M. L., Berry, S. L., and Noble, I. R. 2000. A framework for understanding the relationship between environment and vegetation based on the surface area to volume ratio of leaves. Funct. Ecol. 14:423–437.

Rogers, A., and Ellsworth, D. S. 2002. Photosynthetic acclimation of *Pinus taeda* (loblolly pine) to long-term growth in elevated $pCO_{(2)}$ (FACE). Plant Cell Environ. 25:851–858.

Sage, R. F. 1994. Acclimation of photosynthesis to increasing atmospheric CO_2—The gas exchange perspective. Photosynth. Res. 39:351–368.

Sage, R. F., and Monson, R. 1999. C_4 Plant Biology. San Diego: Academic Press.

Sage, R. F., Sharkey, T. D., and Seemann, J. R. 1989. Acclimation of photosynthesis to elevated CO_2 in five C_3 species. Plant Physiol. 89.

Saxe, H., Ellsworth, D. S., and Heath, J. 1998. Tree and forest functioning in an enriched CO_2 atmosphere. New Phytologist 139:395–436.

Sellers, P. J., Dickinson, R. E., Randall, D. A., Betts, A. K., Hall, F. G., Berry, J. A., Collatz, G. J., Denning, A. S., Mooney, H. A., Nobre, C. A., Sato, N., Field, C. B., and Henderson-Sellers, A. 1997. Modeling the exchanges of energy, water, and carbon between continents and the atmosphere. Science 275:502–509.

Sinclair, T. R., Murphy, C. E. J., and Knoerr, K. R. 1976. Development and evaluation of simplified models for simulating canopy photosynthesis and transpiration. J. Appl. Ecol. 11:617–636.

Stitt, M. 1991. Rising CO_2 levels and their potential significance for carbon flow in photosynthetic cells. Plant Cell Environ. 14:741–762.

Stitt, M. and Krapp, A. 1999. The interaction between elevated carbon dioxide and nitrogen nutrition: The physiological and molecular background. Plant Cell Environ. 22:583–621.

Tissue, D. T., Griffin, K. L., Turnbull, M. L., and Whitehead, D. 2001. Canopy position and needle age affect photosynthetic response in field-grown *Pinus radiata* after five years of exposure to elevated carbon dioxide partial pressure. Tree Physiol. 21:915–923.

Walcroft, A., Whitehead, D., Silvester, W. B., and Kelliher, F. M. 1997. The response of photosynthetic model parameters to temperature and nitrogen concentration in *Pinus radiata* D. Don. Plant Cell Environ. 20:1338–1348.

Waring, R. H., Landsberg, J. J., and Williams, M. 1998. Net primary production of forests: A constant fraction of gross primary production? Tree Physiol. 18:129–134.

Woodward, F. I., Smith, T. M., and Emanuel, W. R. 1995. A global land primary productivity and phytogeography model. Global Biogeochem. Cycles 9:471–490.
Wright, I. J., and Westoby, M. 2002. Leaves at low versus high rainfall: Coordination of structure, lifespan and physiology. New Phytologist 155:403–416.

Appendix 1.

References for CO_2 flux data and associated leaf-level measurements. Numbers in parentheses correspond to numerical data sources listed in Table 8.2.

Anthoni, P. M., Law, B. E., and Unsworth, M .H. 1999. Carbon and water vapor exchange of an open-canopied ponderosa pine ecosystem. Agric. For. Meteorol. 95:51. (13)
Arneth, A., Kelliher, F. M., McSeveny, T. M., and Byers, A. N. 1999. Assessment of annual carbon exchange in a water-stressed *Pinus radiata* plantation: An analysis based on eddy covariance measurements and an integrated biophysical model. Global Change Biol. 5:531–545. (7)
Aubinet, M., Chermanne, B., Vandenhaute, M., Longdoz, B., Yernaux, M., and Laitat, E. 2001. Long term carbon dioxide exchange above a mixed forest in the Belgian Ardennes. Agric. For. Meteorol. 108:293–315. (14)
Baldocchi, D. D., and Harley, P. C. 1995. Scaling carbon-dioxide and water-vapor exchange from leaf to canopy in a deciduous forest. 2. Model testing and application. Plant Cell Environ. 18:1157. (23)
Bassow, S. L., and Bazzaz, F. A. 1997. Intra- and inter-specific variation in canopy photosynthesis in a mixed deciduous forest. Oecologia 109:507–515. (17)
Berbigier, P., Bonnefond, J. M., and Mellmann, P. 2001. CO_2 and water vapour fluxes for 2 years above Euroflux forest site. Agric. For. Meteorol. 108:183–197. (11)
Black, T. A., DenHartog, G., Neumann, H. H., Blanken, P. D., Yang, P. C., Russell, C., Nesic, Z., Lee, X., Chen, S. G., Staebler, R., and Novak, M. D. 1996. Annual cycles of water vapour and carbon dioxide fluxes in and above a boreal aspen forest. Global Change Biol. 2:219. (5)
Carswell, F. E., Meir, P., Wandelli, E. V., Bonates, L. C. M., Kruijt, B., Barbosa, E. M., Nobre, A. D., Grace, J., and Jarvis, P. G. 2000. Photosynthetic capacity in a central Amazonian rain forest. Tree Physiol. 20:179–186. (25)
Chen, W. J., Black, T. A., Yang, P. C., Barr, A. G., Neumann, H. H., Nesic, Z., Blanken Z., Novak M. D., Eley, J., Ketler, R. J., and Cuenca, A. 1999. Effects of climatic variability on the annual carbon sequestration by a boreal aspen forest, Global Change Biol. 5:41. (5)
Chen, J., Falk, M., Euskirchen, E., Paw, U. K.-T., Suchanek, T. H., Ustin, S. L., Bond, B. J., Brosofske, K. D., Phillips, N., and Bi, R. 2002. Biophysical controls of carbon flows in three successional Douglas-fir stands based on eddy-covariance measurements. Tree Physiol. 22:169–177. (15)
Clark, K. L., Gholz, H. L., Moncrieff, J. B., Cropley, F., and Loescher, H. W. 1999. Environmental controls over net exchanges of carbon dioxide from contrasting Florida ecosystems. Ecol. Appl. 9:936. (9)
Dolman, A. J., Moors, E. J., and Elbers, J. A. 2002. The carbon uptake of a mid latitude pine forest growing on sandy soil. Agric. For. Meteorol. 111:157–170. (12)
Flanagan, L. B., Wever, L. A., and Carlson, P. J. 2002. Seasonal and interannual variation in carbon dioxide exchange and carbon balance in a northern temperate grassland. Global Change Biol. 8:599–615. (2)

Goulden, M. L., Munger, J. W., Fan, S. M., Daube, B. C., and Wofsy, S. C. 1996. Exchange of carbon dioxide by a deciduous forest: Response to interannual climate variability. Science 271:1576. (17)

Grace, J., Lloyd, J., Mcintyre, J., Miranda, A., Meir, P., Miranda, H., Moncrieff, J., Massheder, J., Wright, I., and Gash, J. 1995a. Fluxes of carbon dioxide and water-vapor over an undisturbed tropical forest in south-west Amazonia. Global Change Biol. 1:1–12. (24)

Grace, J., Lloyd, J., Mcintyre, J., Miranda, A. C., Meir, P., Miranda, H. S., Nobre, C., Moncrieff, J., Massheder, J., Malhi, Y., Wright, I., and Gash, J. 1995b. Carbon-dioxide uptake by an undisturbed tropical rain-forest in southwest amazonia, 1992 to 1993. Science 270:778–780. (24)

Granier, A., Ceschia, E., Damesin, C., Dufrene, E., Epron, D., Gross, P., Lebaube, S., le Dantec, V., le Goff, N., Lemoine, D., Lucot, E., Ottorinin, J. E., Pointailler, J. Y., and Saugier, B. 2000. The carbon balance of a young beech forest. Funct. Ecol. 14:312–325. (18,21)

Greco, S., and Baldocchi, D. D. 1996. Seasonal variations of CO_2 and water vapour exchange rates over a temperate deciduous forest. Global Change Biol. 2:183. (23)

Harley, P. C., and Baldocchi, D. D. 1995. Scaling carbon-dioxide and water-vapor exchange from leaf to canopy in a deciduous forest. 1. Leaf model parametrization. Plant Cell Environ. 18:1146. (23)

Hollinger, D. Y., Goltz, S. M., Davidson, E. A., Lee, J. T., Tu, K., and Valentine, H. T. 1999. Seasonal patterns and environmental control of carbon dioxide and water vapor exchange in an ecotonal boreal forest. Global Change Biol. 5:891. (10)

Katul, G., Oren, R., Ellsworth, D., Hsieh, C. I., Phillips, N., and Lewin, K. 1997. A Lagrangian dispersion model for predicting CO_2 sources, sinks, and fluxes in a uniform loblolly pine (*Pinus taeda* L.) stand. J. Geophys. Res.-Atmos. 102:9309. (8)

Lai, C. T., Katul, G., Butnor, J., et al. 2002. Modelling night-time ecosystem respiration by a constrained source optimization method. Global Change Biol. 8:124–141. (8)

Law, B., Williams, M., Anthoni, P., Baldocchi, D. D., and Unsworth, M. H. 2002. Measuring and modeling seasonal variation of carbon dioxide and water vapor exchange of a *Pinus ponderosa* forest subject to soil water deficit. Global Change Biol. 6:613. (13)

Lee, X. H., Fuentes, J. D., Staebler, R. M., and Neumann, H. H. 1999. Long-term observation of the atmospheric exchange of CO_2 with a temperate deciduous forest in southern Ontario, Canada. J. Geophys. Res.-Atmos. 104:15975. (16)

Lindroth, A., Grelle, A., and Moren, A. S. 1998. Long-term measurements of boreal forest carbon balance reveal large temperature sensitivity. Global Change Biol. 4:443–450. (4)

Malhi, Y., Nobre, A. D., Grace, J., Kruijt, B., Pereira, M. G. P., Culf, A., and Scott, S. 1998. Carbon dioxide transfer over a Central Amazonian rain forest. J. Geophys. Res.-Atmos. 103:31593–31612. (25)

Malhi, Y., Baldocchi, D. D., and Jarvis, P. G. 1999. The carbon balance of tropical, temperate and boreal forests. Plant Cell Environ. 22:715. (3,23)

Monson, R. K., Turnipseed, A. A., Sparks, J. P., Harley, P. C., Scott-Denton, L. E., Sparks, K., and Huxman, T. E. 2002. Carbon sequestration in a high-elevation, subalpine forest. Global Change Biol. 8:459–478. (1)

Pilegaard, K., Hummelshøj, P., Jensen, N. O., and Chen, Z. 2001. Two years of continuous CO_2 eddy-flux measurements over a Danish beech forest. Agric. For. Meteorol. 107:29–41. (20)

Randerson, J. T., Chapin, F. S., III, Harden, J. W., Neff, J. C., and Harmon, M. E. 2002. Net ecosystem production: A comprehensive measure of net carbon accumulation by ecosystems. Ecol. Applic. 12:937–947. (3,23)

Schmid, H. P., Cropley, F., Su, H.-B., Offerle, B., and Grimmond, C. S. B. 2000. Measurements of CO_2 and energy fluxes over a mixed hardwood forest in the midwestern United States. Agric. For. Meteorol. 103:355. (22)

Valentini, R., Matteucci, G., Dolman, A. J., et al. 2000. Respiration as the main determinant of carbon balance in European forests. Nature. 404:861–865. (11,12,14,18,19,20)

Vygodskaya, N. N., Milyukova, I., Varlagin, A., Tatarinov, F., Sogachev, A., Kobak, K. I., Desyatkin, R., Bauer, G., Hollinger, D. Y., Kelliher, F. M., and Schulze, E.-D. 1997. Leaf conductance and CO_2 assimilation of *Larix gmelinii* growing in eastern Siberian boreal forest. Tree Physiol. 17:607–615. (6)

Wallin, G., Linder, S., Lindroth, A., Rantfors, M., Flemberg, S., and Grelle, A. 2001. Carbon dioxide exchange in Norway spruce at the shoot, tree and ecosystem scale. Tree Physiol. 21:969–976. (4)

Wilson, K. B., and Baldocchi, D. D. 2001. Comparing independent estimates of carbon dioxide exchange over 5 years at a deciduous forest in the southeastern United States. J. Geophys. Res.-Atmos. 106:34167–34178. (23)

Wilson, K. B., Baldocchi, D. D., and Hanson, P. J. 2000. Spatial and seasonal variability of photosynthetic parameters and their relationship to leaf nitrogen in a deciduous forest. Tree Physiol. 20:565. (23)

Wofsy, S. C., Goulden, M. L., Munger, J. W., Fan, S. M., Bakwin, P. S., Daube, B. C., Bassow, S. L., and Bazzaz, F. A. 1993 Net exchange of CO_2 in a mid-latitude forest. Science. 260:1314–1317. (17)

Part 6

Environmental Constraints

9
Chloroplast to Leaf

FRANCESCO LORETO, NEIL R. BAKER, AND DONALD R. ORT

Introduction: Photosynthesis in Hostile Environments

Photosynthesis is highly responsive to environmental changes. Even so, fundamental modifications of the photosynthetic processes during the evolution of plant life have been relatively limited in comparison to the enormous variations in climatic conditions that have occurred during this period. This is evidence of the remarkable plasticity within the photosynthetic process allowing plants to adapt to different life conditions and environmental changes. Currently, the earth is experiencing a series of rapidly developing environmental changes, often collectively referred to as global climate change, and predominantly caused by anthropogenic activities. These may positively or negatively affect photosynthesis as well as trigger yet further adaptive responses. The rapidity of these environmental changes is exemplified by the increases in CO_2 and CH_4 over the last century (Fig. 9.1). It is essential to know if plants have the capacity to adapt to such rapid environmental changes and thereby mitigate the impact on photosynthetic productivity of crops as well as natural plant communities. The effects of global change on photosynthesis can be extremely complex, reflecting not only natural plant biodiversity but also microclimate diversity. As an example, rising atmospheric CO_2 rise per se can enhance photosynthetic carbon fixation and incremental plant growth, but this may be counteracted by the associated temperature increase. Higher temperatures might exceed the optimal temperature for photosynthesis as well as enhance photorespiration, an energetically wasteful process that competes with photosynthetic carbon fixation. These temperature increases, along with suppression of evapotranspiration by elevated CO_2 may also modify climatic conditions and precipitation patterns, increasing the likelihood of severe drought episodes and salt accumulation in littoral environments.

Global change in temperate regions portends a predominantly more extreme climate, leading to the tropicalization in many boreal regions, and to the desertification of large areas previously characterized by a semiarid climate. This is expected to cause plants of traditional agricultural and forest areas to become increasingly constrained by prevailing environmental conditions. How photosynthesis has been and will be impacted is difficult to assess because of the practical difficulties in

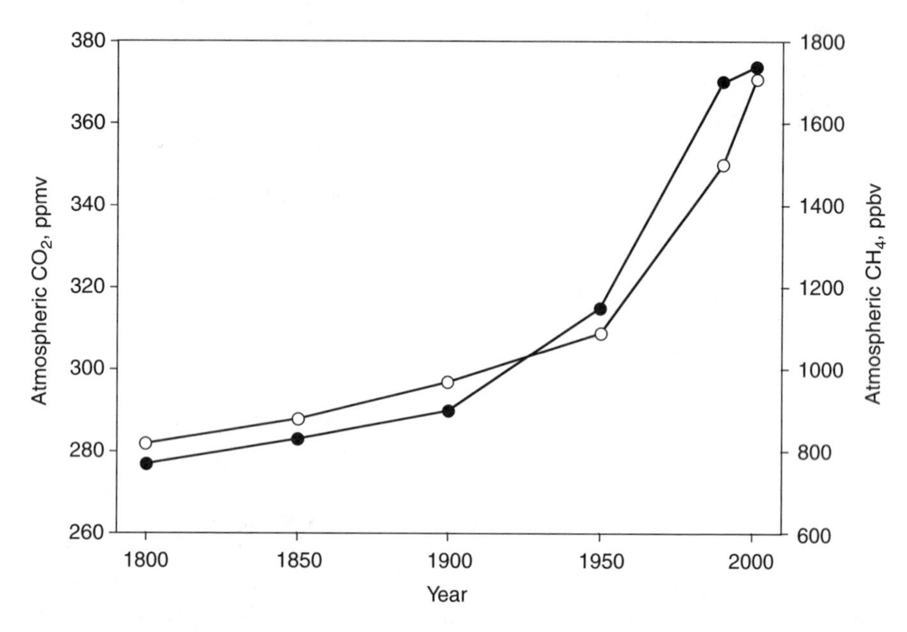

FIGURE 9.1. Atmospheric concentrations of carbon dioxide (●) and methane (○) over the last century (data from [IPCC 94]: Climate Change 1994, and Keeling Whorf 2002).

conducting field experiments with controlled environmental variables. It is an enormous challenge to precisely and concurrently measure the physiological parameters that reliably indicate photosynthesis functionality (e.g., CO_2 and H_2O exchanges between the leaf and air, mesophyll resistances, chlorophyll fluorescence, emission of secondary compounds of photosynthetic origin) under realistic field conditions. Also, it is extremely difficult to simulate the often combined effects of environmental changes on whole plants and plant communities, and for periods sufficiently long to allow adaptation of the plants to take place. The urgent need to understand plant responses to global change has led to the development of sophisticated techniques for growing plants in field conditions under continuously enhanced CO_2. Similar facilities are also being developed to examine the effect of temperature increase, or acute and chronic exposures to ozone or other pollutants.

In this chapter, effects on photosynthesis by some environmental constraints associated with climate change are examined. Emphasis is placed on effects occurring at chloroplast to single leaf scale, and the factors that could compromise photosynthesis of plants growing in hostile environments as a consequence of global climate change.

Photosynthesis in Excessive Light

The light environment in the majority of natural as well as agricultural habitats is extremely variable and plants are often subjected to rapid changes in light intensities over several orders of magnitude (Pearcy, 1990). At high light levels leaves often absorb considerably more light than can be utilized for photosynthesis, creating a potentially hazardous situation in which the photosynthetic apparatus could sustain photodamage (see Chapter 3) with a consequent loss in photosynthetic productivity. Photodamage can also occur at light intensities below those normally required to saturate photosynthesis when other environmental factors, such as low temperatures or drought, limit the ability of leaves to acquire and assimilate CO_2. Plants have evolved mechanisms to protect the photosynthetic apparatus from photodamage in all but the most severe situations. Some plants can reduce the amount of incident light that they absorb through strategic leaf movements and chloroplast movements, and the accumulation of nonphotosynthetic pigments, such as anthocyanins (Long et al. 1994, see also Chapters 2 and 4). However, such strategies for reducing light absorption by the photosynthetic apparatus in leaves appear only to play a minor role in coping with excess light. Mechanisms for rapidly dissipating excitation energy as heat and through photochemistry to electron acceptors other than CO_2 are ubiquitous in higher plant chloroplasts and play central roles in preventing photodamage. Operation of these photoprotective mechanisms unavoidably results in reductions in leaf photosynthetic efficiency and productivity (see Chapter 3). While both photodamage to the photosynthetic apparatus and the development of protective mechanisms to prevent such photodamage both cause decreases in photosynthetic productivity, they differ significantly in that photoprotection permits the restoration of efficient photosynthesis as soon as permissible conditions return. Recovery from photodamage is slow, often incomplete, and can have long-term effects of growth, productivity, and competitiveness.

Mechanisms of Photodamage

The primary site of photodamage of the photosynthetic apparatus is the reaction center of photosystem II (PSII). The largest proteins of the PSII multisubunit complex (Fig. 9.2) are the chlorophyll-binding primary antenna proteins CP47 and CP43 (ca. 47 and 43 kDa), a 33-kDa manganese-stabilizing protein and two additional chlorophyll-binding proteins of 32 and 34 kDa, known as D1 and D2. There are six chlorophyll molecules associated with the PSII reaction center, two of which are bound to the D1 and D2 proteins and function as the primary electron donor. This chlorophyll special pair, known as P680, upon excitation reduces pheophytin, which is an intermediate, metastable electron acceptor on the D1/D2 heterodimer between P_{680} and the two primary, stable quinone electron acceptors, Q_A and Q_B.

Reduced pheophytin transfers its electron to the primary quinone acceptor, Q_A, which in turn reduces the secondary quinone acceptor, Q_B. The semiquinone, Q_B^-, is subsequently further reduced by a second turnover of the reaction center, producing the doubly reduced Q_B^{2-}. Once protonated Q_BH_2 dissociates from the reaction center producing unbound plastoquinol (PQH_2), which is mobile within the lipid phase of the thylakoid membrane. Oxidized P_{680} (P_{680}^+) has an exceptionally high oxidizing potential that normally oxidizes the neighboring Tyr_z residue on the D1 protein, which is in turn reduced by electron transfer from water via the Mn cluster.

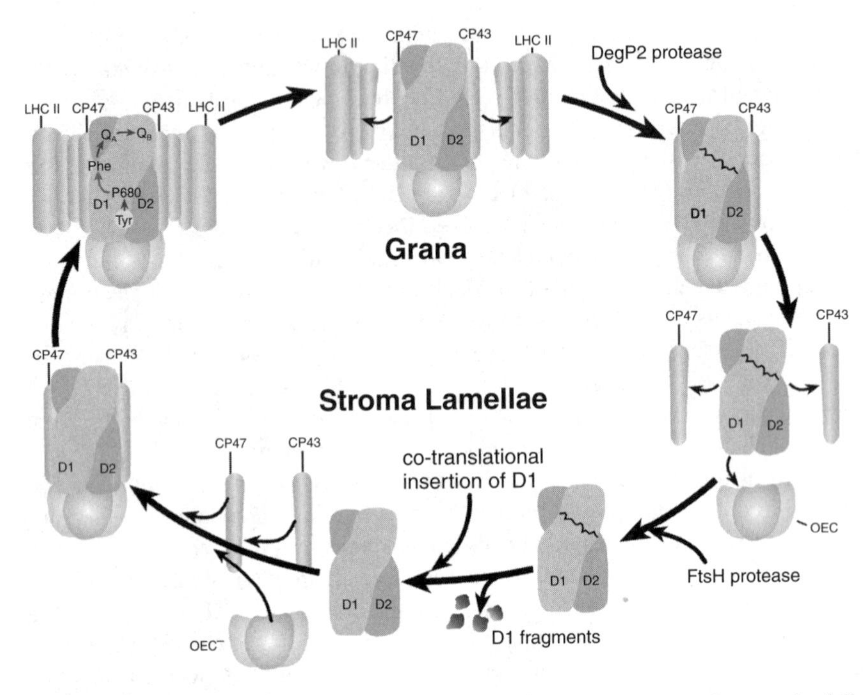

FIGURE 9.2. Cycle of photodamage and repair of the PSII reaction center. See text for full explanation of cycle. CP43, CP47, chlorophyll-binding primary antenna proteins; D1, D2, chlorophyll-binding proteins of the core of the PSII reaction center; LHCII, light-harvesting chlorophyll-protein complexes; $^1O_2$, singlet oxygen; OEC, oxygen evolving complex; P_{680}, reaction center chlorophyll of PSII; $^3P_{680}$, triplet state of reaction center chlorophyll of PSII; Phe, phaeophytin; Q_A, primary quinone electron acceptor of PSII; Q_B, secondary quinone acceptor of PSII (bound plastoquinone).

Photoinactivation of the PSII reaction center can occur by at least two independent mechanisms, associated with either the acceptor or donor side of PSII. Both result in inhibition of electron transfer through PSII and cause degradation and programmed turnover of the D1 protein. Acceptor-side inhibition will occur under high light conditions when the plastoquinone pool is fully reduced causing the Q_B site on D1 to be empty for lack of unbound oxidized plastoquinone. Acceptor-side inhibition can occur in leaves when CO_2 assimilation is restricted (see Chapter 3). In this state, Q_A may become abnormally doubly reduced on a second turnover of the reaction center to form $Q_A{}^{2-}$, as $Q_A{}^-$ cannot transfer an electron to the vacant Q_B. $Q_A{}^{2-}$ then becomes protonated to form Q_AH_2 and is released from the Q_A-binding site on the D1 protein. With the Q_A site vacated, excitation of P_{680} will result in the formation of the radical pair, $P_{680}{}^+$ Pheophytin$^-$. Recombination of these radicals can result in the formation of the triplet state of P_{680}, which reacts with oxygen to form singlet oxygen. Singlet oxygen is potentially damaging to proteins, and is thought to react with the D1 protein thus triggering its degradation (see Fig. 9.2).

Donor-side photoinhibition of PSII is initiated by the dysfunction of water oxidation causing highly reactive P_{680}^{+} and/or Tyr_z^{+} to accumulate. Oxidation of neighboring molecules, such as chlorophylls and carotenoids, occurs and damage to D1 leads to inactivation of PSII. Although donor-side photoinhibition has been frequently observed in in vitro experiments, it is has not yet been demonstrated to be a feature of photoinhibition in leaves operating under physiological conditions.

Under severely excess light, photodamage has also been reported at other sites in the thylakoid membranes as a result of singlet oxygen generation in the light-harvesting antennae. When a chlorophyll molecule absorbs a photon it enters the singlet excited state and normally the excitation energy is rapidly transferred to neighboring chlorophyll molecules. Under high light conditions the probability for the formation of the chlorophyll triplet state increases. The triplet state is long lived compared with the singlet state and can interact with oxygen to produce singlet oxygen, which will readily react with compounds containing carbon-carbon double bonds, such as carotenoids, chlorophylls, and lipids, to form peroxides. Peroxidation of membrane components results in membrane dysfunction and disturbed organization. The peroxidation of chlorophylls leads to bleaching, a commonly observed phenomenon in leaves exposed to severe environmental stresses and high light.

Degradation and Repair of D1 Photodamage

When D1 is photodamaged, the PSII complex undergoes a carefully orchestrated process of disassembly and repair (Fig. 9.3) (Melis, 1999). PSII complexes are normally associated with light-harvesting complexes (LHCII) and are located primarily in the appressed regions of the granal stacks of thylakoid membranes (see Fig. 9.2). Following photodamage, the D1 protein is cleaved by a protease into 22- and 10-kDa polypeptide fragments and a thylakoid protein kinase phosphorylates the associated LHCII complexes, causing their detachment from the PSII complex. Thereafter, the damaged complex migrates from the appressed granal membrane region to a nonappressed, stromal membrane region. After migration to a stromal membrane region, CP43 and CP47 chlorophyll-containing proteins and the oxygen-evolving complex detach from the damaged PSII complex. A new D1 is o-insertionally synthesized on thylakoid-bound chloroplast polysomes and associates with the D2 protein. A functional PSII complex results from reassembly of the CP43, CP47, and the oxygen-evolving complex with the D1/D2 complex. LHCII complexes then reassociate with the PSII complex, which migrates back into an appressed membrane region of the grana as a functional complex to complete the repair cycle.

Even under low light, photodamage to PSII complexes occurs, although at low rates, and the repair cycle is able to keep pace so that inactive centers do not accumulate. In leaves under the majority of physiological conditions, the rate of D1 photodamage is less than the capacity for repair and consequently no significant loss of photosynthetic capacity accumulates. However, if the rate of D1

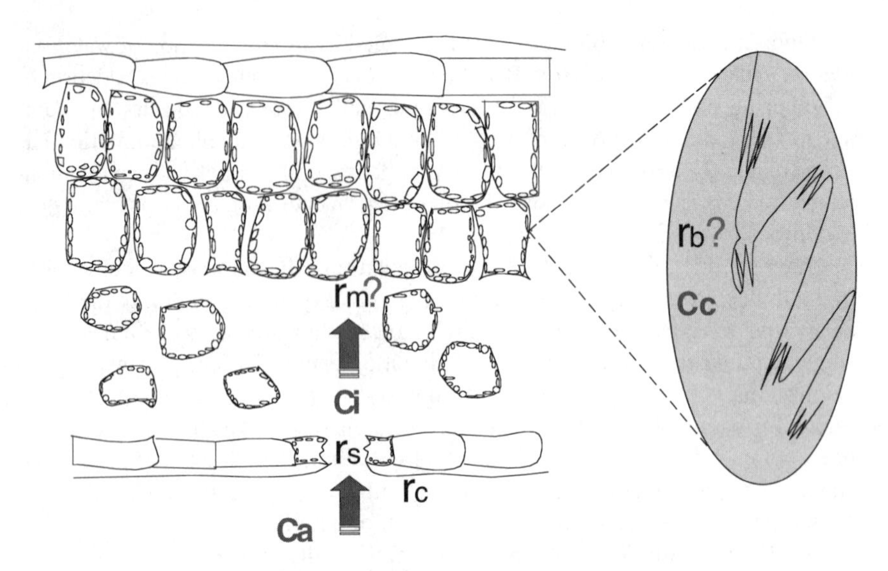

FIGURE 9.3. Pathway of CO_2 acquisition by leaves: from outside leaves (Ca) to intercellular spaces (Ci), to chloroplasts (Cc). This flux is regulated by physical resistances, such as cuticular (rc), stomatal (rs), and mesophyll (rm), and biochemical resistances mainly attributed to the capacity of CO_2 fixation by Rubisco (rb). The effect of mild drought and salinity stress on stomatal resistance is well documented, while the effect on mesophyll and biochemical resistances is debated.

photodamage exceeds the rate of D1 repair, then there will be an accumulation of damaged PSII complexes that will result in a decrease in photosynthetic efficiency and capacity.

Photoinhibition in the Natural Environment

A considerable body of information demonstrates that light-induced inhibition of photosynthesis occurs regularly in the field. Perhaps the most convincing demonstrations of photoinhibition in the natural environment have been made from comparisons of photosynthetic performance in exposed and shaded leaves. Shading of both crop and tree leaves in canopies experiencing low temperatures prevented light-dependent decreases in the maximum quantum efficiency of photosynthesis that occurred at high light (Long et al. 1994). However, it is not yet clear whether such photoinhibition is due to photodamage to PSII reaction centers or photoprotection such as light-induced increases in antennae quenching. This important issue has proved to be difficult to resolve, primarily due to the practical difficulties of studying the detailed characteristics of photoinhibition under field conditions. A number of field studies on a range of species in different habitats have reported decreases in photosynthetic efficiency due to light-induced increases in the rate of thermal dissipation of excitation energy with increasing light during the morning, followed by recovery as light decreased in the afternoon (Long et al. 1994). Although there is no doubt that both the rates of photodamage to and repair of PSII complexes increases when leaves are exposed to environmental stresses in high light, it is not clear to what extent photodamaged PSII complexes accumulate during midday and result in transient depressions of photosynthetic efficiency routinely observed in field situations. It may be expected that accumulation of photodamaged PSII complexes would be more probable when plants are at their geographical limits of their distribution. Alternatively, accumulation of photodamaged PSII complexes in many leaves may only be a brief stage prior to rapid death, with the probability of recovery being extremely low.

Photoinhibition of PSII electron transport is associated with decreases in CO_2 assimilation in field situations. However, it is often difficult to assess exactly what the consequences of such photoinhibition are for the overall photosynthetic productivity of a plant or canopy. The linear relationship between the absorption of light and the production of dry matter that has been observed for a number of crops indicates that crop photosynthetic productivity is normally limited by light, and supports the argument that the majority of crop photosynthesis is performed at light levels below those required to saturate photosynthesis (Baker and Ort 1992). Consequently even moderate levels of photoinhibition, which decrease the maximum quantum efficiency but not in the light-saturated rate of photosynthesis, would be expected to measurably influence canopy carbon gain. Modeling studies of leaf and canopy photosynthesis have suggested that even in the absence of any significant environmental stresses, approximately 10% of the potential carbon gain can be lost due to photoinhibition on clear days (Long et al. 1994). Perhaps the most compelling evidence that photoinhibition does significantly influence plant productivity is the correlation between decreased conversion of dry matter production and the amount of intercepted light in developing crop canopies during periods of chilling. These events are also accompanied by decreased maximum quantum yields of CO_2 assimilation by the leaves (Long et al. 1994).

Photosynthesis in a Water-Scarce and Salt-Rich World

Drought

Due in large part to land-use decisions, an increasing proportion of the earth's land surface suffers recurrent or enduring water scarcity. This situation is exacerbated by multiple and concurrent factors associated with climate change, particularly temperature increase. Mesophytic plants of temperate regions will be faced with more frequent and prolonged drought episodes and it is vital to understand how this will impact photosynthetic productivity. Because plant growth is affected by drought considerably before effects on photosynthesis are observed (Nonami and Boyer 1990) and because the inhibition of photosynthesis in drought-stressed leaves is often relieved when these leaves are exposed to very high CO_2 concentration (e.g., Cornic et al. 1989) it is widely held that the photosynthetic apparatus is resistant to drought (Cornic and Massacci 1996). However, when relative water content (RWC) falls below ~70%, a cascade of physiological processes can be affected resulting in irreversible or slowly reversible damage to the photosynthetic apparatus (e.g., Kaiser 1987; Cornic and Massacci 1996). The occurrence of mild drought stress (i.e., RWC > 70%) is the most common circumstance in nature and our focus exclusively here.

The entry of CO_2 into leaves is determined by a series of physical and biochemical resistances (Evans and Loreto 1999). The primary physical resistances are at cuticular, stomatal, and mesophyll level, while "biochemical resistances" are often those involving CO_2 fixation or RuBP carboxylation (Laisk and Loreto 1996) (see Fig. 9.3). A fully developed cuticle has an extremely high resistance to CO_2 flux and thus only when the other resistances to CO_2 movement are very high (i.e., when stomata are tightly closed), does cuticular conductance need to be considered (Meyer and Genty 1998). Stomatal aperture is the primary change affecting leaf conductance when plants sense a reduction of soil water content (Davies and Zhang 1991) or low atmospheric humidity (Schulze et al. 1987). The physiological mechanism leading to stomatal closure involves stress perception and the activation of a signal transduction pathway in which ABA biosynthesis and translocation play an integral role. Recent findings indicate that while ABA-dependent stomatal closure dominates slowly developing drought stress, when water loss occurs rapidly due to decreased air humidity, the stomatal closure is apparently not primarily mediated by ABA accumulation and may be a direct response to osmotic changes (Assmann et al. 2000). Perception and response of stomata to leaf water status is a very active field of research and exciting progress has been made in identifying receptors and genes in the cellular pathways involved in stomatal response to stress (Ingram and Bartels 1996). It is important to recognize, however, that while reduced stomatal conductance is an effective barrier against water loss under drought conditions, stomatal closure simultaneously restricts CO_2 entry in the leaves and therefore photosynthesis.

Gas-exchange measurements allow calculation of the CO_2 concentration in the intercellular spaces of the leaf (C_i) and modeling of stomatal limitation to CO_2 entry (Farquhar and Sharkey 1982). The standard procedure that has evolved to assess the individual stomatal and nonstomatal contributions to the overall control of net photosynthesis has been to examine the dependence of assimilation on the C_i. That is, the extent to which net photosynthesis can be stimulated by increases in the C_i has been taken to be a quantitative measure of the control exerted on photosynthesis by the stomata. However, a good deal of evidence is now emerging that shows that there are significant interactions between these two levels of control, making their individual evaluation more difficult and perhaps less informative than previously recognized. Since the exchange of water vapor and CO_2 between the leaf and the surrounding atmosphere occur along a common diffusion pathway, the complexity and subtlety of this interaction between stomatal and nonstomatal controls is especially relevant when water availability is being considered. Indeed, since the driving force for water vapor loss is most often much greater than that for CO_2 uptake into the leaf, mesophytic crop plants (particularly C_3 plants) typically lose a great deal of water in exchange for comparatively modest acquisition of CO_2. From this viewpoint it would seem remarkable if there were not well-developed biochemical controls in place to help balance the demand of photosynthesis for CO_2 with the restricted supply arising from the stomatal closure that accompanies decreasing leaf water potential. Since the actual relationship between assimilation and C_i is strongly influenced by conductance, multiple linear regression procedures can be useful to factor out the effect of conductance and reveal any underlying response of assimilation to C_i in some cases revealing that the response of photosynthesis to C_i is unambiguously negative (e.g., Wise et al. 1992). Thus, although the rate of light-saturated net photosynthesis was strongly dependent on leaf conductance, this analysis reveals that there was also an underlying negative dependence of the rate of photosynthesis on C_i, indicating that the biochemical demand for CO_2 had been downregulated in response to declining CO_2 availability associated with water-deficit-induced stomatal closure. A significant contribution to this issue was the development by Lauer and Boyer (1992) of an elegant method for direct measurement of CO_2 within leaves. Their study demonstrated that an increase in C_i accompanied water-deficit-induced decreases in net photosynthesis of sunflower, soybean, and bush bean. Apparently contradictory data were reported by Cornic and Briantais (1991) based on a clever application of chlorophyll fluorescence methods, which enable them to obtain qualitative information about changes in CO_2 availability with the chloroplast (C_c) in response to water deficit. Their data indicated a decrease in C_c of slowly dehydrated French bean leaves which in turn resulted in an increase in photorespiratory turnover of Rubisco with O_2. This result is not necessarily contradictory to Lauer and Boyer (1992) since it seems likely that although significant downregulation of CO_2 demand occurred, it was nevertheless not sufficient to prevent a decline in CO_2 availability. These data of Cornic and Briantais (1991) also remind us that changing ratios of photorespiration to photosynthesis as well as other alternative electron sinks can be a factor assisting to balance CO_2 demand with declining CO_2 supply (Ort and Baker 2002) during drought.

In addition to the intricacies introduced by the interaction between stomatal and nonstomatal factors, analysis of the limitation of photosynthesis by water deficit was further complicated when evidence of inhomogeneities in stomatal apertures across the leaf surface began to emerge (e.g., Daley et al. 1989, Downton et al. 1988, Laisk et al. 1980, Loreto and Sharkey 1990, Terashima et al. 1988). It was suggested that severe inhomogeneities in conductance within a leaf could lead to an overestimation of C_i owing to an underestimation of leaf conductance and thereby creating the illusion of nonstomatal limitations when none actually exist (Farquhar et al. 1987, Terashima et al. 1989). The rationale behind this argument has been strongly challenged (Cheeseman 1991). Although such inhomogeneities do develop upon dehydration of nonacclimated chamber-grown plants or in uprooted field plants, patchiness probably seldom occurs under natural conditions (Wise et al. 1992).

Whereas the pathway shared between water loss and CO_2 acquisition is basic to models of photosynthesis it often does not account for additional resistances to CO_2 diffusion that may also occur from the intercellular spaces to the chloroplasts and contribute to reductions in the CO_2 concentration at the carboxylation sites. So-called mesophyll resistances have been theoretically enumerated (Nobel 1971). Their calculation with independent methods has shown that in non-stressed leaves mesophyll resistances are proportional to both photosynthesis and stomatal conductance (Loreto et al. 1992). However, mesophyll resistances are not often introduced as a possible factor contributing to photosynthesis limitation. In the case of drought-stressed leaves the intercellular CO_2 concentration is low because of stomatal closure, and mesophyll resistances likely contribute to further reducing the CO_2 concentration and, consequently, to limiting photosynthesis. This was clearly argued by Cornic and Massacci (1996) on the basis of an increased discrepancy between the electron transport rate and photosynthesis in a number of studies with drought-stressed leaves. This discrepancy is likely to be due to a low CO_2 concentration in the chloroplasts, which results in an increase in the rate of O_2, compared with CO_2, reduction by photosynthetic electron transport. Unfortunately, under drought conditions mesophyll resistance measurements are difficult and imprecise although a few studies have calculated the C_c and indicate that mesophyll resistance contributes to the limitation of photosynthesis under drought (Lauteri et al. 1997, Delfine et al. 2001). Thus, it appears that the diffusion resistances limiting photosynthesis in leaves subjected to mild drought-stress have both stomatal and mesophyll components.

A diverse array of photochemical and biochemical changes have been described in mildly drought-stressed leaves. On occasion, such leaves may show symptoms of photoinhibition (Powles 1984), reduced nirate reductase activity (Smirnoff et al. 1985), and reduction of sucrose phosphate synthase (Vassey and Sharkey 1989). Photoinhibition occurs if the rate of photosynthesis decreases and leaves cannot make full use of the absorbed light (see previous section Photosynthesis in Excessive Light). Consequently, photoinhibition is often observed when drought-stressed leaves are also exposed to high light intensities. Shade leaves may also be susceptible to photoinhibition caused by sunflecks under drought conditions (Valladares and Pearcy 2002). Consequent to, or more often immediately preceding photochemical damage, energy dissipation mechanisms are activated as monitored by nonphotochemical fluorescence quenching and xanthophyll deepoxidation (Saccardy et al. 1998) and alternative electron transport possibly feeding direct O_2 photoreduction (Loreto et al. 1995). The question of whether metabolic or diffusional factors limit photosynthesis in drought-stressed leaves is still a matter of debate (e.g., Cornic 2000). To reconcile the different opinions it could be argued that it is just a matter of stress severity and species-specific sensitivity, as there is consensus that slowly reversible or irreversible damage of photosynthesis associated with metabolic perturbations does occur when the drought stress becomes severe (i.e., RWC $< 70\%$, as reviewed by Kaiser 1987). But data of Tezara et al. (1999) suggest that metabolic limitations are present whenever a stress effect on photosynthesis is measurable, independently of stress onset velocity. The most recent view (Lawlor and Cornic 2002) maintains that plants may respond to mild drought stress first by diffusional followed by metabolic limitation of photosynthesis. But in our view, this sequential model does not adequately incorporate the importance of the interaction of these two levels of limitation.

Salinity

Concurrently with water scarcity, an increasing amount of cultivated soil is becoming heavily salinized. Salinization occurs because of seawater infiltration consequent to lowering of the table water in littoral soils, salt resurgence consequent to soil drying, or agricultural practices such as heavy fertilization and irrigation (Szabolcs 1994). Despite its importance, there is not agreement concerning the most significant physiological limitations caused by salt accumulation in plants. Munns (1993) pointed out that many studies on salinity contradict previous inferences that growth of salt-exposed plants is limited by photosynthesis or turgor. Rather, salt accumulation may cause early leaf senescence and restrain assimilation availability for sink leaves. Growth is initially reduced because of the salt-induced decrease of soil water potential. This is an effect that mimics water stress and is mainly caused by accumulation of salt in the soil. Salt is also accumulated within leaves (starting with the oldest and less-active leaves) thereby exerting direct inhibitory effects. As a result, these leaves rapidly senesce and lose their function as a carbon source for younger leaves.

One important prediction of this biphasic view of salinity effects is that plant breeding for tolerance to salt accumulation by leaves would alone be ineffective in dealing with the full scope of problems encountered by the plants during the initial phase of salt accumulation. However, this notion has been challenged by Neumann (1997), who argued that salt accumulation in the soil induces changes at the cellular level, eventually leading to differentiating growth response in susceptible and tolerant genotypes. If correct, selecting genotypes solely for tolerance to moderate and external salt accumulation could result in varieties with overall improved tolerance to saline conditions.

Recent findings (Delfine et al. 1998, 1999) seem to reconcile these two contrasting views showing that a moderate Na accumulation (about 3% of leaf dry mass) in spinach leaves does not impede the biochemical and photochemical reactions, and, thus, does not have direct toxic effects on photosynthesis. However, salt accumulation, by reducing both stomatal and mesophyll conductance to CO_2, reduces the CO_2 at the site of carboxylation within the chloroplast thereby indirectly limiting photosynthesis. This is an effect similar to that found in drought stress experiments, further suggesting a mechanistic tie between these two stresses. Perhaps the cascade of events leading to the reduction of stomatal conductance by salt stress involves activation of an overlapping perception-signal transduction pathway, and possibly the synthesis of the same stress-related compounds, as occurs with drought stress. But, contrary to the belief that this response only occurs once salt accumulation in the soil increases the osmotic potential, which is sensed as a drought stress by plants with the consequent stomatal closure (Munns 1993, Blumwald et al. 2000), a drought/salt syndrome can be extended to the early and mild accumulation of Na in leaves (probably compartmentalized to vacuole). This early, nonosmotic effect of salt accumulation in leaves is therefore somehow in between the two phases of the Munns model (i.e., between the decrease of soil water potential and the direct toxic effects). Interestingly, Delfine and coworkers (1999) showed that salinity-induced reductions in conductance (stomatal and mesophyll) are reversed by suspending saline irrigation, and that photosynthesis recovers in proportion to conductance. The first important consequence of this finding is confirmation that mild salt accumulation within leaves is unlikely to affect the biochemistry and photochemistry of carbon assimilation. Still, growth is curbed and photosynthesis at ambient CO_2 is reduced. Consistent with this finding is the work of Centritto et al. (2002), showing a peak of photosynthesis in salt-stressed leaves after a long exposure to low CO_2 to induce stomata to open. This transient increase of photosynthesis was found in several cultivars of olive with different genotypic sensitivity to salt, indicating that the early phase of the stress is initiated by diffusional limitation for CO_2 and that this response is common to both tolerant and susceptible genotypes. A second outcome of the Delfine and coworkers (1999) study is the demonstration that transient salt accumulation, which is sufficient to inhibit photosynthesis and retard growth, can do so without triggering events that lead to accelerated senescence.

In the past, mesophyll resistance has been attributed to physical constraints to CO_2 movement within the intercellular space of leaves, mostly due to scherophyllous anatomy and pathway tortuosity (Loreto et al. 1992, Syvertsen et al. 1992, Miyazawa and Terashima 2001). As opposed to stomatal resistance, increases of mesophyll resistance were assumed to be developmental and irreversible. For example, an irreversible increase of mesophyll resistance has been shown in ageing leaves (Loreto et al. 1994). The experiments of Delfine and coworkers (1999) suggest that mesophyll resistance of salt- and perhaps also of drought-stressed leaves may be as rapidly reversed as stomatal resistance. The most attractive hypothesis for how this might occur is that CO_2 diffusion into mesophyll cells is facilitated by aquaporins. Aquaporins are hydrophobic proteins that passively channel water (Preston 1992), and are suspected to exert the same functions for dissolved CO_2 though contrasting evidence has been presented (see review by Tyerman et al. 2002). In plants, aquaporins similar to those of animals have been found and their role may also be similar. The possibility that leaf aquaporins allow CO_2 transport was tested by Terashima and Ono (2002) who reported a 30 to 40% reduction of mesophyll conductance in leaves treated with $HgCl_2$, which inhibits aquaporins but also reacts with a large number of other proteins. A very recent experiment, conducted over expressing tobacco NtAQP1 membrane aquaporin, has further clarified that aquaporins may indeed facilitate CO_2 transport in leaves (Uehlein et al. 2003). Aquaporins may change in leaves exposed to drought stress (Smart et al. 2001) and may therefore be a very important determinant of CO_2 diffusion in stressed leaves, although they are likely to exert a larger effect when the CO_2 gradient is low (Uehlein et al. 2003) such as when photosynthesis is limited by biochemistry rather than by gas diffusion. Carbonic anhydrase, the enzyme catalyzing the CO_2 – HCO_3^- conversion could also be affected by stresses and this could also influence the mesophyll conductance of the leaves. But most of the studies addressing this problem have shown that strong reductions of carbonic anhydrase activity, as induced by genetic engineering (Price et al. 1994) or inhibitors (Peltier et al. 1995) do not have appreciable consequence on the internal resistance.

As in the case of drought stress, a wealth of biochemical changes occur when salt stress is prolonged and salt accumulates in leaves to high levels. Besides increasing levels of ABA, plants accumulate various compatible solutes (polyols, hydrocarbons, amino acids, and betaines) probably in part serving as osmoprotectants and preserving the cellular turgor. A protective role for compatible solutes has been elucidated in numerous studies (e.g., Bohnert et al. 1995) and has received further experimental support from genetically engineered plants with deficient levels of these compounds. However, compatible solutes generally accumulate only after the photosynthetic apparatus has already begun to suffer irreversible damage such as reduction of Rubisco activity and RuBP regeneration capacity (Delfine et al. 1999). Compatible solutes are important compounds allowing survival of glycophytic plants under severe salt stress but may not be involved in early stages of adaptation or when mesophytic plants are exposed to low and transient salt-stress levels.

Photosynthesis in a Polluted World

Air pollution is increasingly becoming a major problem for human health and crop production in both developing and developed regions of the world. The increases in CO_2 and ozone in the lower atmosphere (troposphere) have major implications for the productivity of crops and sustainability of ecosystems. Also, the potential for further increases in the level of ultraviolet-B (UV-B) radiation reaching the earth's surface due to ozone depletion in the upper atmosphere (stratosphere) portends additional new environmental challenges for many plants.

Elevated CO_2

The amount of CO_2 in the atmosphere is on the rise. The increase is proportional to the amounts of CO_2 humans have released into the global atmosphere by burning fossil fuels and forests. In 1800, the concentration of CO_2 in the atmosphere was about 250 parts per million parts of air (ppm), by 2000 it was 365 ppm and by 2100 it is expected to be between 600 and 700 ppm. Photosynthesis can provide some protection against atmospheric change. Each year photosynthesis removes about 110 Gt of carbon from the atmosphere and each year the respiration of plants, animals and microorganisms releases about 110 Gt back into the atmosphere. This balance had been stable for centuries, until humans started to release significant additional carbon to the atmosphere from fossil fuel use. Today fossil fuel combustion adds another 5.5 Gt and forest destruction 1.5 Gt to the atmosphere. Since this anthropogenic emission of 7 Gt of carbon is about 1% of the total atmospheric concentration, the atmospheric concentration would be expected to rise at about 1% per year. However, photosynthesis is halving the rate of rise in CO_2 that would otherwise occur as the actual rate of rise is about 0.5% per year. Thus at present, photosynthesis is protecting us from a more rapid rate of atmospheric and climate change. If rising CO_2 were the only change occurring in the atmosphere, this might be expected to unequivocally benefit photosynthetic production. A doubling of CO_2 concentration would roughly halve photorespiratory losses in C_3 crops. In addition to the carboxylation of RuBP by CO_2, Rubisco will also catalyze its oxidation by atmospheric O_2 (oxygenation) to yield one molecule of a 3-carbon compound (3-phosphoglyceric acid; PGA) and a molecule of a 2-carbon compound. This oxygenation reaction creates a significant inefficiency in the photosynthetic process because the 2-carbon compound cannot enter the C_3 cycle. In a typical C_3 crop (e.g., wheat or soybean), the rate of Rubisco-catalyzed oxygenation is about 20% of the rate of Rubisco-catalyzed CO_2 fixation in the current atmosphere. The inability of Rubisco to distinguish between molecular O_2 and CO_2 appears to be an unavoidable consequence of having evolved in an oxygen-free atmosphere. Due to this competition between O_2 and CO_2, the proportion of oxygenation to carboxylation will continue to decline as the CO_2 concentration in the atmosphere continues to increase. Photosynthesis of C_3 plant species would be further increased because the current CO_2 concentration is insufficient to saturate Rubisco, with its surprisingly low affinity for CO_2. Additionally, it is clear that elevated CO_2 has the potential to impact water-use efficiency by inducing stomatal closure in both C_3 and C_4 species. It should be recognized however that, while reduced transpiration at elevated CO_2 has the potential to reduce drought stress, most precipitation within continental areas is driven by evapotranspiration. Therefore, CO_2-induced inhibition of transpiration may reduce water stress in that location, but simultaneously reduce precipitation in downwind continental areas. Less evaporation from leaves will also reduce the sensible heat loss, increasing vegetation temperature. The potential implications of such CO_2-induced alterations of transpiration on future climate, land-use patterns, and ecosystems in arid and other regions is profound. Accordingly, the effect of rising CO_2 on vegetation water fluxes are included in the current third-generation global atmospheric general circulation models (AGCM), such as SiB2-GCM (Sellers et al. 1996,1997). The coupled photosynthesis-conductance submodel used in AGCM, based on controlled environment studies, predicts a 14% reduction in leaf conductance in response to a 50% increase in CO_2 (Sellers et al. 1996). However, had this model assumed that plants also downregulate photosynthesis in response to elevated CO_2, the reduction in stomatal conductance would triple to over 40%. Downregulation of photosynthesis to growth in elevated CO_2 has been often observed, especially when plants are grown in small pots or have other limitations to their sink capacity (Arp 1991). This acclimation to elevated CO_2 often involves a reallocation of plant nitrogen away from Rubisco toward RuBP regeneration capacity.

Elevation of CO_2 concentration in greenhouses has shown large increases in yields of the major crops. However, enclosing crops such as wheat in a greenhouse affects growth, production, and appearance. Such plants can respond very differently to environmental treatments than do plants in the field. A new series of experiments being conducted at various locations around the world takes advantage of the natural air movement to enrich crops with CO_2 in the open air in a precisely controlled manner using a new technology: free air carbon dioxide enrichment (FACE). The world's first such system, in Arizona, examined wheat crops over 4 years. Increase of CO_2 concentration to 550 ppm, the level expected for the second half of this century, resulted in a 28% increase in rates of photosynthesis, however, grain yield was only increased by about 10%. One urgent task will be selecting cultivars of crops that are able to fully translate the increased photosynthesis allowed by rising CO_2 into increased yield.

Rising atmospheric CO_2 concentration also has several indirect negative effects on agriculture and will almost certainly alter natural plant communities. Crops grown in elevated CO_2 throughout their life often show decreased nitrogen and protein contents per unit mass, so increased quantity may be gained at the expense of nutritional value. Also, plants differ substantially in their ability to use the additional CO_2, some show large production increases while others show none. In natural ecosystems, this may alter competitive balance and lead to widespread changes in natural communities. Because photosynthesis in C_4 plants is CO_2 saturated in the current atmosphere, this group of plants may lose competitive advantage with C_3 plants where photosynthesis has the potential to be stimulated substantially by increased CO_2. Carbon dioxide is a potent "greenhouse gas," that is, a gas that traps the longwave infrared radiation emitted by earth's surface. This trapping warms the atmosphere. As a result, average global temperatures are expected to rise by at least 2 °C, and probably more, this century. As mentioned earlier, this may allow crop production at higher latitudes than at present, but at the same time is expected to depress yields in warm and tropical climates. It will also alter rainfall patterns, portending increased drought in some areas of the globe.

Ozone

Pollution of the troposphere by ozone is a major problem in many industrialized regions of the world. Ozone, besides causing severe respiratory problems in humans, produces major damage to crops and natural vegetation. Ozone pollution has been linked to large declines in crop productivity and damage to natural vegetation, even at sites distant from large, polluting urban areas. Automobile exhausts and industrial emissions release a family of nitrogen oxide gases (NO_x) and volatile organic compounds (VOC) resulting from the burning of gasoline and coal. Biogenic VOC are also released by vegetation. During periods of high light and high temperatures that frequently occur during late spring, summer, and early fall, NO_x and VOC readily combine with oxygen to form ozone, often surprisingly distant from the urban sites of production of these pollutants. The highest levels of ozone are usually encountered in the heat of the afternoon and early evening, and decrease to undetectable levels during the night as light levels and temperatures drop. Exposure of plants to can ozone result in decreased rates of photosynthesis, reduced root and shoot growth, accelerated senescence, leaf chlorosis and necrosis, and reduced yield.

Ozone can affect plant tissues in three ways; (1) acute stress, (2) chronic stress, and (3) accelerated senescence (Heath 1996). *Acute stress* results from exposure to high concentrations of ozone for short periods of time. Cell and tissue death rapidly occur and produce visible areas of injury. Such damage is so severe that it cannot be prevented or repaired by plant physiological processes. *Chronic stress* occurs when plants are exposed to lower concentrations of ozone for long periods of time. The damage to tissues is considerably less and often does not result in any visual signs. However, cell metabolism is altered and decreases in photosynthesis and productivity occur. *Accelerated senescence* occurs when plants receive very low doses of ozone for extended periods of time. Even at very low concentrations, ozone induces plant tissues to prematurely begin the ageing process (senescence). This appears to be due to genes involved in the control the onset of normal senescence being switched on in response to the cells being affected by ozone. Consequently leaf fall and flowering occur much earlier than they would normally, and productivity declines.

Ozone induces many responses in plants (Bray et al. 2000). Entry of ozone into leaves is by diffusion from the atmosphere through stomata, where it reacts immediately with cells in the substomatal cavities (Fig. 9.4, Heath 1996). This results in ozone initially having a heterogeneous effect on photosynthesis across a leaf. Ozone is highly reactive and will react rapidly with a range of molecules associated with cell walls and membranes, for example, lipids and proteins. Consequently, it is unlikely to move into the cytoplasm of cells in significant amounts and directly react with internal cellular components (Heath 1996). The intercellular concentration of ozone has been estimated at very close to zero in leaves exposed to a very high level of 1500 nmol mol^{-1} (Laisk et al. 1989), which is an order of magnitude greater than ozone levels found in heavily polluted urban atmospheres. The reactions of ozone with cell wall and plasma membrane components produce a range of oxidative products, for example, hydrogen peroxide, hydroxyl radicals, and superoxide anions, which if not detoxified rapidly by antioxidants and enzymes, will diffuse into the cell and react with and damage cell components, especially lipids, proteins and nucleic acids.

Photosynthesis is particularly sensitive to ozone, with decreases in stomatal conductance, CO_2 assimilation, and the Rubisco activity among the earliest responders on exposure of leaves to ozone (Heath1996). Decreases in leaf chlorophyll content will also occur during prolonged exposures to ozone. Decreases in stomatal conductance are associated with ozone-induced modifications of the plasma membranes of guard cells of the stomata that will affect the ionic relations of the cells. Closure of stomata will restrict diffusion of CO_2 into the leaf,

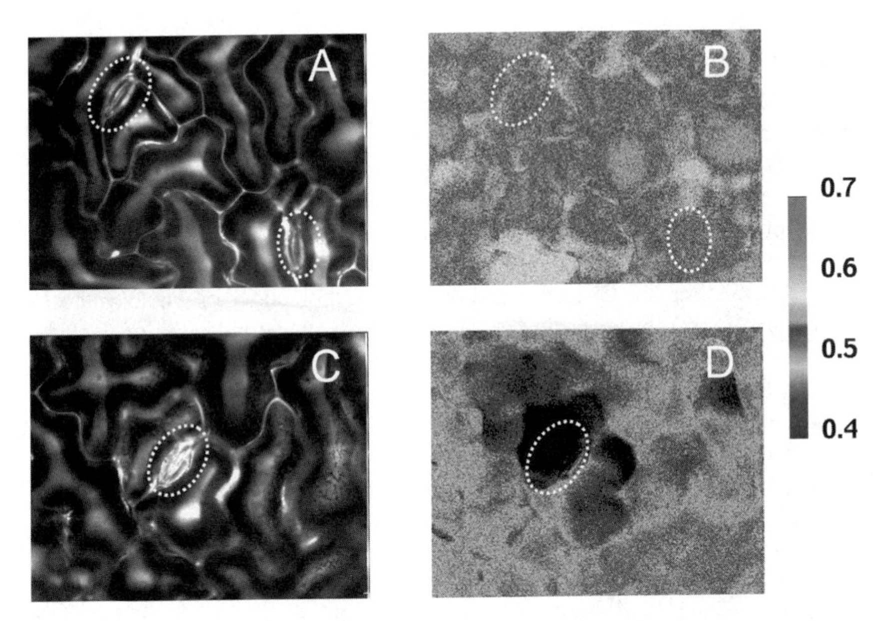

FIGURE 9.4. The effect of ozone on photosynthetic electron transport in a French bean leaf. Images of the quantum efficiency of electron transport in a control leaf (B) and a leaf that had been exposed to 250 nmol mol^{-1} ozone for 3 h (D) demonstrate the severe inhibition of photosynthetic activity induced by ozone in cells around the stomatal pore. The colored bar at the right of the images indicates the relative rate of electron transport. The surface of the leaves are shown in A and C, respectively, for control and ozone-treated leaves; dotted white lines indicate the position of stomata.

which may limit photosynthesis. Similarly, further ozone diffusion into the leaf will also be restricted and this can limit the damage caused by exposure to ozone. Damage to the plasma membranes of mesophyll cells can disrupt the normal ionic balance of the cells, which will inevitably result in perturbation of many metabolic reactions, including photosynthetic metabolism (Heath 1996). Decreases in Rubisco activity are associated with a loss of both the large and small subunits of Rubisco (Heath 1996). The loss of these proteins cannot be attributed to direct effects of ozone on the enzyme since it is highly unlikely that any ozone will reach the chloroplasts. This loss of Rubisco is often considered to be a symptom of premature leaf senescence. Characteristic features of the onset of leaf senescence are a loss of chlorophyll and soluble protein, with Rubisco being the predominant protein lost during the early stages of senescence. It is likely that exposure to ozone will induce changes in cell signaling that result in changes in gene expression, some of which are associated with switching on the degradative events involved in senescence. Ozone-induced modifications of gene expression also result in alterations in a plant's developmental sequence, in some cases large morphological differences are apparent in plants exposed to ozone pollution.

An important consequence of ozone damage to leaves is the increase in the potential for photoinhibition of photosynthesis and photodamage to PSII. As the capacity to dissipate the products of photosynthetic electron transport [adenosine triphosphate (ATP) and reduced nicotinamide-adenine dinucleotide phosphate (NADPH)] decreases due to ozone-induced closure of stomata, perturbation of photosynthetic metabolism, and loss of Rubisco, increases in the rate of thermal dissipation of excitation energy in the light-harvesting antennae and photodamage to PSII reaction centers will increase and result in further decreases in photosynthetic carbon assimilation (see Chapter 3 and previous section above on Photosynthesis in Excessive Light). Increases in the rate of photosynthetic reduction of O_2 may also occur and increase the potential for damage by the reactive oxygen species produced (see Chapter 3). However, in many plants exposure to ozone, like other oxidative stresses, will stimulate synthesis of antioxidant metabolites and increase antioxidant enzyme activities, and consequently reduce the possibility of damage by reactive oxygen species and increase tolerance to ozone (Bray et al. 2000).

Clearly ozone can influence the physiology of plants in many ways and inhibit productivity. Decreases in productivity are difficult to predict accurately because of the complexities of the responses to ozone and the variation in responses depending on plant species and environmental conditions (e.g., temperature, light intensity). However, it is evident that ozone pollution is having, and will continue to have, for the foreseeable future, a considerable impact on crop productivity and natural vegetation.

Ultraviolet-B Radiation

Ozone is the primary UV-B absorbing component in the atmosphere with approximately 90% of atmospheric ozone being in the stratosphere (the upper layer of the atmosphere above the troposphere, which is immediately adjacent to the earth's surface). Increases in the release of pollutants into the stratosphere have resulted in a severe depletion of ozone in this region. The primary culprits for this ozone depletion are man-made chlorine and bromine compounds, such as chlorofluorocarbons (CFCs) and halogens. Industrial activity and traffic have produced large increases in ozone in the troposphere, however this increase has little effect on the overall total ozone concentration in the atmosphere (both troposphere and stratosphere). Tropospheric warming as a result of greenhouse gas emissions has resulted in a cooling of the stratosphere, which has further exacerbated ozone destruction. Decreases of ozone in the stratosphere have resulted in the formation of ozone holes near the poles and a generally thinning of the ozone layer elsewhere, through which increased levels of UV-B radiation reach the earth's surface. UV-B radiation is absorbed strongly by the nucleic acids, proteins, and lipids in cells of biological systems (Jordan 1996) and damages the structural integrity and function of these molecules resulting perturbations of many activities, including photosynthesis.

It has long been recognized that exposure to high doses of UV-B radiation can inhibit photosynthesis in many plants, however there is a wide range of intra-

and interspecific sensitivity differences (Teramura and Ziska 1996). Approximately half of all plant species tested are negatively affected by increases in UV-B radiation, indicating that many species are well adapted to dealing with exposure to high levels of UV-B. In order for UV-B to have any direct or developmental effects on photosynthetic productivity, UV-B must penetrate into the leaf to a depth to be absorbed by chromophores associated with the photosynthetic apparatus or associated genes and gene products. Leaves absorb over 90% of incident UV-B. Leaves contain water-soluble phenolic pigments, such as flavonoids, which strongly absorb UV-B radiation while not absorbing photosyntheticly active radiation. Consequently, in many leaves UV-B, penetration to sites where it might be damaging to the photosynthetic apparatus may be minimal.

When UV-B penetrates into the photosynthetic cells of leaves it is evident that many photosynthetic processes can be affected (Teramura and Zizka 1996, Allen et al. 1998). UV-B has been reported to reduce stomatal conductance, inhibit PSII photochemistry, and decrease Rubisco and sedoheptulose bisphosphatase (SBPase) contents (Allen et al. 1998). While UV-B will induce stomatal closure in many plants, the direct effects of decreases in stomatal conductance are not sufficiently great to account for the large decreases in CO_2 assimilation in the leaves (Allen et al. 1998). The major factor determining the decrease in CO_2 assimilation when leaves are exposed to high levels of UV-B appears to be the loss of Rubisco, SBPase, and possibly other C_3 carbon reduction cycle enzymes (Allen et al. 1998). The process by which UV-B irradiation induces this loss of C_3 carbon reduction cycle enzymes is not well understood. The selective loss of specific enzymes suggests that this mechanism is not a nonspecific destruction of leaf proteins resulting from UV-B absorption by these proteins, or a UV-B-induced production of damaging reactive oxygen species. An alternative mechanism may be the induction by UV-B of proteases specific to the proteins that are degraded. It is possible that UV-B is inducing such a senescence-like response, as has been implicated in plant responses to tropospheric ozone where specific proteins are degraded that are spatially separated from the site of primary oxidative damage (see previous section on Mechanisms of Photodamage).

Numerous investigations have demonstrated that PSII is the most sensitive component of the thylakoid membrane photosynthetic apparatus on exposure to UV-B irradiation (Allen et al. 1998). Consequently, in many reviews PSII damage has often been implicated as the major potential limitation to photosynthesis in UV-B–irradiated leaves, as is the case in the photoinhibition of photosynthesis by excess photosyntheticly active radiation (400 to 700 nm) (see previous section on Photosynthesis in Excessive Light). However, UV-B exposure initially induces decreases in the light-saturated CO_2 assimilation rate in the absence of any major inhibition of the quantum efficiency of PSII photochemistry, demonstrating that no significant photodamage to PSII complexes occurred in parallel with inhibition of CO_2 assimilation and that UV-B inhibition of PSII photochemistry is not a ubiquitous primary limitation to photosynthesis, as is frequently claimed (Allen et al. 1998).

Besides the direct effects of UV-B on leaf photosynthesis discussed above, UV-B can also have indirect effects on leaf photosynthetic performance by modifying developmental processes. Early stages of seedling development appear to be particularly sensitive to UV-B. Plant photosynthetic productivity is a product of leaf area and the photosynthetic rate per unit area. UV-B–induced reductions in leaf area have been observed in many species (Allen et al. 1998). While some studies demonstrated a decline in both leaf area and photosynthetic competence, reductions in leaf area have been observed in the absence of UV-B–induced inhibition of photosynthesis (Allen et al. 1998). The impact of UV-B on leaf area has received much less attention than the effect of UV-B irradiation on photosynthetic competence, despite the importance of both of these components in determining plant productivity.

Ultraviolet-B radiation reduces the total leaf area of plants by reducing the size of leaves, not by reducing leaf number, indicating an inhibition of cell division and/or cell expansion in developing leaves. A primary cause of reduced area of leaves exposed to UV-B is an inhibition of cell division, and in particular inhibition of epidermal cell division since these cells receive a significantly higher incident UV-B flux than the other leaf cells. Leaf growth is also regulated by the rate of expansion of the epidermis and UV-B inhibition of leaf growth may also involve an inhibition of the growth of the adaxial epidermis (Allen et al. 1998). UV-B inhibition of cell expansion has also been observed and could be due to changes in turgor pressure or cell wall extensibility (Allen et al. 1998).

The majority of studies on UV-B inhibition of photosynthesis have involved exposure of leaves in controlled environment cabinets and glasshouses to levels of UV-B that are considerably in excess of both current UV-B fluxes found in natural environments and the predicted increased UV-B levels that will occur in the next 50 years due to depletion in stratospheric ozone. Also the visible light levels used for plant growth in the majority of these studies were considerably below the light levels occurring in natural environments. This is problematical since the sensitivity of plants to a given UV-B exposure is far greater when the plants are grown under low photosyntheticly active photon flux densities (Allen et al. 1998). Consequently, it is difficult to predict from these data with confidence the effects of the future increases in UV-B on the productivity of crops. Field experiments using lamps to produce a constant supplementary UV-B have reported decreases in plant productivity and photosynthesis (Allen et al. 1998). However, the inhibitions observed are generally less pronounced than similar UV-B treatments in glasshouses with lower photosyntheticly active photon flux densities and some studies found no inhibition of photosynthetic productivity (Allen et al. 1998). A limited number of field experiments using modulated UV-B supplementation to ensure a constant percentage increase in UV-B as photosyntheticly active photon flux densities fluctuated throughout the day and seasons have indicated that UV-B enhancements produce no significant changes in photosynthetic parameters (Allen et al. 1998). The weight of evidence emerging from these modulated UV-B field experiments suggests that current predictions for ozone depletion will not adversely affect plant productivity and photosynthesis in crops directly.

Photosynthesis at Limiting Temperatures

Native temperate species respond to low-temperature episodes, or sometimes to secondary environmental signals that are normally the harbinger of cold weather, by enacting a genetically programmed adaptive response. Some species can survive temperatures below −40° C but they cannot tolerate anything close to these extreme temperatures during much of their annual life cycle. There are many aspects involved in cold acclimation that conspire for full development of low-temperature tolerance. For example, sucrose and other simple sugars accumulate during the acquisition of cold tolerance in most species (Anchordoguy et al. 1987). Although accumulation of these sugars is a requisite component of cold acclimation and is believed to be instrumental in protecting membranes during freezing, this alone is not sufficient to confer full tolerance. This fact is presaged by the bewildering complexity of changes in gene expression that accompanies the acquisition of cold and freezing tolerance in species capable of cold acclimation (Thomashow 1999). Although the actual function of most of these cold-induced genes in the development of cold tolerance is unknown, it is known that many cold-regulated genes contain a drought-responsive DNA element (DRE) that interacts with the transcriptional activator CBF1 (Thomashow 1999). Constitutive expression of CBF1 confers considerable cold/freezing tolerance to unacclimated plants confirming its central role the acclimation process (Jaglo-Ottosen et al. 1998, Liu et al. 1998). While native temperate plants respond to repeated exposures to cool temperature by progressively acquiring greater tolerance through intricate if only partially understood acclimation processes, many warm-climate species lack the capacity to acclimate and thus respond to cool temperature excursions by accumulating damage and progressively becoming physiologically unfit. The modest level of genetic variation for chilling tolerance that does exist in these warm climate species appears to reside in their recovery ability rather than in their acclamatory capacity.

Different growing regions represent different challenges for warm-climate, chilling-sensitive plants. Photosynthetic metabolism is among the most chilling-sensitive processes in these plants and the chilling sensitivity of photosynthesis plays a critical role both in limiting the geographical range where these plant species grow as well as contributing to the annual variation in their success particularly for those growing near the northern border of their natural range or agricultural cultivation. Thus it is very clear that the chilling sensitivity of species has far-reaching consequences concerning where it is found or where it can be cultivated. However, it less well appreciated that the underlying mechanism of chilling sensitivity can be quite different depending upon the prevailing climatic conditions of the target growing region. This point is perhaps best illustrated by considering the very different issues concerning low temperature that need to be confronted in growing maize in the cool, wet climates of northern Europe versus the warm, humid continental climate of the Midwestern corn belt of the United States. The agronomic challenge for maize in northern Europe is early growing season temperatures that routinely hover at the lower limit of the permissive range with frequent episodes, particularly at night and early morning periods, when temperatures fall into the chilling sensitivity range. Thus, in this growing environment, issues concerning the effects of low temperature on development and particularly on chloroplast development are centrally important in understanding the mechanism and in devising amelioration strategies (Baker and Nie 1994).

In contrast to cold-tolerant crop species such as winter rye or wheat, maize plants grown at 14 °C develop small, chlorotic leaves with a reduced abundance of photosynthetic components on a leaf area basis and perform photosynthesis less efficiently. This cold exposure phenotype arises from a number of quite striking aberrations during chloroplast development. For example, it has been shown that the accumulation of the chloroplast-encoded thylakoid proteins of the PSI and PSII reaction center complexes, cytochrome f and the ‡ heterodimer of the chloroplast ATPsynthase were preferentially suppressed compared with the nuclear-encoded chloroplast proteins LHCII and LHCI (Nie and Baker 1991). In tomato, a highly chilling-sensitive species, even a single cool night (i.e., 4 °C for 16 h in the dark) strongly suppresses the synthesis of chloroplast-encoded proteins relative to those that are synthesized in the cytoplasm and imported into the chloroplast. The observed reduction in accumulation of these chloroplast proteins and the associated anomalies in thylakoid membrane deployment undoubtedly contribute directly to the decreased photosynthetic efficiency and capacity in maize grown at cool temperatures (Nie et al. 1995).

The agricultural issues pertinent to low-temperature sensitivity of maize are distinctly different in the Midwestern corn belt of the United States than for the cool, wet climates of northern Europe. In the major corn growing regions of the Midwest, air temperatures are for the vast majority of the growing season solidly within the permissive range. Chilling episodes are infrequent, are confined almost exclusively to quite early or very late in growing season, and occur as transient low-temperature excursions followed by rapid return to permissive warm temperatures. Thus, in contrast to the northern Europe condition, low-temperature-induced chloroplast developmental changes have little if any role in explaining the chilling-induced dysfunctions in photosynthetic metabolism that are relevant to the U.S. corn belt. It should be evident that the experimental systems and treatments that are appropriate to studying these two circumstances are also quite different. It follows that while there may be some overlapping aspects between these two chilling environments, the underlying inhibitory mechanisms and therefore the suite of genes and gene products that are involved are likely to be quite different. A survey of recent research supports this view even though these different outcomes have sometimes led to controversy when the importance of differences in the chilling environment failed to be recognized.

In addition to the climatic circumstances under which a chilling-sensitive plant encounters low-temperature exposure, another factor that has a strong effect on the underlying inhibitory mechanism is whether or not the low-temperature exposure coincides with high light exposure. The combination of cool temperatures and high light inhibit photosynthesis by various photooxidative mechanisms. Most warm-climate plant species are sensitive to brief exposures to low, nonfreezing temperatures. Low-temperature exposure in combination with even moderate irradiance levels causes rapid, often very severe, inhibition of photosynthesis in a broad range of warm-climate plants, including maize (Baker et al. 1983), cucumber (Peeler and Naylor 1988), and tomato (Martin and Ort 1985). Numerous potentially contributing elements to the inhibition have been identified and all may ultimately arise from the photosynthetic production of active oxygen species (Wise 1995). While the reduction of oxygen is a highly effective strategy to sustain electron flow and thereby avoid dangerous pigment states and overreduction of NADP, it nevertheless carries the hazards inherent in formation of reduced oxygen and hydrogen peroxide (Asada 1996). Superoxide radicals generated by the one-electron reduction of molecular oxygen by PSI/ferredoxin are rapidly converted within the chloroplast to hydrogen peroxide by CuZn-superoxide dismutase. Whereas hydrogen peroxide associated with the photorespiratory pathway is generated and detoxified by catalase in peroxisomes (Ogren 1994), detoxification of hydrogen peroxide produced in the chloroplast relies almost exclusively on the activity of ascorbate peroxidase bound to the thylakoid membranes and localized in the vicinity of PSI (Miyake and Asada 1992). Because monodehydroascorbate reductase catalyzes the regeneration of ascorbate in the chloroplast at the expense of NAD(P)H, monodehydroascorbate reduction must be stoichiometric with oxygen reduction as a photosynthetic electron sink in order to ensure sustained capacity to detoxify reactive oxygen species. The triplet ground state of molecular oxygen allows it to participate in unusual chemistry that generates an additional, potentially hazardous, form of reactive oxygen within chloroplasts. The triplet excited state of the primary donor of PSII, $^3P_{680}^*$, can form during charge recombination reactions within the PSII reaction center (Vass and Styring 1993). Although reaction center carotenoids generally efficiently quench chlorophyll triplets, when the rate of triplet formation is increased by reduced flux of electrons from the reaction center, molecular oxygen can interact with the excited triplet state of the pigment generating strongly oxidizing and highly reactive singlet oxygen, $^1O_2$.

Under most circumstances, the aggregate effect of these protective and detoxification processes coupled with considerable capacity for repair, effectively prevents the inhibition of photosynthesis by excessive irradiance. However, there is plentiful evidence that illumination of chilling-sensitive plants at low temperature causes an enhanced production of photosyntheticly and photochemically generated reactive oxygen species that can outpace the protective and detoxification process of the chloroplast (e.g., Wise 1995, Asada 1996) thereby expending a substantial portion of the chloroplast's reductant reserves causing a lowering of the redox poise of the chloroplast (Hutchison et al. 2000). There is convincing evidence that the lowered redox poise of the chloroplast interferes with the reductive activation of certain reductively activated enzymes of the carbon reduction cycle (Sassenwrath et al 1990, Hutchison et al. 2000) which in turn causes a significant and persistent inhibition of photosynthesis. Low-temperature episodes disrupt many cellular processes in chilling-sensitive species from nutrient transport to photosynthesis (e.g., Martin et al. 1981, Monroy et al. 1997,

Ort et al. 1989). Much work has focused on the effects of low temperature on carbon assimilation, particularly exposure to chilling under high light conditions as discussed above and reviewed more extensively in Allen and Ort (2001). In addition to decreased thioredoxin-dependent reductive activation of stromal bisphosphatases (Sassenrath et al. 1987, 1990, Hutchinson et al. 2000) direct damage to the photosynthetic apparatus by enhanced formation of oxygen radicals and damage to the oxidative side of PSII are also well documented in many thermophilic species (e.g., Martin et al. 1981, Wise 1995). Thus, credible mechanistic links have been established between high light chill damage and dysfunctions in photosynthetic metabolism.

The mechanistic connection between low temperature and photosynthetic metabolism is more tenuous in circumstances in which chilling-sensitive species encounter chilling temperatures at night (i.e., in the dark). Nevertheless, the deleterious effects that a cool night (<10 °C) can have on whole-plant photosynthesis the following day are substantial and well documented in a wide range of warm climate species. *Phaseolus vulgaris, Sorghum vulgare, Oryza sativa, Medicabo sativa, Digitaria decumbens* and *Lycopersicon esculentum* all exhibit a substantial depression of photosynthesis following a night chill exposure (Chatterton et al. 1972, Crookston et al. 1974, Izhar and Wallace 1967, Kishitani and Tsunoda 1974, Pasternak and Wilson 1972, Peoples and Koch 1978, Martin et al. 1981). For example, in tomato a 16-h dark chill (4 °C) causes a $\sim$60% decline in photosynthesis the next day after permissive temperatures are restored. This substantial inhibition is only in small part accounted for by changes in stomatal conductance that accompany such conditions (Martin et al. 1981). Rates of thylakoid membrane electron transfer, although depressed after the chill, are ample to support control rates of photosynthetic carbon fixation and thus cannot account for the decline in net photosynthesis (Kee et al. 1986). Nor are there changes in the affinity of Rubisco for CO_2 under chilling conditions that would explain the chill-induced dysfunction seen at the whole-plant level. In fact, direct chill-induced damage to the photosynthetic apparatus does not appear to be a major contributor in these chill-induced effects on photosynthetic performance.

One of the more promising clues concerning potential factors underlying the altered photosynthetic metabolism due to low temperatures at night was the discovery that endogenous circadian rhythms in chilling-sensitive species are impacted by exposure to dark chills (Martino-Catt and Ort 1992). Catalase, PSII chlorophyll a/b binding protein (cab), phosphoenolpyruvate carboxylase (PEPcase), and Rubisco activase (rca) all exhibit endogenous rhythms in transcript levels (Carter et al. 1991, Martino-Catt and Ort 1992, Zhong et al. 1994). Rhythms in many cellular processes are maintained even when plants are held under constant (free-running) conditions, with a period of approximately 24 h. These circadian rhythms, by definition, have temperature compensation to maintain the same cycle period over a range of temperatures. However, circadian rhythms in cab and rca mRNA expression in tomato and other warm climate plants that were investigated are stalled by chilling in the dark. Although these enzymes are so abundant in photosynthetic cells of mature leaves that transient changes in de novo synthesis do not have a significant impact on activity, this work raised the intriguing notion that some of the depression of photosynthesis in thermophilic plants following a chill is the result of the mistiming of multiple circadian-regulated processes. In the photosynthetic cells of a leaf, as with all cells, there are a wide variety of reactions and pathways that must be separated from one another to avoid futile cycles and other types of biochemical interference. We normally think about the compartmentalization that achieves the requisite separation in physical terms involving organelles, substrate channelling within enzyme supercomplexes, and other forms of physical separation. However, in principle, compartmentation could be achieved temporally and circadian control of gene transcription and enzyme regulation would be expected to be the central controlling element of this mechanism for separating potentially competing reactions and metabolic processes. As was the case with the endogenous rhythms of cab and rca gene transcription, changes in the rhythm of sucrose phosphate synthase (SPS) activity in tomato are stalled during a 12-h dark chill resuming upon rewarming, resulting in oscillations that are 12 h out of phase with the actual diurnal time. The low-temperature induced shift in SPS activity rhythm appears in turn to be driven by low-temperature interference with the circadian transcriptional regulation of a type 2A protein phosphatase (Jones and Ort 1997, Jones et al. 1998). This mistiming of SPS appears to be detrimental to a plant by causing feedback inhibition of photosynthesis due to a lack of available inorganic phosphate in the stroma (Sharkey 1990, Laporte and Sharkey 1995). However, chilling does not delay all circadian rhythms in chill-sensitive species as illustrated by the lack of effect on chilling on the endogenous oscillation in mango leaf stomatal conductance (Allen et al. 2000) and the different mode of disruption of chilling on nitrate reductase regulation in tomato (Tucker and Ort 2002).

Summary

Human activity has immensely impacted over a very short time span the environmental conditions that confront plants. Through the burning of fossil fuels and tropical forests we have caused a 50% increase in atmospheric CO_2, polluted the troposphere with ozone, diluted the globe's atmospheric UV-B screen, and set in motion a greenhouse warming event that will dramatically impact global climate and rainfall patterns. Our agricultural and land-use practices have resulted in widespread water depletion, the salinization of soils, and the introduction of plant species far outside their evolutionary habitat. Largely through photosynthesis, crop and natural plant communities have been substantially affected by these recent anthropogenic changes to the environment and at the same time are themselves impacting the change. In many cases, the basis for understanding response of plants to their new environment is to be found in the response of chloroplast function within a leaf. Plant responses to the stresses are difficult to predict, due to the large variety and species specificity of adaptive mechanisms optimizing and defending photosynthesis in plants. In mesophyllous plants growing in agricultural and forest areas of temperate regions, different components of the photosynthetic apparatus may be impaired, depending on the stress. Photosynthesis is primarily limited by diffusive resistances in plants exposed to drought and salinity stress, although biochemical limitation may contribute. Stomata also help mediate plant tolerance to ozone and the response to rising CO_2 concentration. Photosystem II emerges as highly sensitive to a wide variety of nonfavorable environments because these varied stress conditions have in common the photochemical generation of reactive compounds that target PSII function. There is diversity not only in the genetically based physiological capacity of different plants to respond to different stresses, but there is equally important diversity in the combinations of stresses that plants encounter even when grown in very close proximity.

References

Allen, D. J., and Ort, D. R. 2001. Impacts of chilling temperatures on photosynthesis in warm-climate plants. Trends Plant Sci. 6:36–42.

Allen, D. J., Nogués, S., and Baker, N. R. 1998. Ozone depletion and increased UV-B radiation: Is there a real threat to photosynthesis? J. Exp. Bot. 328:1775–1788.

Allen, D. J., Ratner, K., Giller, Y. E., Gussakovsky, E. E., Shahak, Y., and Ort, D. R. 2000. An overnight chill induces a delayed inhibition of photosynthesis at midday in mango (*Mangifera indica* L.). J. Exp. Bot. 51:1893–1902.

Anchordoguy, T. J., Rudolph, A. S., Carpenter, J. F., and Crowe, J. H. 1987. Modes of interaction of cryoprotectants with membrane phospholipids during freezing. Cryobiology. 24: 324–331.

Arp, W. J. 1991. Effects of source-sink relations on photosynthetic acclimation to elevated CO_2. Plant Cell Environ. 14:869–75.

Asada, K. 1996. Production and scavenging of radicals. In: Advances in Photosynthesis: Photosynthesis and the Environment, Vol. 5. N. R. Baker (ed.), pp. 123–150. Dordrecht, The Netherlands: Kluwer Academic Publishers.

Assmann, S. M., Snyder, J. A., and Lee, Y. R. I. 2000. ABA-deficient (aba1) and ABA-insensitive (abi1-1, abi2-1) mutants of Arabidopsis have a wild-type stomatal response to humidity. Plant Cell Environ. 23:387–395.

Baker, N. R., and Nie, G. Y. 1994. Chilling sensitivity of photosynthesis in maize. In: Biotechnology of Maize. Y. P. S. Bajaj (ed.), pp. 465–481. Berlin: Springer-Verlag.

Baker, N. R., and Ort, D. R. 1992. Light and crop photosynthetic performance. In: Crop Photosynthesis: Spatial and Temporal Determinants, N. R. Baker and H. Thomas (eds.), pp. 289–312. Amsterdam: Elsevier Science Publishers.

Baker, N. R., East, T. M., and Long, S. P. 1983. Chilling damage to photosynthesis in young *Zea mays*. II. Photochemical function of thylakoids in vivo. J. Exp. Bot. 34:189–197.

Blumwald, E., Aharon, G. S., and Apse, M. P. 2000. Sodium transport in plant cells. Biochim. Biophys. Acta 1465:140–151.

Bray, E. A., Bailey-Serres, J., and Weretilnyk, R. 2000. Responses to abiotic stresses. In: Biochemistry and Molecular Biology of Plants. B. Buchanan, W. Gruissem, and R. Jones (eds.), Rockville: American Society of Plant Physiologists.

Bohnert, H. J., Nelson, D. E., and Jensen, R. G. 1995. Adaptations to environmental stresses. Plant Cell 7:1099–1111.

Carter, P. J., Nimmo, H. G., Fewson, C. A., and Wilkins, M. B. 1991. Circadian rhythms in the activity of a plant protein kinase. EMBO J. 10:2063–2068.

Chatterton, N. J., Carlson, G. E., Hungerford, W. E., and Lee, D. R. 1972. Effect of tillering and cool nights on photosynthesis and chloroplast starch in *Pangola*. Crop Sci 12:206–208.

Cheeseman, J. M. 1991. PATCHY: Simulating and visualizing the effects of stomatal patchiness on photosynthetic CO_2 exchange studies. Plant Cell Environ. 14:593–599.

Cornic, G. 2000. Drought stress inhibits photosynthesis by decreasing stomatal aperture—not by affecting ATP synthesis. Trends Plant Sci. 5:187–188.

Cornic, G., and Briantais, J.-M. 1991. Partitioning of photosynthetic electron flow between CO_2 and O_2 in a C_3 leaf (*Phaseolus vulgaris* L.) at different CO_2 concentrations and during drought stress. Planta 183:178–184.

Cornic, G., and Massacci, A. 1996. Leaf photosynthesis under drought stress. In: Photosynthesis and the Environment. N. Baker (ed.), pp. 347–366. Dordrecht, The Netherlands: Kluwer Academic Publishers.

Cornic, G., Le Gouallec, J. L., Briantais, J.-M., and Hodges, M. 1989. Effect of dehydration and high light on photosynthesis of two C3 plants [*Phaseolus vulgaris* L. and *Elatostema repens* (Lour.)]. Hall. Planta 177:84–90.

Crookston, R. K., O'Toole, J., Lee, R., Ozbun, J. L., and Wallace, D. H. 1974. Photosynthetic depression in beans after exposure to cold for one night. Crop Sci. 14:457–464.

Daley, P. F., Raschke, K., Ball, J. T., and Berry, J. A. 1989. Topography of photosynthetic activity of leaves obtained from video images of chlorophyll fluorescence. Plant Physiol. 90:1233–1238.

Davies, W. J., and Zhang, J. 1991. Root signals and the regulation of growth and development of plants in drying soil. Annu. Rev. Plant Physiol. 42:55–76.

Delfine, S., Alvino, A., Zacchini, M., and Loreto, F. 1998. Consequences of salt stress on conductance to CO2 diffusion, Rubisco characteristics and anatomy of spinach leaves. Aust. J. Plant Physiol. 25:395–402.

Delfine, S., Alvino, A., Villani, M. C., and Loreto, F. 1999. Restrictions to CO2 conductance and photosynthesis in spinach leaves recovering from salt stress. Plant Physiol. 119:1101–1106.

Delfine, S., Loreto, F., and Alvino, A. 2001. Drought-stress effects on physiology, growth and biomass production of rainfed and irrigated bell pepper plants in the Mediterranean region. J. Am. Soc. Hort. Sci. 126:297–304.

Downton, W. J. S., Loveys, B. R., and Grant, W. J. R. 1988. Stomatal closure fully accounts for the inhibition of photosynthesis by abscisic acid. New Phytologist 108: 263–266.

Evans, J. R., and Loreto, F. 1999. Acquisition and diffusion of CO_2 in higher plant leaves. In: Photosynthesis: Physiology and Metabolism. R. C. Leegood, T. D. Sharkey, and S. von Caemmerer (eds.), pp. 322–351. Dordrecht, The Netherlands: Kluwer Academic Publishers.

Farquhar, G. D., and Sharkey, T. D. 1982. Stomatal conductance and photosynthesis. Annu. Rev. Plant Physiol. 33:317–345.

Farquhar, G. D., Hubick, K. T., Terashima, I., Condon, A. G., and Richards, R.A. 1987. Genetic variation in the relationship between photosynthetic CO_2 assimilation and stomatal conductance to water loss. In: Progress in Photosynthetic Research, Vol. IV. J. Biggins (ed.), pp. 209–212. Dordecht, The Netherlands: Martinus Nijhoff Publishers.

Heath, R. L. 1996. The modification of photosynthetic capacity induced by ozone exposure. In: Photosynthesis and the Environment. N. R. Baker (ed.), pp 469–476. Dordrecht, The Netherlands: Kluwer Academic Publishers.

Hutchison, R. S., Groom, Q., and Ort, D. R. 2000. Differential effects of chilling-induced photooxidation on the redox regulation of photosynthetic enzymes. Biochemistry 39:6679–6688.

Kaiser, W. M. 1987. Effect of water deficit on photosynthetic capacity. Physiol. Plant. 71:142–149.

Kee, S. C., Martin, B., and Ort, D. R. 1986. The effects of chilling in the dark and in the light on photosynthesis of tomato: Electron transfer reactions. Photosynth. Res. 8:41–51.

Kishitani, S., and Tsunoda, S. 1974. Effect of low and high temperature pretreatment on leaf photosynthesis and transpiration in cultivars of *Oryza sativa*. Photosynthetica 8:161–167.

Ingram, J., and Bartels, D. 1996. The molecular basis of dehydration tolerance in plants. Annu. Rev. Plant Physiol. Plant Mol. Biol. 47:377–403.

Izhar, S., and Wallace, D. H. 1967. Effect of night temperature on photosynthesis of *Phaseolus vulgaris* L. Crop Sci. 7:546–547.

Jaglo-Ottosen, K. R., Gilmour, S. J., Zarka, D. G., Schabengerger, O., and Thomashow, M. F. 1998. Arabidopsis CBF1 overexpression induces COR genes and enhances freezing tolerance. Science 280:104–106.

Jones, T. L., and Ort, D. R. 1997. Circadian regulation of sucrose phosphate synthase activity in tomato by protein phosphatase activity. Plant Physiol. 113:1167–1175.

Jones, T. L., Tucker, D. E., and Ort, D. R. 1998. Chilling delays circadian pattern of sucrose phosphate synthase and nitrate reductase activity in tomato. Plant Physiol. 118:149–158.

Jordan, B. R. 1996. The effects of UV-B on plants: A molecular perspective. Adv. Bot. Res. 22:97–162.

Laisk, A., and Loreto, F. 1996. Determining photosynthetic parameters from leaf CO_2 exchange and chlorophyll fluorescence: Rubisco specificity factor, dark respiration in the light, excitation distribution between photosystems, alternative electron transport rate and mesophyll diffusion resistance. Plant Physiol. 110:903–912.

Laisk, A., Oja, V., and Kull, O. 1980. Statistical distribution of stomatal apertures of *Vicia faba* and *Hordeum vulgare* and the Spannungs-phase of stomatal opening. J. Exp. Bot. 31:49–58.

Laisk, A., Kull, O., and Moldau, H. 1989. Ozone concentration in leaf intercellular air spaces is close to zero. Plant Physiol. 90:1163–1167.

Lauer, M. J., and Boyer, J. S. 1992. Internal CO_2 measured directly in leaves. Abscisic acid and low leaf water potential cause opposing effects. Plant Physiol. 98:1310–1316.

Lauteri, M., Scartazza, A., Guido, M. C., and Brugnoli, E. 1997. Genetic variation in photosynthetic capacity, carbon isotope discrimination and mesophyll conductance in provenances of *Castanea sativa* adapted to different environments. Funct. Ecol. 11:675–683.

Laporte, M. M., and Sharkey, T. D. 1995. Effects of temperature on photosynthesis, partitioning, and growth in SPS transformed tomato plants. Plant Physiol. 108S:56.

Lawlor, D. W., and Cornic, G. 2002. Photosynthetic carbon assimilation and associated metabolism in relation to water deficits in higher plants. Plant Cell Environ. 25:275–294.

Liu, Q., Kasuga, M., Sakuma, Y., Abe, H., Miura, S., Yaaguchi-Shinozaki, K., and Shinozaki, K. 1998. Two transcriptional factors, DREB1 and DREB2, with an EREBP/AP2 DNA binding domain separate two cellular signal transduction pathways in drought- and low temperature-responsive gene expression, respectively, in Arabidopsis. Plant Cell. 10:1391–1406.

Long, S. P., Humphries, S., and Falkowski, P. G. 1994. Photoinhibition of photosynthesis in nature. Annu. Rev. Plant Physiol. Plant Mol. Biol. 45:633–662.

Loreto, F., and Sharkey, T. D. 1990. Low humidity can cause uneven photosynthesis in olive (*Olea europea* L.) leaves. Tree Physiol. 6:409–415.

Loreto, F., Harley, P. C., Di Marco, G., and Sharkey, T. D. 1992. Estimation of the mesophyll conductance to CO_2 flux by three different methods. Plant Physiol. 98:1437–1443.

Loreto, F., Di Marco, G., Tricoli, D., and Sharkey, T. D. 1994. Measurements of mesophyll conductance, photosynthetic electron transport and alternative electron sinks of field grown wheat leaves. Photosynth. Res. 41:397–403.

Loreto, F., Tricoli, D., and Di Marco, G. 1995. On the relationship between electron transport rate and photosynthesis in leaves of the C_4 plant *Sorghum bicolor* (L.) Moench exposed to water stress, temperature changes and carbon metabolism inhibition. Aust. J. Plant Physiol. 22:885–892.

Martin, B., and Ort, D. R. 1985. The recovery of photosynthesis subsequent to chilling exposure. Photosynth. Res. 6:121–132.

Martin, B, Ort, D. R., and Boyer, J. S. 1981. Impairment of photosynthesis by chilling-temperatures in tomato. Plant Physiol. 68:329–334.

Martino-Catt, S., and Ort, D. R. 1992. Low temperature interrupts circadian regulation of transcriptional activity in chilling-sensitive plants. Proc. Natl. Acad. Sci. USA 89:3731–3735.

Melis, A. 1999. Photosystem-II damage and repair cycle in chloroplasts: What modulates the rate of photodamage in vivo? Trends Plant Sci. 4:130–135.

Meyer, S., and Genty, B. 1998. Mapping intercellular CO_2 mole fraction (c_i) in Rosa rubiginosa leaves fed with abscissic acid by using chlorophyll fluorescence imaging. Plant Physiol. 116:947–957.

Miyake, C., and Asada, K. 1992. Thylakoid-bound ascorbate peroxidase in spinach chloroplasts and photoreduction of its primary oxidation product monodehydroascorbaste radicals in thylakoids. Plant Cell Physiol. 33:541–553.

Miyazawa, S. I., and Terashima, I. 2001. Slow chloroplast development in the evergreen broad-leaved tree species: Relationship between leaf anatomical characteristics and photosynthetic rate during leaf development. Plant Cell Environ. 24:279–291.

Monroy, A. F., Labbe, E., and Dhindsa, R. S. 1997. Low temperature perception in plants: Effects of cold on protein phosphoylation in cell-free extracts. FEBS Lett. 410:206–209.

Munns, R. 1993. Physiological processes limiting plant growth in saline soils: Some dogmas and hypotheses. Plant Cell Environ. 16:15–24.

Munns, R., Schachtman, D. P., and Condon, A. G. 1995. The significance of the two-phase growth response to salinity in wheat and barley. Aust. J. Plant Physiol. 22:561–569.

Neumann, P. 1997. Salinity resistance and plant growth revisited. Plant Cell Environ. 20: 1193–1198.

Nie, G.-Y., and Baker, N. R. 1991. Modifications to thylakoid composition during development of maize leaves at low growth temperatures. Plant Physiol. 95:184–191.

Nie, G.-Y., Robertson, E. J., Fryer, M. J., Leech, R. M., and Baker, N. R. 1995. Response of the photosynthetic apparatus in maize leaves grown at low temperature on transfer to normal growth temperature. Plant Cell Environ. 18:1–12.

Nobel, P. S. 1991. Physicochemical and environmental plant physiology. San Diego: Academic Press.

Nonami, H., and Boyer, J. S. 1990. Primary events regulating stem growth at low water potentials. Plant Physiol. 93:1601–1609.

Ogren, W. L. 1994. Energy utilization by photorespiration. In: Regulation of atmospheric

CO_2 and O_2 by photosynthetic carbon metabolism. N. E. Tolbert, and J. Preiss (eds.), pp. 115–125. Oxford: Oxford University Press.

Ort, D. R., and Baker, N. R. 2002. Photoprotection: The role of electron sinks. Curr. Opin. Plant Biol. 5:193–198.

Ort., D. R., Martino, S., Wise, R. R., Kent, J., and Cooper, P. 1989. Changes in protein synthesis induced by chilling and their influence on the chilling sensitivity of photosynthesis. Plant Physiol. Biochem. 27:785–793.

Pasternak, D., and Wilson, G. L. 1972. After-effects of night temperatures on stomatal behavior and photosynthesis of Sorghum. New Phytol. 71:683–689.

Pearcy, R. W. 1990. Sunflecks and photosynthesis in plant canopies. Annu. Rev. Plant Physiol. Plant Mol. Biol. 41:421–453.

Peeler, T. C., and Naylor, A. W. 1988. A comparison of the effects of chilling on thylakoid electron transfer in pea (*Pisum sativum* L.) and cucumber (*Cucumis sativus* L.). Plant Physiol..86:147–151.

Peoples, T. R., and Koch, D. W. 1978. Physiological response of three alfalfa cultivars to one chilling night. Crop Sci. 18:255–258.

Powles, S. B. 1984. Photoinhibition of photosynthesis induced by visible light. Annu. Rev. Plant Physiol. 35:15–44.

Preston, G. M., Carroll, T. P., Guggino, W. B., and Agre, P. 1992. Appearance of water channels in Xenopus oocytes expressing red cell CHIP28 protein. Science 256:385–387.

Saccardy, K., Pineau, B., Roche, O., and Cornic, G. 1998. Photochemical efficiency of photosystem II and xanthophylls cycle components in *Zea mays* leaves exposed to water stress and high light. Photosyn. Res. 56:57–66.

Sassenrath, G. F., Ort, D. R., and Portis, A.R., Jr. 1987. Effect of chilling on the activity of enzymes of the photosynthetic carbon reduction cycle. In: Progress in Photosynthesis Research, Vol. 4, J. Biggins (ed.), pp. 103–106. Dordrecht, The Netherlands: Martinus Nijihoff.

Sassenrath, G. F., Ort, D. R., and Portis, A. R., Jr. 1990. Impaired reductive activation of stromal bisphosphatases in tomato leaves following low-temperature exposure at high light. Arch. Biochem. Biophys. 282:302–308.

Schulze, E. D., Turner, N. C., Gollan, T., and Schakel, K. A. 1987. Stomatal responses to air humidity and to soil drought. In: Stomatal Function. E. Zeiger, G. D. Farquhar, and I. R. Cowan (eds.), pp. 311–321. Stanford University Press, Stanford, CA.

Sellers, P. J., Bounoua, L., Collatz, G. J., Randall, D. A., Dazlich, D. A., Los, S. O., Berry, J. A., Fung, I., Tucker, C. J., Field, C. B., and Jensen, T. G. 1996. Comparison of radiative and physiological effects of doubled atmospheric CO_2 on climate. Science 271:1402–1406.

Sellers, P. J., Dickinson, R. E., Randall, D. A., Betts, A. K., Hall, F. G., Berry, J. A., Collatz, G. J., Denning, A. S., Mooney, H. A., Nobre, C. A., Sato, N., Field, C. B., and Henderson-Sellers, A. 1997. Modeling the exchanges of energy, water, and carbon between continents and the atmosphere. Science 275:502–509.

Sharkey, T. D. 1990. Feedback limitation of photosynthesis and the physiological role of ribulose bisphosphate carboxylase carbamylation. Bot. Mag. Tokoyo (special issue 2) 87–105.

Smart, L. B., Moskal, W. A., Cameron, K. D., and Bennett, A. B. 2001. MIP genes are down-regulated under drought stress in *Nicotiana glauca*. Plant Cell Physiol. 42:686–693.

Smirnoff, N., Windslow, M. D., and Stewart, G. R. 1985. Nitrate reductase activity in leaves of barley (*Hordeum vulgare*) and durum wheat (*Triticum durum*) during field and rapidly applied water deficits. J. Exp. Bot. 36:1200–1208.

Syvertsen, J. P., Lloyd, J., McConchie, C., Kriedemann, P. E., and Farquhar, G. D. 1995. On the relationships between leaf anatomy and CO_2 diffusion through the mesophyll of hypostomatous leaves. Plant Cell Environ. 18:149–157.

Szabolcs, I. 1994. Soils and salinization. In: Handbook of plant and crop stress. M. Pessarakli (ed.), pp. 3–11. New York: Marcel Dekker.

Teramura, A. H., and Ziska, L. H. 1996. Ultraviolet-B radiation and photosynthesis. In: Photosynthesis and the Environment. N. R. Baker (ed.), pp. 435–450. Dordrecht, The Netherlands: Kluwer Academic Publishers.

Terashima, I., Wong, S. C., Osmond, C. B., and Farquhar, G. D. 1988. Characterization of non-uniform photosynthesis induced by abscissic acid in leaves having different mesophyll anatomies. Plant Cell Physiol. 29:385–395.

Terashima, I., and Ono, K. 2002. Effects of $HgCl_2$ on CO_2 dependence of leaf photosynthesis: Evidence indicating involvement of aquaporins in CO_2 diffusion across the plasma membrane. Plant Cell Physiol. 43:70–78.

Tezara, W., Mitchell, V. J., Driscoll, S. P., and Lawlor, D. W. 1999. Water stress inhibits plant photosynthesis by decreasing coupling factor and ATP. Nature 401:914–917.

Thomashow, M. F. 1999. Plant cold acclimation: Freezing tolerance genes and regulatory mechanisms. Annu. Rev. Plant Physiol. Mol. Biol. 50:571–599.

Tucker, D. E., and Ort, D. R. 2002. Low temperature induces expression of nitrate reductase in tomato that temporarily overrides circadian regulation of activity. Photosynth. Res. 72: 285–293.

Uehlein, N., Lovisolo, C., Siefritz, F., and Kaldenhoff, R. 2003. The tobacco aquaporin NtAQP1 is a membrane CO_2 transporter with physiological functions. Nature 425: 734–737.

Valladares, F., and Pearcy, R. W. 2002. Drought can be more critical in the shade than in the sun: A field study of carbon gain and photoinhibition in a Californian shrub during a dry El Nino year. Plant Cell Environ. 25:749–759.

Vass, I., and Styring, S. 1993. Characterization of chlorophyll triplet promoting states in photosystem II sequentially induced during photoinhibition. Biochemistry. 32:3334–3341.

Vassey, T. L., and Sharkey, T. D. 1989. Mild water stress of *Phaseolus vulgaris* plants leads to reduced starch synthesis and extractable sucrose phosphate synthase activity. Plant Physiol. 89:1066–1070.

Wise, R. R., Ortiz-Lopez, A., and Ort, D. R. 1992. Spatial distribution of photosynthesis during drought in field-grown and acclimated and nonacclimated growth chamber-grown cotton. Plant Physiol. 100:26–32.

Wise, R. R. 1995. Chilling-enhanced photooxidation: The production, action and study of reactive oxygen species produced during chilling in the light. Photosynth. Res. 45: 79–97.

Zhong, H. H., Young, J. C., Pease, E. A., Hangarter, R. P., and McClung, C. R. 1994. Interactions between light and the circadian clock in the regulation of CAT2 expression in *Arabidopsis*. Plant Physiol. 104:889–898.

10
Leaf to Landscape

Stanley D. Smith, Elke Naumburg, Ülo Niinemets,
and Matthew J. Germino

Introduction

One fundamental "problem" for maximizing carbon gain at the leaf and higher organizational levels entails the link between light capture and leaf energy budgets. The balance between the two processes, however, depends on the environment. For example, shade environments limit carbon gain due to low light levels, and so we would expect plants to display traits that maximize light interception and traits that facilitate survival in a low-resource environment (relative to photosynthesis). These topics have received wide attention and have been reviewed numerous times (for the most relevant contributions in the context of this book, see Ackerly 1996, Poorter 1999, Poorter and Werger 1999, Valladares 1999, Walters and Reich 1999; see also Chapter 2). Conversely, in high-light environments, leaf energy balance becomes the driving force for structural traits. This is particularly true for environments that are at the same time limited by water availability. Similarly, although in the opposite direction, leaf energy balance is a major factor in carbon gain in cold environments. A third significant environmental axis is nutrient availability which, when limiting, can result in evergreen, often sclerophyllous, leaves which have potentially low photosynthetic capacity but high efficiency of resource utilization.

Dry Environments

General Adaptations

Water-limited environments impose multiple limitations on the ability of plants to gain carbon. Although water stress per se has often been viewed as the primary determinant of plant structural features in water-limited environments, there is ample evidence that structural and physiological adaptations of leaves and stems are for maximizing carbon gain and regulating energy budgets (Gibson 1998). Furthermore, a lack of water, high light, and high temperature all interact to increase plant stress in a nonadditive way. Nevertheless, plant carbon balance is as important a selective factor for plant distributions across mesic-to-xeric gradients as is physiological stress tolerance (Smith et al. 1997). Physiological water stress is well known to affect photosynthesis and plant growth at the cellular level (Tardieu and Davies 1993), but it also has profound effects on the structure of plant canopies. As discussed above, photosynthetic canopies tend to be optimized to capture light (Küppers 1989), but in the absence of water, light can fundamentally change from a resource to a stress factor. As a result, canopies are structured based on tradeoffs between light capture for photosynthesis, energy balance, and drought-induced photoinhibition. It is therefore not surprising that leaf energy balance and avoidance of photoinhibition may become the most important selective factors in water-limited environments.

As in any environment, the presumed evolutionary "goal" of the plant canopy is to maximize carbon gain per unit investment in photosynthetic surface area. When exposed to water limitation, canopy morphology may respond in several ways. In general, water-limited environments also tend to be high-light environments, so selection for light capture may be less pronounced as in more mesic environments. High-light, water-limited environments also tend to have higher temperatures and lower relative humidities. For a plant, high light during the midday period may become stressful, particularly if water limitations and/or dry air cause partial to complete stomatal closure. High temperatures, particularly in concert with water stress (and therefore a lack of latent heat flux), may also limit the ability of plants to construct canopies with high leaf area or large leaves, both of which limit convective heat exchange between leaves and the air. In such environments, leaf and canopy morphologies may therefore be adapted to (1) minimize solar interception during midday hours or during drought periods, and (2) maintain leaf temperatures at or near air temperature. Traits that may minimize solar interception include vertical leaf orientation, midday wilting, paraheliotropism (leaf cupping), and pubescence (Smith et al. 1997a). Because light interception is an important driver of leaf energy balance, all of these traits also affect leaf temperature. However, leaf size can control leaf temperature independently of light interception and/or latent heat loss, and so many dryland plants are often microphylls that maintain leaf temperature near air temperature at all times of the day. Additional traits that are commonly observed in dryland plants, and that can affect both light interception and canopy temperature, include seasonal leaf polymorphism and deciduousness, and stem/twig photosynthesis (Ehleringer 1985).

Leaf Size, Density, and Thickness

Adaptive patterns in canopy form and function can be analyzed by either examining trends in overall floras or traits in individual species. Decreases in site water availability lead to dramatic changes in leaf structure and chemical composition. In particular, increasing aridity results in thicker and more lignified plant cell walls, and smaller and more tightly packed mesophyll cells. Several case studies have shown that these changes collectively lead to a negative relationship between leaf density (D, dry mass per unit volume) and water availability (Witkowski and Lamont 1991, Groom and Lamont 1997) as well as across all major earth biomes (Figure 10.1).

Modifications in D have important implications for leaf water relations, because increases in tissue density are compatible with enhanced bulk leaf elastic modulus (ϵ, change in hydrostatic leaf pressure per unit change in leaf symplastic water content) (Fig. 10.2A). Larger ϵ results in greater change in leaf water potential, $\Delta\Psi$, for a certain leaf water loss (see inset in Fig. 10.2A). Thus, increases in ϵ as a result of structural leaf modifications allow water extraction from drying soil with a lower leaf water loss. An alternative way to attain large $\Delta\Psi$ is via increases of the concentration of cell osmotica, either by decreases in cell volume in the case of elastic cell walls (low ϵ) or by net accumulation of osmotica (Morgan 1984, Abrams 1990). However, osmotic regulation of $\Delta\Psi$ may be inherently limited because of the damaging effects of high symplastic ionic activities on all cellular processes (Kaiser 1987a,b). At the global scale, leaf elastic modulus changed more than 10-fold (see Fig. 10.2A), but the minimum foliar osmotic potentials only 4-fold (Niinemets 2001), suggesting that adjustment of ϵ through changes in foliar internal architecture plays a primary role in developing a high water potential gradient between the soil and leaves. Although increases in D provide an important way to improve plant water status, there is a negative relationship between foliage photosynthetic capacity and D (see Fig. 10.2B). This negative relationship is associated with lower leaf nitrogen con-

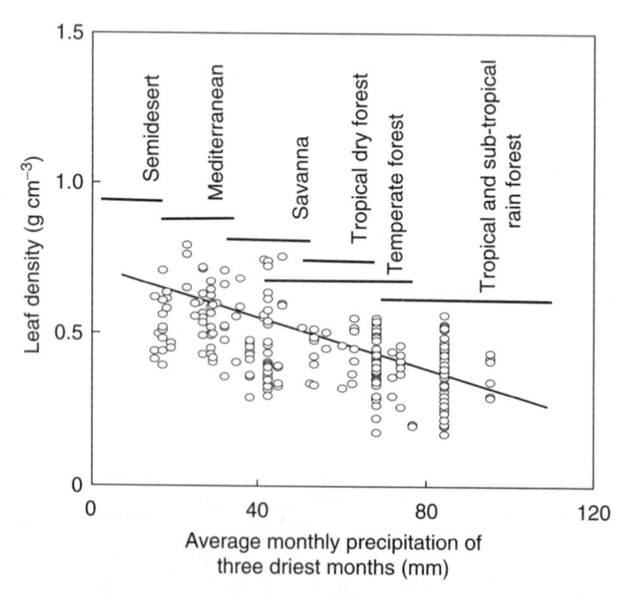

FIGURE 10.1. Leaf density (dry mass per unit volume) in relation to site aridity in 401 broad-leaved trees and shrubs sampled across all major earth biomes with woody vegetation (modified from Niinemets 2001).

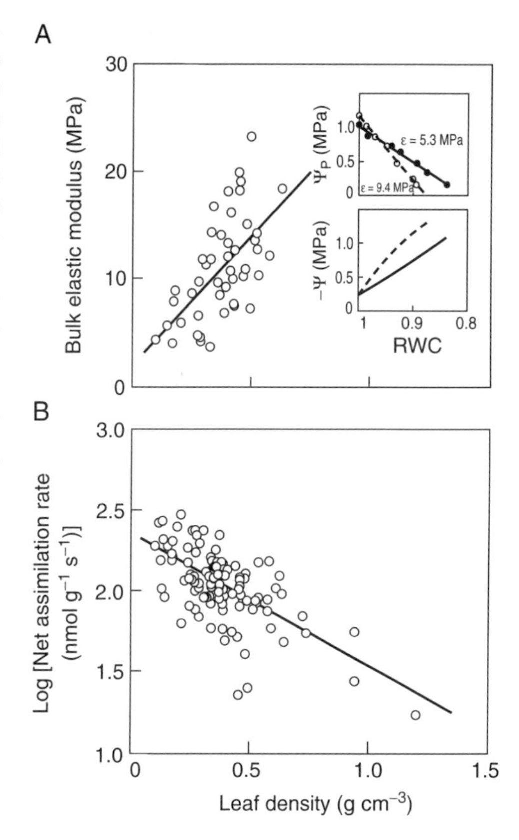

FIGURE 10.2. Leaf bulk elastic modulus (A) and net assimilation rate per unit dry mass (B) in relation to leaf density in a global leaf structure versus function data set (modified from Niinemets 1999, 2001). Inset in A demonstrates representative leaf pressure (Ψ_P) and water potential (Ψ) versus relative leaf symplastic water content (RWC) curves for a drought-stressed (- - ○ - -) and control (- - ● - -) plant of a broad-leaved temperate tree *Populus tremula* (unpublished data of Aasamaa and Niinemets). Leaf bulk elastic modulus (ϵ) is given as the slope of the Ψ_P versus RWC relationship.

centrations (Niinemets 1999), but also apparently with a greater internal gas- and liquid-phase diffusion resistance from substomatal cavities to carboxylation sites in the chloroplast in leaves with higher D (Syvertsen et al. 1995, Evans and Loreto 2000). Accordingly, a tradeoff occurs between adjustments in leaf water relations and photosynthetic capacity. An improvement of leaf water extraction capacity inherently leads to lower foliar photosynthetic capacities.

Leaf age is also an important factor in these relationships, as plant communities are usually made up of many different life forms, ranging from annuals to woody evergreens. Vendramini et al. (2002) observed leaf dry mass per unit area (M_A) to vary widely across growth forms in a semiarid community in Argentina, ranging from low values in herbaceous species to high values in evergreen shrubs (over a 3-fold difference in M_A between growth forms). In chaparral, Ackerly et al. (2002) observed leaf size at the community level to decline and M_A to increase with increasing insolation (Fig. 10.3), and therefore openness of the vegetation and aridity of the habitat. However, these traits were not correlated across species, suggesting that these two traits are decoupled and associated with different aspects of performance across the gradient. Closer inspection showed M_A to mainly differ between deciduous and evergreen species—evergreens increase in M_A as the gradient becomes more arid but deciduous species do not—and therefore M_A varies as a function of the abundance of these two functional types along the gradient, whereas the *variance* in leaf size increased with insolation (see Fig. 10.3), suggesting a greater diversity of water use and energy balance strategies under drier, more-exposed conditions (Ackerly et al. 2002).

Berry and Roderick (2002) proposed that much of the variability in carbon assimilation and leaf longevity between plants can be related to variations in the surface:volume ratio of leaves. They proposed three basic global leaf types, in order of declining surface:volume ratio (Table 10.1): (1) "turgor" leaves, represented by seasonal herbaceous plants and characteristic of disturbed habitats; (2) "mesic" leaves, represented by broadleaf trees and characteristic of habitats with

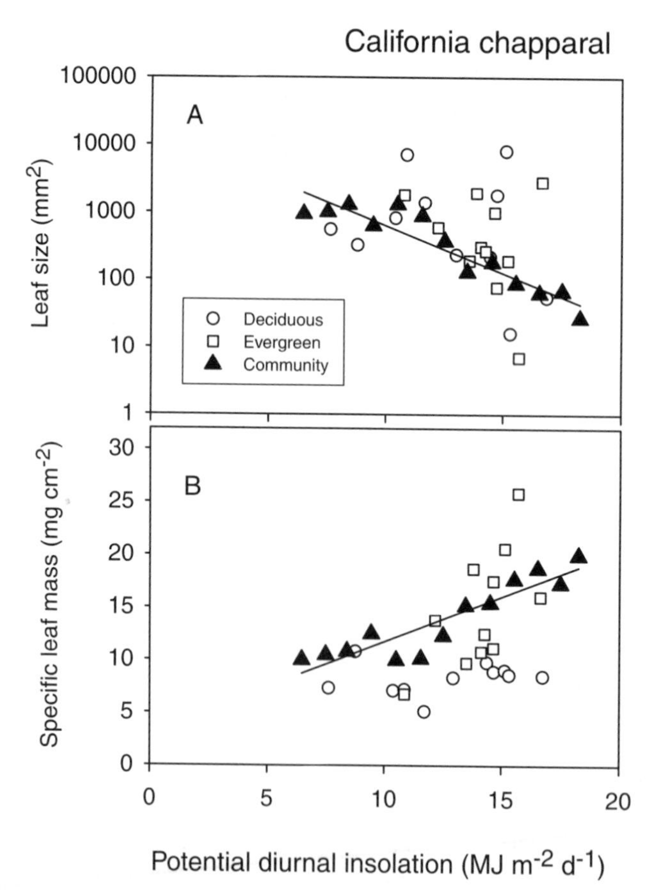

FIGURE 10.3. Response of leaf size (A) and specific leaf mass (B) to potential diurnal insolation in California chaparral vegetation. Responses are for individual species of deciduous (○) and evergreen (△) shrubs, and for the community-wide mean value of all species (▲); the regression line is for the community-wide response in each case (data from Ackerly et al. 2002).

TABLE 10.1. Characteristics of turgor (T), mesic (M) and sclerophyll (S) leaves (from Berry and Roderick 2002). Parameter terms: SLA = specific leaf area ($m^2\ g^{-1}$); V = leaf volume ($m^3\ m^{-2}$); $A_{max,\ v}$ = photosynthetic capacity per unit leaf volume; C:N = leaf carbon/nitrogen ratio.

Parameter	T	M	S
SLA/V (m^{-1})	> 10,000	4000–10,000	< 4,000
Thickness ($m \times 10^{-6}$)	< 200	200–500	> 500
$A_{max,\ v}$ (kg CO_2 $m^{-3}\ s^{-1}$)	0.002–0.004	0.001–0.002	0.0005–0.001
C:N (mass basis)	< 25	25–42	> 42
Leaf longevity (yr)	< 0.5	0.5–2	> 2

optimum light, nutrients and water; and (3) "sclerophyll" leaves, represented by evergreen shrublands and forests and characteristic of water- or nutrient-limited environments. Roderick and coworkers (2000) provided a conceptual model for how these leaf types sort along environmental gradients. Thin "turgor" leaves typically have a high photosynthetic capacity, a low C:N ratio, and are short lived, whereas "sclerophyll" leaves have low photosynthetic capacity, high C:N ratio, and are long lived. "Mesic" leaves are intermediate. From an adaptation perspective, leaves in nonstressful environments differ primarily in N content and photosynthetic capacity, whereas in leaves from more stressful environments, leaf longevity is more variable than photosynthetic rates. Indeed, Reich and coworkers (1997,1998a) have documented considerable variation in leaf longevity among plant species with thick leaves. This conceptual model provides a useful framework for examining leaf structure-function relationships, with photosynthetic capacity and leaf longevity inversely related along nutrient- and water-availability gradients.

Finally, although microphylly is considered to be the norm in desert plants due to their convective properties, it is important to note that small leaves are usually clumped on shoots, even in desert species (e.g., *Banksia ericacea*). This may often result in a single boundary layer for the entire shoot (later, in the cold environment section, this aspect of small leaves is considered in detail). What may actually drive lower temperatures in some microphylls is that the upper leaves are shading the lower ones such that overall light interception is lower for the shoots of microphylls than for broad-leaved species. Although this may appear on the surface to be contradictory, the final effect may very much depend on surface properties such as pubescence or wax coatings, which may lead to multiple reflections and more uniform light distribution within the shoot.

Leaf Angle

A thin, horizontally oriented leaf with well-developed palisade and spongy mesophyll layers is considered optimum for maximizing photosynthetic capacity (Smith et al. 1997b) and light-use efficiency (Ustin et al. 2001). Yet many plant species have primarily vertically oriented leaves, leaflets, or phyllodes. Although many plant canopies with high leaf area have vertical sun leaves in the upper canopy that give way to horizontal shade leaves in the lower canopy, many other species have vertical leaves throughout the canopy. Smith et al. (1998) found that decreases in annual precipitation (from 15 to 3 cm y^{-1}) and increases in total daily sunlight (4 to 29 mol photons m^{-2}) corresponded strongly to an increase in the number of species in a given community with more inclined ($> 45°$) leaves; in environments with less than 8 cm y^{-1} precipitation, $> 70\%$ of the species had vertical, amphistomatous leaves with distinct adaxial/abaxial palisade layers. Similarly, the leaflets of the evergreen desert shrub *Larrea* show an increasing vertical orientation as their distribution extends into drier environments, both for *Larrea* spp. in Argentina (Ezcurra et al. 1991) and for *Larrea tridentata* in North America (Neufeld et al. 1988). Furthermore, *L. tridentata* can vertically close their bifoliate leaflets with respect to direct solar radiation, and field data from different Mexican deserts showed a significant correlation between leaflet aperture and the mean inclination of leaf pairs (Ezcurra et al. 1992), with plants from the driest habitats having vertically oriented leaflets that close during the midday period to further reduce the interception of solar radiation.

Erect leaves are not common, and a model analysis indicates that when averaged over a whole day's course of solar angles, canopy photosynthesis is fairly insensitive to leaf angle, and at low leaf area index (LAI), leaf erectness would be at a clear disadvantage (Gutschick and Wiegel 1988). However, leaf erectness reduces leaf heat load and hence temperature and vapor pressure deficit (VPD), which would in turn increase water-use efficiency (WUE) by shifting maximum photosynthesis to earlier in the day when VPD is lowest (Gibson 1998). An additional explanation is the avoidance of high-light stress, which has been shown to be the case in the evergreen chaparral shrub *Heteromeles arbutifolia. Heteromeles* has a distinct sun/shade canopy, with steeply inclined sun leaves in the upper canopy and horizontal shade leaves in the lower canopy. Although sun leaves have higher M_A and photosynthetic capacity than do shade leaves, they absorb only 3 times more solar radiation than shade shoots (despite having 7 times more solar energy available) and exhibit daily carbon gain only double those of shade shoots (Valladares and Pearcy 1998). However, when vertical leaves were horizontally restrained so that they received maximum midday solar radiation, they showed significantly lower photosynthetic rates and chlorophyll fluorescence (F_v/F_m) values, indicative of photoinhibition of the photosynthetic apparatus (Fig. 10.4). Valladares and Pearcy (1997) concluded from this experiment that the vertical orientation of sun leaves in this species provided an efficiency compromise between maximizing carbon gain while minimizing the time that leaf surfaces are exposed to light levels in excess of those required for light saturation of photosynthesis (and therefore potentially photoinhibitory).

Leaf Pubescence

Leaf pubescence is a clear adaptation to high-light environments (see Chapter 2), and is observed in a wide range of growth forms from annuals to deciduous shrubs and succulents. It has been extensively studied in the drought-deciduous shrub

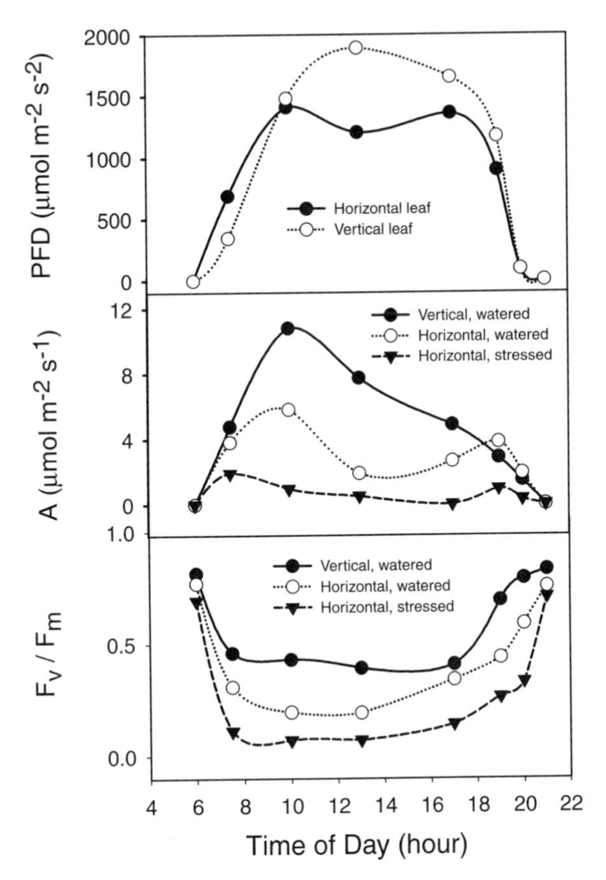

FIGURE 10.4. Mean leaf interception of photosynthetic photon flux density (PFD) (top), leaf CO_2 assimilation (*A*) (middle), and chlorophyll fluorescence (F_v/F_m) (bottom) as a function of time of day in the California chaparral shrub *Heteromeles arbutifolia*. Experimental treatments were naturally occurring vertical leaves versus leaves artificially held at a horizontal angle when well watered or after water stress (data from Valladares and Pearcy 1997).

Encelia farinosa and several other congeneric species. Pubescence in *E. farinosa* has been shown to increase seasonally as large, glabrous winter leaves give way to smaller, pubescent leaves in late spring prior to leaf fall in the summer (Ehleringer 1982). Across *Encelia* species, progressively more pubescent, reflective leaves are found along aridity gradients in both North and South America (Ehleringer et al. 1981), a phenomenon that has also been observed within *E. farinosa* in North American deserts (Sandquist and Ehleringer 1998). Common garden experiments have shown the desert-inhabiting *E. farinosa* to perform better (higher photosynthesis and more rapid growth) under desert conditions than do the nonpubescent desert species *E. frutescens* or the coastal species *E. californica*, particularly in the late spring and summer when pubescence should have its greatest effect (Ehleringer and Cook 1990). Within *E. farinosa*, there is a clear genetic basis for the pubescence-rainfall relationship in the western U.S., as common garden experiments have shown pubescence and physiological performance attributes to be retained (Sandquist and Ehleringer 1997). Pubescence in broadleaf shrubs such as *Encelia* thus appears to maximize water-use efficiency and minimize photoinhibition through seasonal leaf dimorphism, with large, glabrous leaves maximizing carbon assimilation in the cool winter months but then pubescent leaves minimizing solar absorption during times of higher light and water stress later in the season. Furthermore, the observation that pubescence increases along aridity gradients appears consistent with a tradeoff scenario in which plants from more mesic habitats trade-off water conservation for higher carbon gain, whereas plants from drier regions reduce water consumption and extend leaf longevity to maintain photosynthetic activity in response to the unpredictable growing conditions that typify arid regions (Monson et al. 1992, Sandquist and Ehleringer 1998).

Photosynthetic Stems

As leaf area and total canopy cover decline along aridity gradients, the potential for supporting stems to contain chlorophyll and to therefore contribute to canopy carbon gain increases. In arid regions, many drought-deciduous shrubs and trees, herbaceous plants, and succulents exhibit photosynthetic stem surfaces. In water-limited environments, reductions in leaf area and increased reliance on stems for photosynthetic activity have been explained as a mechanism to conserve water (Gibson 1983), but an alternative hypothesis is that when stems are the primary photosynthetic organ of the plant, leaves have been reduced or eliminated because they block sunlight from striking the stem surface (Gibson 1998). Indeed, many desert shrubs and trees (e.g., *Cercidium* spp.) are almost completely leafless, depending largely on stems for their annual carbon balance (Szarek and Woodhouse 1978). A second alternative hypothesis may be that photosynthetic stems are advantageous because of their high volume: surface area ratio, and thus high heat storage capacity, which should result in lower variation in temperature over 24-h cycles. In large succulents, photosynthetic tissues can reach extremely high temperatures (Smith et al. 1984), but such temperatures may be advantageous in colder months when soil resources are more available for photosynthesis (Nobel 1988).

Comstock et al. (1988) surveyed a number of desert shrub species and found a wide variety of phenological patterns occur in the balance between seasonal occurrence and activity of leaf and stem surfaces. Three primary groups were observed based on the relative contributions of leaves and stems to seasonal carbon gain (Fig. 10.5). The groups were observed to be: (1) species, mostly suffrutescent subshrubs, with a low relative contribution by stem photosynthesis but with a high level of drought deciduousness (see Fig. 10.5A); (2) species (e.g., *Chrysothamnus, Gutierrezia, Hymenoclea*) that exhibit primarily leaf carbon gain during the growing season but then maintain a considerable fraction of green canopy and therefore comparable contributions of leaves and stems to canopy carbon gain through

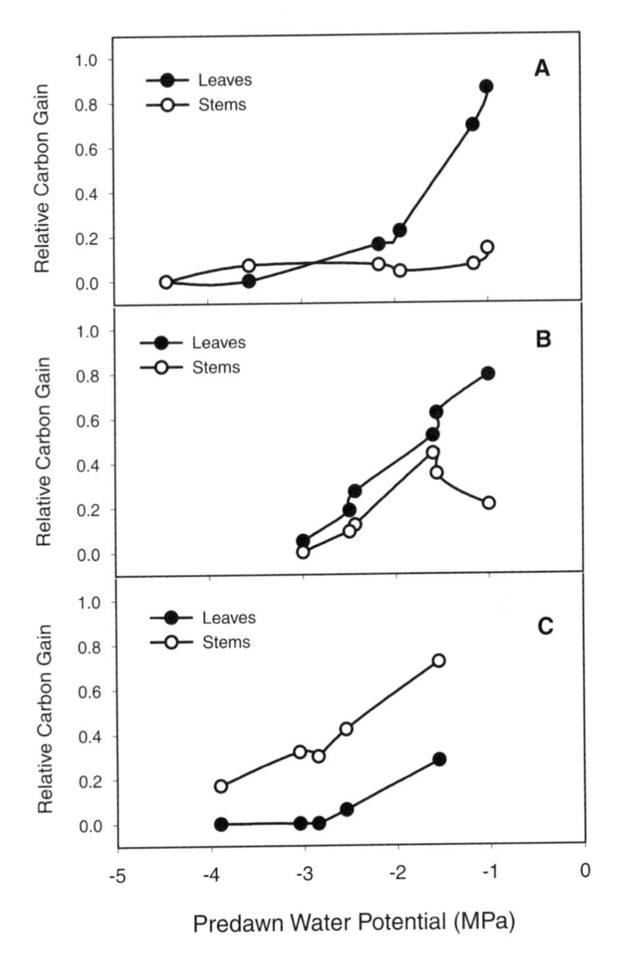

FIGURE 10.5. Three primary groups of desert plants with deciduous leaves, based on the relative contribution of leaves and stems to whole-plant carbon gain as plant water potential declines from winter to summer in the Sonoran Desert. A: Suffrutescent subshrubs with small stem contribution; B: shrubs with approximately equal contribution from leaves and stems; and C: trees/shrubs with a majority of carbon gain from stems (data from Comstock et al. 1988).

drought periods (see Fig. 10.5B); and (3) deciduous species (e.g., *Cercidium, Salazaria*) with reduced leaf area in the primary growing season, and therefore a carbon budget dominated by stems and showing very little reduction in green twig area during drought (see Fig. 10.5C). Clearly, a wide range of phenological patterns are observed when comparing the seasonal contributions of leaves and stems to whole-plant carbon gain across different species and growth forms.

A significant amount of research has been conducted to determine the advantages of stem versus leaf photosynthesis in response to aridity. Studies of the herbaceous perennial *Eriogonum inflatum* (Osmond et al. 1987), the drought-deciduous shrub *Hymenoclea salsola* (Comstock and Ehleringer 1988), and the tree *Psorothamnus spinosus* (Nilsen et al. 1989) all indicate that stems have lower rates of nitrogen and chlorophyll contents, and therefore lower photosynthetic capacities, than do leaves, although in all cases much of the mass-based or 2-dimensional area-based rate differences disappear when projected areas of the cylindrical stems are used. Additionally, stems appear to have higher temperature optima and their stomata are less sensitive to high VPD conditions (Osmond et al. 1987, Nilsen et al. 1989), consistent with their larger relative role in plant carbon gain under the warmer, drier conditions that prevail later in the growth season. As a result, the primary attribute of stems may be that they operate at higher water-use efficiency. In a survey of 14 species with both leaf and stem photosynthesis, Ehleringer et al. (1987) observed a mean $\delta^{13}C$ (a proxy for photosynthetic WUE) of -26.5% for leaves and -25.0% for stems, and in all cases stems operated at higher (less negative) carbon isotope ratio than did leaves. Smith and Osmond (1987) also found stems in *Eriogonum inflatum* to operate at higher WUE, and in a population from Death Valley, photosynthetic surface area integrated over a full growth cycle was observed to be 34% leaves and 66% stems, but only 40% of whole-plant water loss was attributed to stems.

There are a number of plant species in dryland environments that do not employ stem photosynthesis as a significant component of canopy carbon gain, so under what conditions do stems become an important adaptation to water and/or high-light stress? Comstock and Ehleringer (1990) addressed this question by comparing desert perennials with and without stem photosynthesis, and their analysis indicated that scaling relationships between leaf and twig dimensions have a strong influence on the kind of photosynthetic activity that develops in twig tissues. Their analysis found that positive net assimilation of atmospheric CO_2 by twigs is generally restricted to species with small leaves, small twig diameters, and low rates of respiratory CO_2 production, and since leaf size tends to decrease along aridity gradients, the abundance of desert species with photosynthetic stems may be a result of allometric relationships between leaves and stems. A more recent analysis by Comstock (2000) found that *Hymenoclea salsola* exhibits an increased reliance on stem photosynthesis as aridity increases, and this is related to the decreased ability of plant hydraulic conductance to support leaf area as water stress intensifies (Fig. 10.6). Therefore, stem photosynthesis may be an important mechanism to increase hydraulic conductance per unit photosynthetic canopy by increasing the allocation of photosynthetic machinery to an organ that simultaneously performs photosynthetic, support, and transport functions.

Scaling Issues

Plants in water-limited environments are exposed to not only a lack of water, but also to high-light, high-temperature, and high-VPD conditions. Because they tend to occur simultaneously in natural conditions, it is often difficult to attribute each

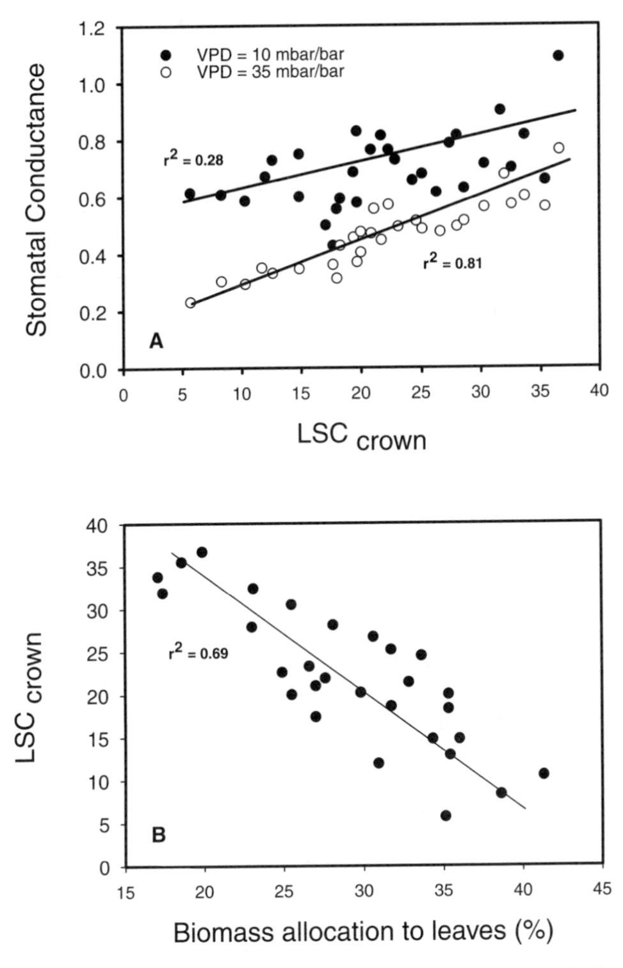

FIGURE 10.6. A: Leaf stomatal conductance as a function of leaf-specific hydraulic conductance of the crown for two desert shrub species (*Ambrosia dumosa* and *Hymenoclea salsola*) at low (10 mbar bar^{-1}) (●) and high (35 mbar bar^{-1}) (○) VPD. B: Leaf-specific hydraulic conductance of the crown of *Ambrosia* and *Hymenoclea* as a function of the amount of biomass allocated to leaves (data from Comstock 2000).

of these stress factors as important drivers for the structural adaptations that were discussed above. However, a controlled-environment experiment with *Encelia farinosa*, a desert shrub with pronounced seasonal polymorphic behavior in leaf structure, provided important insights into potential driving variables for morphological change as climate seasonally dries. This study showed that irradiance had its greatest effect on leaf absorptance (pubescence) and thickness, plant water potential had the greatest effect on leaf size, and temperature per se did not appear to have a controlling effect on leaf structural properties (Smith and Nobel 1977, 1978). As discussed above, it appears that irradiance may be a primary driver for leaf orientation, as vertical leaves are an important adaptation to avoid photoinhibition as soils dry (Valladares and Pearcy 1997), and plant moisture status—specifically hydraulic limitations—may be a controlling factor in the shift of plant canopies from leaf to stem photosynthesis (Comstock 2000). In a simulation analysis, Smith and Nobel (1978) assessed the importance of leaf size (length), pubescence (absorptance), and thickness (A^{mes}/A) for whole-canopy photosynthesis, transpiration and WUE for three desert broadleaf shrubs: *Encelia farinosa*, a highly polymorphic species; *Hyptis emoryi*, a moderately polymorphic species with distinct sun/shade leaves; and *Mirabilis tenuiloba*, a nonpolymorphic species with large, thick leaves (Table 10.2). In all cases, the "natural" condition of leaf structure during the growing season resulted in greatest carbon gain. In the polymorphic species, simulation of a year-round canopy of early leaves with high A_{max} resulted in reduced carbon gain and increased transpiration, and thus strong reductions in WUE. In contrast, simulation of a year-round canopy of drought-adapted leaves resulted in slightly higher WUEs than observed in the natural condition, but at significantly lower carbon gain. This analysis suggests that canopy structure is not driven by stress factors alone, but that the sea-

TABLE 10.2. Influence of leaf parameters on simulated photosynthesis, transpiration and water-use efficiency (WUE) for three desert shrub species: *Encelia farinose*, a deciduous shrub with seasonal leaf polymorphism; *Hyptis emoryi*, a deciduous shrub with distinct sun-shade leaves; and *Mirabilis tenuiloba*, an evergreen sclerophyll. Simulations use constant growing-season conditions (i.e., early spring when soils are moist). "Natural" designates the yearly pattern of leaf morphology from actual field observations, "Combined" refers to maintenance of leaf morphology throughout the year as they occurred at peak growing season ("wet") or late in the growing season ("dry") (data from Smith and Nobel 1977b).

Taxa	Parameter	Photosynthesis	Transpiration	WUE
Encelia	Natural	1.00	1.00	1.00
	Maximum length	0.96	1.15	0.83
	Maximum absorptance	0.93	1.18	0.77
	Minimum A^{mes}/A	0.84	1.00	0.84
	Combined: wet	0.73	1.40	0.52
	dry	0.76	0.72	1.04
Hyptis	Natural	1.00	1.00	1.00
	Maximum length	0.95	1.09	0.86
	Maximum absorptance	0.95	1.12	0.85
	Minimum A^{mes}/A	0.86	1.00	0.86
	Combined: wet	0.80	1.33	0.60
	dry	0.83	0.79	1.04
Mirabilis	Natural	1.0	1.0	1.0
	Maximum length	1.08	1.06	1.01
	Maximum absorptance	1.00	1.00	1.00
	Minimum A^{mes}/A	0.97	1.00	0.97
	Combined: wet	1.03	1.06	0.97
	dry	0.98	1.00	0.98

sonal changes in leaf and canopy morphology in response to drying conditions also optimize carbon gain by the canopy.

We can now ask two critical questions: (1) How do plants change leaf, branch, and canopy form to adapt to water-limited environments? and (2) At which organizational level of form is the greatest apparent advantage possible? These are difficult questions to answer given a lack of specific mechanistic studies that have addressed issues of scale in plants from water-limited ecosystems. Dryland plants exhibit an incredible diversity of canopy structures, from (1) highly polymorphic deciduous species such as *Encelia*, to (2) shrubs such as *Heteromeles* and rosette plants such as *Yucca brevifolia* (Smith et al. 1978) with high leaf area and a systematic gradation from horizontal leaves in the lower canopy to vertical leaves in the upper canopy, to (3) evergreens such as *Larrea* with open canopies and primarily vertically oriented leaves. Each of these functional types exhibits specific combinations of leaf absorptance, leaf angle, and leaf thickness, as well as trade-offs between leaf and stem photosynthetic surfaces. Adaptations such as leaf pubescence clearly have their greatest impact at the level of the individual leaf, but pubescence is also important in an *Encelia* plant in the transition phase between wet and dry seasons, and thus may possess approximately equal amounts of glabrous and pubescent leaves. Leaf angle also has important implications for individual leaves as an adaptation to avoid photoinhibitory damage, but also has important effects on light penetration into canopies. And finally, the transition from leaf to stem photosynthesis has clear branch- and canopy-level mechanistic origins, particularly if hydraulic limitations help drive the process.

At the canopy scale, there are indications that foliage is not randomly distributed within canopies even though most aridland plants have relatively open canopies in a high-light environment. For example, Neufeld et al. (1988) found that foliage clusters in *Larrea* canopies are predominantly oriented toward the southeast (SE). Within SE sectors of the crown, branches were shorter, resulting in reduced self-shading early in the morning when maximum gas exchange typically occurs. This effect was not apparent at the summer solstice (Neufeld et al. 1988), again suggesting that maximization of carbon gain is the driving force for this nonrandom pattern rather than stress avoidance per se. Similar results have been observed in very different plant species, both morphologically and phylogenetically. Specifically, rosettes in the Joshua tree (*Yucca brevifolia*) occur in a nonrandom distribution with more rosettes occurring in the SE quadrant of the tree "canopy" (Rasmuson et al. 1994), and side branches of the saguaro cactus (*Carnegiea gigantea*) also occur nonrandomly toward the SE quadrant (Geller and Nobel 1986). In both cases, the interpretation was to reduce self-shading between these modular photosynthetic "units" and to enhance whole-canopy carbon assimilation primarily in the winter-spring period when soil resources are most available. In conclusion, drought is an important determinant of CO_2 fluxes in water-limited ecosystems, but scaling relationships from leaves to the ecosystem remain complicated by canopy structural characteristics such as nonrandom distribution of foliage on branches, branching structure, and vertical structure of canopies and communities (Rambal et al. 2003).

Cold Environments

As detailed in the previous chapter, freezing temperatures, especially if followed by high-light conditions, can severely decrease A_{net} during subsequent favorable temperature regimes. Some of the damaging effects can be mitigated by physiological adjustments within the leaf. Here, we discuss how adjustments in the display of leaves and shoots affect leaf temperature (T_l) and cold-induced photoinhibition. Unfortunately, in comparison to the abundance of studies addressing the within-leaf physiology of low-temperature effects on A_{net}, relatively few studies have addressed how larger-scale plant properties affect carbon gain in cold environments.

As in water-limited environments, architectural features of leaves and crowns in cold environments may reflect optimization of leaf energy balances as much as for sunlight interception for photosynthesis. Cool habitats in temperate and tropical latitudes occur often due to elevation effects, which also lead to lower ambient pressure and more rapid diffusion rates. More rapid diffusion, combined with lower water vapor holding capacities of cool air, cause water stress to be a common, but frequently unappreciated, stress that affects photosynthesis in cold environments, irrespective of water supply (Smith and Geller 1979, Leuschner 2000). Low-temperature stress has well-known effects on photosynthesis and plant growth at the cellular level, and can also affect canopy structure. Thus, canopy structure in cold environments probably reflects optimization for photosynthesis, energy balance, and avoidance of photoinhibition.

A major difference for plants in cold compared with dry environments is that leaf displays may reflect avoidance (instead of enhancement) of convective cooling during days, and can lead to appreciable warming of leaves for enhanced carbon gain (Smith and Carter 1988). However, leaf and crown adaptations in cold environments tend to be less consistent in how they affect energy balances. For example, many plants from cool environments can have leaves as small or smaller than dryland plants that would normally be close to air temperature. On the other hand, these small leaves are often densely arranged into crowns that are within a few cm of the ground, as in "cushion" plants that can warm many degrees above air temperature during the day (Körner 1999). Therefore, morphological factors that affect radiative or convective heat exchange during the day also substantially affect these energy balance considerations at night. Finally, low-temperature stress need not occur together with light to affect photosynthesis—nighttime frost can have appreciable effects on subsequent carbon gain.

Microclimate Effects on Nighttime Leaf Temperature and Freezing

Leaf temperature (T_l) is a function of radiant energy exchange (only longwave radiation at night) between a leaf and the sky/ground as well as the exchange of sensible and latent energy with the surrounding air. A fourth factor, heat storage, is small for most leaves and is generally ignored in the energy balance. Based on these factors, freezing T_l can occur by sensible heat exchange with subfreezing air temperatures (T_{air}) and/or via radiative cooling—longwave energy exchange of the leaf with the cold sky during clear nights. Subfreezing air temperatures are more likely in depressions and at the base of slopes via cold-air drainage. Also, during still and clear nights, temperature inversions result in the lowest air temperatures closest to the ground (Leuning and Cremer 1988, Jordan and Smith 1995a). As a consequence, leaves and plants located in cold-air drainage depressions or close to the ground experience lower nighttime temperatures and more subfreezing hours (Leuning and Cremer 1988, Germino and Smith 2001). Lower windspeeds also occur close to the ground and thus reduce convective warming for leaves that cool below air temperature on clear nights as a result of longwave radiation loss to clear, cold skies. Herbaceous canopies in alpine meadows can further decrease windspeeds and minimize convective warming for radiatively cooled leaves (Germino et al. 2002). Further, air temperatures close to the ground are dependent on how well the surface conducts stored heat. Vegetation and plant litter are comparatively poor conductors and, as a consequence, air and leaf temperatures are 1 to 2 °C lower above grass than moist bare soil, which is a good heat conductor (Leuning and Cremer 1988, Ball et al. 1997). These differences are significant for plant carbon gain and survival. For example, *Eucalyptus pauciflora* (snow gum) seedlings growing in grassland displayed more photoinhibition, from which they recovered more slowly, than those growing in bare soil patches (Ball et al. 1997). Further, seedlings with the greatest biomass gain through the winter and spring experienced less photoinhibition (Fig. 10.7).

The second factor that impacts nighttime T_l is radiative cooling. Microsites with a high likelihood of radiation freezing have an unobstructed "view" of the sky, a condition that allows the free exchange of longwave radiation between leaves and the sky. As a consequence, plants growing in clearings experience dramatically more frost events and lower leaf temperatures than plants located at the clearing edge or underneath a plant canopy (Ball et al. 1991, Jordan and Smith 1995a). Because radiation frost is most likely during clear still nights, which are generally followed by clear sunny days, plants in clearings are more

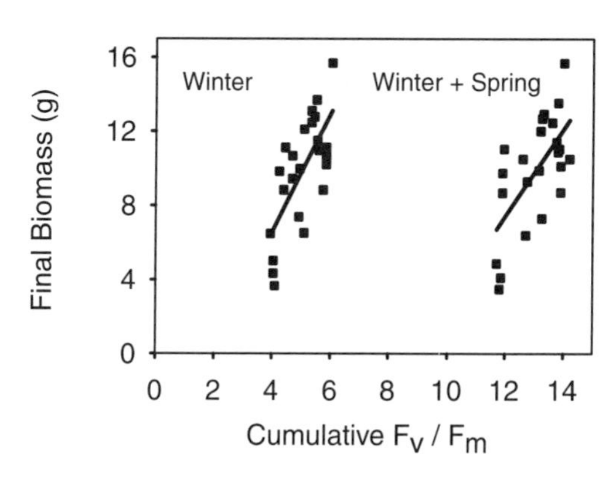

FIGURE 10.7. Final spring biomass of *Eucalyptus pauciflora* as a function of cumulative weekly midday F_v/F_m over winter (1 June to 31 August; $r^2 = 0.51$) or winter + spring (1 June to 30 November; $r^2 = 0.41$) time periods (data from Ball et al. 1997).

likely to experience persistent photoinhibition when temperatures are low (Ball et al. 1991, Germino and Smith 1999, Egerton et al. 2000). This, in turn, may influence the spatial distribution of plants—species that are sensitive to cold-induced photoinhibition are preferentially located in microsites that shield them from nighttime radiation loss and morning sun (Ball et al. 1991, Germino and Smith 1999). In another example, seedling survival of *Carnegiea gigantea* (saguaro cactus) was greater in rocky microsites and decreased with increasing distance from "nurse rocks" due to a loss of frost protection and shading (Steenbergh and Lowe 1969).

Plant Structural Effects on Microclimate and Nighttime Leaf Temperature and Freezing

The above discussion focused on the effects of microsites on plant performance in frost-prone environments. However, plants can ameliorate the negative effects of their microsite via physiological adjustments (see Chapter 9) or via structural adjustments that directly affect the microclimate they experience. Among these structural adjustments are leaf size, leaf orientation, and canopy display.

Because small leaves have a higher boundary-layer conductance, they track ambient T_{air} better than larger leaves. This adaptation is advantageous in dry, high-light environments (see above), but may also hold in frost-prone environments. The 4-fold wider leaves of *Eucalyptus pauciflora* were 1 to 2 °C cooler at night than the smaller *E. viminalis* leaves when placed horizontally under identical microclimatic conditions (Leuning and Cremer 1988). Although leaf orientation may have compounded the leaf size effect, similar results were obtained from comparisons of alpine herbs and conifer seedlings (Jordan and Smith 1995a, Germino and Smith 2001). In a within-species comparison, leaf width and length of the alpine forb *Erigeron peregrinus* was greater at the margin of forest clearings than in random sites (Jordan and Smith 1995b). This difference could not be explained by daytime photon flux density, because the forest margin plants were growing at a southern exposure and thus had similar daily values as plants in clearings. Similarly, leaves of the forb *Taraxacum officinale* decreased in size with elevation, and leaf size showed a strong negative correlation with longwave sky radiation (Figure 10.8). Further, in a common-garden experiment with the Hawaiian tree *Metrosideros polymorpha*, seed sources from high elevations had smaller leaves than those from lower elevations (Fig. 10.8). Thus, these studies lend support to the hypothesis that smaller leaf size is adaptive in cold environments. As leaves get smaller, they may also increase in density, as Hultine and Marshall (2001) reported increasing leaf mass (M_A) of conifers with elevation.

Leaf orientation affects the degree of radiative cooling because vertical leaves are less exposed to the sky than horizontal ones. Moreover, convective heat exchange under conditions of low windspeeds can be nearly twice as great for vertical compared with horizontal leaves (Germino and Smith 2001). When leaves of *Eucalyptus pauciflora* and *E. viminalis* were displayed vertically rather than horizontally, their nighttime minimum temperatures were 0.2 to 1.5 °C warmer

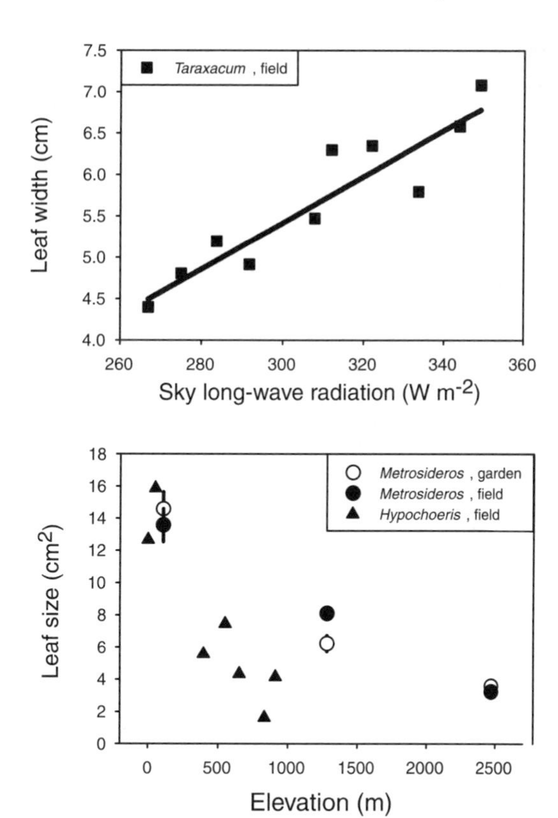

FIGURE 10.8. (Top) Leaf width of *Taraxacum officinale* to average sky longwave radiation and (bottom) leaf size of garden and field populations of *Metrosideros polymorpha* as a function of the elevation of the seed source (data from Jordan and Smith (1995b) for *Taraxacum* and from Cordell et al. (1998) for *Metrosideros*).

(Leuning and Cremer 1988). Similarly, horizontal leaves of two alpine herbaceous species were 1.0 to 1.6 °C lower than leaves in their natural positions (Germino and Smith 2001). This phenomenon led King (1997) to hypothesize that *Eucalyptus* species differences in leaf angle may contribute to their geographical distribution. Indeed, species with steep leaf angles dominate both dry and low-latitude/high-elevation regions, the latter an environment where very cold nighttime temperatures can be followed by intense midday sun. Similarly, two cold-tolerant species of *Rhododendron* diurnally respond to decreasing temperatures with steeper leaf angles (Nilsen 1991). While this may ameliorate minimum leaf temperatures, more likely protection from photoinhibition caused by high-light exposure of cold leaves is the driving force behind variable leaf angles in *Rhododendron* (Bao and Nilsen 1988). Further, frost-tolerant *Rhododendron* spp. have a strong leaf-curling response (Nilsen 1991), which would decrease the amount of leaf area exposed to high light. The *Rhododendron* example suggests that while leaf angle can modify nighttime leaf temperatures, reduced light interception during morning hours may be of greater physiological benefit.

Further evidence of the complex interplay of microsite properties, low temperature, and light comes from a set of studies at the alpine treeline in Wyoming

FIGURE 10.9. *Erythronium grandiflorum* (left) and *Caltha leptosepala* (right) are alpine herbaceous species that differ in their microsite preferences as well as leaf display. Both factors contribute to *Caltha* experiencing significantly lower nighttime leaf temperatures.

(USA). Four different species, representing a range of leaf sizes and canopy displays, were compared with respect to their degree of photoinhibition and depression of A_{net} following freezing events. *Caltha leptosepala* and *Erythronium grandiflorum* (Fig. 10.9) grow in close proximity to snowbanks but have very different morphologies and microsite preferences—*Caltha* has broader, more horizontally inclined leaves that occur in clusters, while *Erythronium* has two opposite, vertical, and partially folded leaves and grows in less frost-prone microsites. As would be expected, *Caltha* experienced more freezing nights over the growing season and lower minimum and early morning T_l than *Erythronium* (Table 10.3). Neither species, however, showed a depression of photosynthetic rates following frost and full sunlight in comparison with nonfrost days or shaded plants (see Table 10.3) (Germino and Smith 2000a).

This finding is in contrast to those for *Abies lasiocarpa* (subalpine fir) seedlings, which because of their small leaf size, track T_{air} closely (Jordan and Smith 1995a). *Abies* seedlings occur in microsites with reduced incidence of radiation frost (Germino and Smith 1999), but experience a 90% reduction in A_{net} after frost and sun exposure compared with nonfrost and shaded conditions (see Table 10.3). *Picea engelmannii* (Engelmann spruce) seedlings that have only slightly greater leaf inclination than *Abies* seedlings, showed no such depression of photosynthesis (Germino and Smith 2000b). As an example of the potential contribution of frost and high sunlight to seasonal carbon gain for entire seedlings of *Abies*, Germino and Smith (1999) found that a 30 W m^{-2} increase in thermal radiation from the sky led to 1 °C warmer needle temperatures and a 50% reduction in the number of

TABLE 10.3. Morphological characteristics of alpine herbs (*Caltha leptosepala* and *Erythronium grandiflorum*) and conifer seedlings (*Abies lasiocarpa* and *Picea engelmannii*) and their effect on leaf temperatures (T_l) relative to air temperature (T_a) and daytime photosynthesis (A_{net}). STAR is a measure of the relative leaf area intercepting direct sunlight for a given sun angle. Not all data were available for 1-y-old conifer seedlings, so we added data for 5- to 10-y-old seedlings in those cases (data from Germino and Smith 1999,2000b,2001).

Trait	*Caltha*	*Erythronium*	*Picea* – 1y	*Abies* – 1y
Leaf length*width (cm)	7.9 ± 0.3 * 6.6 ± 0.2	12.3 ± 0.1 * 2.5 ± 0.1	~1.0 * 0.06	~1.0 * 0.13
Number of leaves/plant	3–6, grow in clusters: >20 leaves/cluster	2 opposite	7–8 cotyledons	7–8 cotyledons
Leaf inclination	38% > 60°	57% > 60°	69.3 ± 2.2° (5–10 yr open)	69.1 ± 2.7° (5–10 yr open)
Midday STAR	0.16 ± 0.02	0.13 ± 0.04	0.25 ± 0.04 0.15 ± 0.02 (5-10 y)	0.33 ± 0.07 0.23 ± 0.03 (5–10 y)
T_l-T_{air}	8 to 14 °C day −5 to −8°C night	2 to 4°C day −2 to −4°C night		
A_{net} (% difference protected vs frost&sun)	+6	−7	−11	−90

nights with frost during the growing season at treeline (from 34 to 14% of nights with frost). This reduction in frost frequency alone might lead to about a 13% increase in seasonal carbon gain for *A. lasiocarpa*. Decreasing sunlight in addition to the occurrence of frost (e.g., with tree- or cloudcover) would lead to an additional 17% increase in seasonal carbon gain, leading to a total 30% increase due to reduced frost, sunlight, and low-temperature photoinhibition (LTP).

These studies highlight that for a given environmental constraint on photosynthesis, different physiological and structural traits can be viable mechanisms for maintaining positive photosynthetic rates. Based on microsite preferences and structural attributes, we would expect *Caltha* to suffer severe depressions in A_{net} due to LTP, for which no evidence exists. Thus, physiological adjustments within *Caltha* leaves must be the major mechanism for maintaining carbon gain under stressful temperature regimes. In contrast, *Abies* and *Eucalyptus pauciflora* seedlings (see above) showed comparatively little evidence for protective physiological adjustments, and consequently are only successful in protected microsites (Ball et al. 1991, Germino and Smith 1999). The structural adjustments—leaf size and orientation—of these species that ameliorate their nighttime leaf temperatures were not as effective as the physiological adjustments that were apparent in *Picea*, *Caltha* and *Erythronium* (Germino and Smith 2001).

Plant Structural Effects on Daytime Leaf Temperature

While plants in cold environments face the challenge of low nighttime temperatures, the same can be said for daytime hours. Depending on latitude and site-specific characteristics, up to 60 °C differences in T_{air} between night and day are possible (Goldstein et al. 1996). Again, physiological shifts in photosynthetic temperature optima and increased nitrogen content and carboxylation efficiencies can maintain A_{net} comparable over an altitudinal gradient (Oleskin et al. 1998, Cordell et al. 1999). Alternatively, leaf and shoot structural properties can elevate leaf temperatures well above air temperatures by decreasing convective heat loss (incidentally, several of the same traits can lead to low nighttime leaf temperatures). Consequently, the alpine herb *Erigeron peregrinus* had T_l values ~5 °C higher than *Abies lasiocarpa* seedlings, which were closely coupled to T_{air} due to its small leaf size (Jordan and Smith 1995a). Similarly, daytime T_l values of *Caltha leptosepala* were up to 10 °C higher than those of *Erythronium grandiflorum* (Germino and Smith 2001), which has more vertically inclined and narrower leaves (see Fig. 10.9). However, due to lower early morning leaf temperatures and greater fractions of oblique and shaded leaf areas, *Caltha* plants during mid-morning were ~3 °C cooler than *Erythronium*.

Clustering leaves and shoots, as displayed by conifer species, elevates daytime T_l by 2 to 8 °C above T_{air} (Smith and Carter 1988, Smith and Brewer 1994), especially in dense krummholz mats (LAI $>$ 20), which can have nearly twice this warming (Hadley and Smith 1988). Furthermore, because soil surface temperatures in sun-exposed sites are well above ambient T_{air}, close contact of leaves with the soil surface results in conductive heat transfer, which can significantly elevate T_l (Smith et al. 1983). Lower wind speeds close to the soil surface can lead to foliar temperatures in sunlight that are >20 °C warmer than nearby air temperatures, leading to potentially excessive temperatures (e.g., 45 °C leaves; Gauslaa 1984). We are not aware of any data to indicate whether conductive warming would counteract the strong radiative cooling at night expected for cushion plants, as a result of their low windspeeds.

In conclusion, the importance of the observed differences in leaf temperature for daily carbon gain strongly depends on the relationship between A_{net} and T_l, and between A_{net} and photosynthetic photon flux density (PFD) if self-shading is substantial. Studies that combine physical and physiological measurements to model consequences of low nighttime temperatures and cold-induced photoinhibition for daily carbon gain of whole plants are scarce (e.g., Germino and Smith 2000b, Valladares and Pearcy 1998 for warm and dry environments), and should be prioritized in the future.

Low-Nutrient Environments

Effects of Low Nutrient Availabilities on Shoot Morphology and Branching Architecture

In conifers, differences in clumping of foliar elements may significantly alter foliar light interception efficiency and thereby photosynthesis rates. The ratio of shoot silhouette to total needle area (STAR) generally increases with decreasing irradiance (Stenberg et al. 1998, Niinemets et al. 2001, Palmroth et al. 2002), indicating that the fraction of exposed needle surface area, and accordingly shoot light interception efficiency, increases at lower irradiance.

Decreases in site nutrient availability result in shorter needles and shoots, and increases needle number per unit shoot length (Roberntz 1999, Niinemets et al. 2002a). These modifications lead to lower shoot light interception efficiency at nutrient-limited sites, especially at lower irradiances (Fig. 10.10). Thus, apart from direct effects of low nutrient availability on foliar photosynthesis, limited light interception efficiency of lower-canopy shoots significantly constrains whole-canopy photosynthetic productivity on infertile sites.

At the crown scale, nutrient shortage leads to arrested leader growth and less frequent branching of upper-canopy laterals. As a result of these changes, as well as limiting competition for light by neighboring trees, the crowns in nutrient-limited sites are often flat-topped, with short internodes and large within-crown shading (Roberntz 1999, Niinemets and Lukjanova 2003). Inherently large self-shading at infertile sites may effectively diminish midday photoinhibitory damage of leaves that have low capacities for photosynthetic excitation energy quenching due to limited nitrogen content of foliage. However, high self-shading also means that photosynthetic nutrient-use efficiency (NUE) is significantly lower on infertile sites during periods of moderate and low quantum flux densities. Thus, shoot, branch and canopy architecture significantly interact with

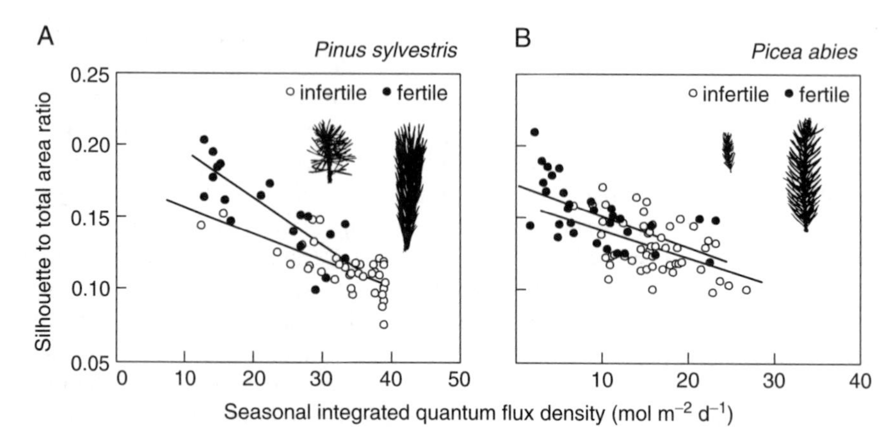

FIGURE 10.10. Mean silhouette to total needle area ratio relative to Q_{int} in *Pinus sylvestris* (left) and *P. abies* (right). Silhouette area was spatially averaged (Niinemets et al. 2002a, Palmroth et al. 2003). Representative silhouettes of upper canopy shoots in infertile and fertile sites are also depicted for both species (Niinemets et al. 2002a and unpublished). The scale for shoot silhouettes is the same for each species.

leaf-level photosynthetic NUE. We conclude from these studies with conifers that nutrient-related modifications in crown architecture significantly constrain the photosynthetic productivity of the entire canopy. Architectural models, including plant plastic adjustments to nutrients, water, light, and temperature, are required to quantify the significance of stress-related architectural constraints on plant photosynthetic productivity (Valladares 1999).

Nitrogen Allocation Within Crowns

Leaf nitrogen contents per unit area (N_A) generally increase with increasing long-term irradiance (Q_{int}) in plant canopies, leading to a strong canopy gradient in foliar photosynthetic capacity (e.g., Ellsworth and Reich 1993). Studies demonstrate that a positive scaling of leaf dry mass per unit area (M_A) with Q_{int} is the primary determinant of canopy variation in N_A, because nitrogen contents per unit dry mass (N_M) are generally relatively invariant in plant canopies (Niinemets and Tenhunen 1997).

Foliar N_M decreases with decreasing site nitrogen availability, and if there were no site effects on M_A, changes in N_A within the plant canopy would be exactly proportional to modifications in N_M (e.g., Rosati et al. 2000). This would imply that only the intercepts of N_A versus Q_{int} relationships are affected by site fertility. However, recent evidence demonstrates that differences in site nutrient availability may affect whole-plant M_A (Meziane and Shipley 1999, Niinemets et al. 2002b) as well as M_A versus light relationships (see Fig. 10.11A) (Niinemets et al. 2001, Palmroth et al. 2002).

In broad-leaved species, decreases in nutrient availability often result in a higher M_A at a common Q_{int} (see Fig. 10.11A) (Niinemets et al. 2002b). Given that $N_A = M_A N_M$, such increases in M_A in response to nutrient limitations partly compensate for the decreases in N_A due to low N_M (see Fig. 10.11B). Consequently, canopy gradients in N_A may differ considerably less between different sites than canopy gradients in M_A (see Fig. 10.11A,B).

In conifers, decreases in site nitrogen availability are associated with decreased M_A (Fig. 10.12A,C) as a result of limited needle thickness growth (Niinemets et

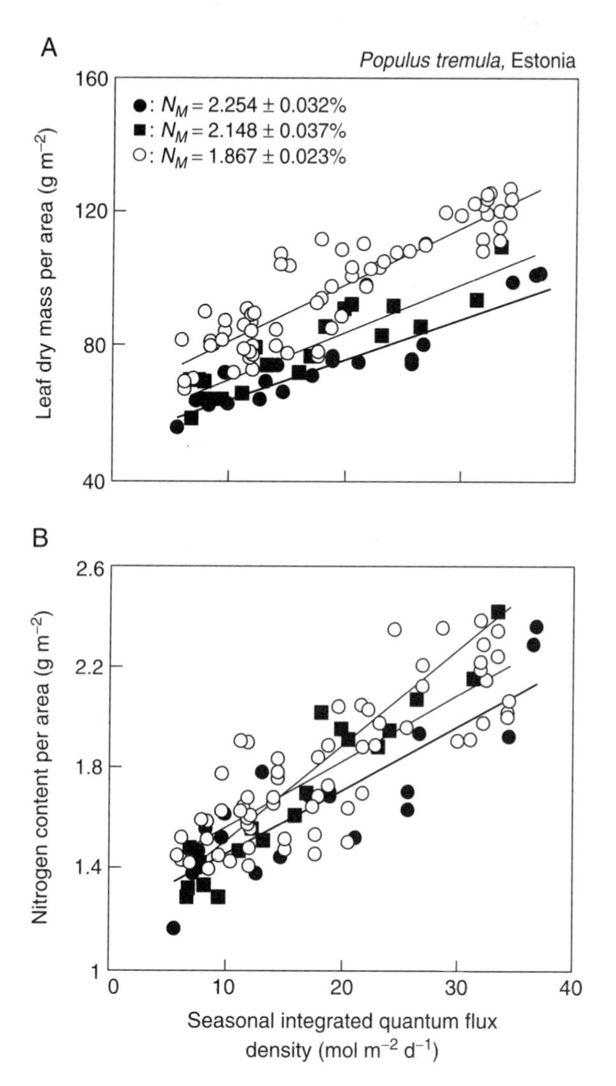

FIGURE 10.11. (A) Leaf dry mass per unit area (M_A) and (B) leaf nitrogen content per unit area (N_A) in relation to within-canopy gradient in seasonal average integrated quantum flux density during leaf growth and development (Q_{int}) in the broad-leaved tree *Populus tremula* from sites of high (●, data from Niinemets et al. 1999), moderate (■, Niinemets and Kull 1998), and low (○; unpublished data of Niinemets et al.) fertility. Average (± SE) nitrogen contents per unit dry mass (N_M) are also given for each site.

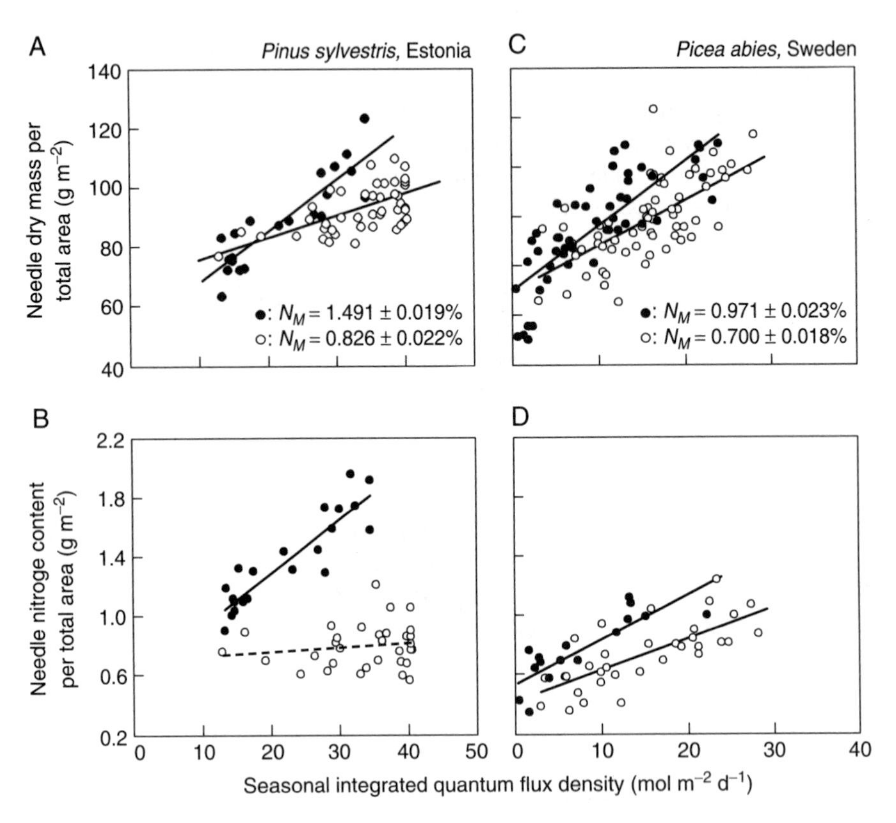

FIGURE 10.12. Needle dry mass per unit total area (A, C) and nitrogen content per area (B, D) in dependence on Q_{int} in the temperate conifers *Pinus sylvestris* (A, B) sampled in a high- (●) and low-fertility (○) site (Niinemets et al. 2001), and *Picea abies* (C, D) sampled in a fertilized (●) and control (○) plot (recalculated from Palmroth et al. 2003). Q_{int} for the Swedish site was determined from the actual measurements of global solar radiation (May–July) (Bergh et al. 1999), and using a representative conversion factor of 1.93 mol MJ^{-1} (Aubinet et al. 2000). N_M is the average (± SE) nitrogen content per unit dry mass.

al. 2001, Palmroth et al. 2002). This further augments the differences in N_A versus Q_{int} relations (see Fig. 10.12B,D), and implies that canopy gradients in photosynthetic capacity differ between sites of contrasting nutrient availability considerably more in conifers than in broad-leaved trees.

Research in Australia has long pointed to limiting nutrients as a key factor that helps determine leaf form and function, and since nutrient availability is correlated to a certain extent with water availability, it has often been difficult to separate these two important factors. Indeed, Beadle's (1966) theory that sclerophylly has a nutritional (especially low phosporus), rather than drought, basis is a long-held paradigm in plant ecology. As a test of this theory, Lamont et al. (2002) examined M_A (as an index of sclerophylly) in evergreen Proteaceae along extensive rainfall gradients in southeastern Australia and the Cape of South Africa. They found a strong inverse curvilinear relationship between mean M_A per site and mean annual rainfall, but no such relationship with mean nitrogen or phosphorus on a mass basis. They concluded that when water and nutrient supply vary independently in the field, M_A is more closely correlated with rainfall (as an index of water status) than with nutrient status, and suggested that M_A may indeed be an important adaptation to water limitation. However, this is not to suggest that nutrients are unimportant in shaping leaf and canopy morphologies, even in water-limited environments.

Summary

This chapter has focused on the interplay between photosynthesis and the tolerance of plants to key environmental stress factors, such as drought, freezing temperatures, and lack of nitrogen. We reinforce the view that plants are faced with tradeoffs between photosynthetic carbon gain on one hand and the avoidance of stress on the other. The adaptations to adverse conditions such as small leaves of higher mass, vertical leaf orientation, pubescence, clustering of leaves around shoots, and stem photosynthesis all can be viewed as improving stress tolerance (or, more precisely, avoiding the intensity of stress events such as temperature extremes and photoinhibition), but at the expense of optimum carbon gain. The reviewed data demonstrate that plants simultaneously maximize their stress tolerance and the efficiency with which plants use limiting resources (e.g., WUE and/or NUE). We also note that adaptations like increased leaf density, sclerophylly, and steeper leaf orientations can potentially be viewed as a genetically linked suite of traits (e.g., stress resistance syndrome) (Chapin et al. 1993), which allow plants to adapt to multiple stress factors such as low water, cold temperatures, and lack of nutrients. However, the consequences of these adaptations for community- or ecosystem-level photosynthesis may be straightforward in the case of leaf density, but may be very difficult to predict in the case of other traits such as leaf orientations. Clearly, this suggests that more empirical evidence for the effects of these adaptations at larger organizational levels is needed.

This review further leads us to the conclusion that the impacts of environmental stresses cannot be analyzed without consideration of interactions between multiple stress factors, especially with sunlight. Moreover, scaling the effects from leaves to greater levels of organization is not trivial because of this plurality of interacting stresses. For example, the chapter emphasized how adaptations for minimizing water stress affect sunlight interception, and any adaptation for sunlight interception is likely also to affect convective heat exchange. Structural adaptations for sunlight interception also affect nighttime microclimate and frost occurrence. Changes in light interception and light-use efficiency also affect photosynthetic nutrient-use efficiencies. These complex interactions of stresses make inferring the adaptive value of photosynthetic properties difficult. Furthermore, plants rarely express a response in only one character to stress; instead, suites of properties tend to change (Chapin et al. 1993). Thus, models such as Y-plant (e.g., Valladares and Pearcy 1998), which enable testing of the sensitivity of whole-plant carbon gain to changes in one or a suite of properties to multiple changes in environmental conditions, are key to scaling multiple stresses from leaves to levels of organization that are more relevant to evolution (the individual) or ecology (communities and ecosystems).

From a perspective of scale we have by necessity, given the data available, focused primarily on leaves and branches, with some modeling projections for whole plants. We note that there has been only a minimal amount of assessment of how water, temperature extremes, and nutrient stress affect photosynthesis at the ecosystem level. Given that the light response of photosynthesis generally exhibits progressively less saturation at higher levels of structural organization, decreases in quantum yield due to stress might lead to greater reductions in carbon gain at the canopy scale, compared with crown, branch, leaf, and chloroplast levels, or shifting the canopy bulk photosynthetic activity from upper layers to middle and lower, more shaded, canopy layers. At larger (ecosystem, landscape) scales data sets are emerging from eddy-flux studies from around the globe that can be analyzed across seasonal and annual changes in environmental conditions, but to our knowledge no studies that have explicitly examined stress as a potential controller of carbon gain from leaf-to-landscape scales. Such studies are needed to evaluate the relative benefit of the range of leaf and canopy structures displayed in plants that inhabit stressful environments.

Acknowledgments. We thank Dene Charlet (UNLV) for valuable assistance with graphics and final formating of the manuscript.

References

Abrams, M. D. 1990. Adaptations and responses to drought in *Quercus* species of North America. Tree Physiol. 7:227–238.

Ackerly, D. D. 1996. Canopy structure and dynamics: Integration of growth processes in tropical pioneer trees. In: Tropical Forest Plant Ecophysiology. S. S. Mulkey, R. L. Chazdon, and A. P. Smith (eds.), pp. 619–658. New York: Chapman and Hall.

Ackerly, D. D., Knight, C. A., Weiss, S. B., Barton, K., and Starmer, K. P. 2002. Leaf size, specific leaf area and microhabitat distribution of chaparral woody plants: Contrasting patterns in species level and community level analysis. Oecologia 130:449–457.

Aubinet, M., Grelle, A., Ibrom, A., Rannik, Ü., Moncrieff, J., Foken, T., Kowalski, A. S., Martin, P. H., Berbigier, P., Bernhofer, C., Clement, R., Elbers, J., Granier, A., Grunwald, T., Morgenstern, K., Pilegaard, K., Rebmann, C., Snijders, W., Valentini, R., and Vesala, T. 2000. Estimates of the annual net carbon and water exchange of forests: The EUROFLUX methodology. Adv. Ecol. Res. 30:113–175.

Ball, M. C., Hodges, V. S., and Laughlin, G. P. 1991. Cold-induced photoinhibition limits regeneration of snow gum at tree-line. Funct. Ecol. 5:663–668.

Ball, M.C., Egerton, J. J. G., Leuning, R., Cunningham, R. B., and Dunne, P. 1997. Microclimate above grass adversely affects spring growth of seedling snow gum *Eucalyptus pauciflora*. Plant Cell Environ. 20:155–166.

Bao, Y., and Nilsen, E. T. 1988. The ecophysiological significance of leaf movements in *Rhododendron maximum*. Ecology 69:1578–1587.

Beadle, N. C. W. 1966. Soil phosphate and its role in molding segments of the Australian flora and vegetation, with special reference to xeromorphy and sclerophylly. Ecology 47:992–1007.

Bergh, J., Linder, S., Lundmark, T., and Elfving, B. 1999. The effect of water and nutrient availability on the productivity of Norway spruce in northern and southern Sweden. For. Ecol. Manage. 119:51–62.

Berry, S. L., and Roderick, M. L. 2002. Estimating mixtures of leaf functional types using continental-scale satellite and climatic data. Global Ecol. Biogeogr. 11:23–39.

Chapin, F. S. III, Autumn, K., and Pugnaire, F. 1993. Evolution of suites of traits in response to environmental stress. Am. Nat. 142:S78–S92.

Comstock, J. P. 2000. Variation in hydraulic architecture and gas-exchange in two desert sub-shrubs, *Hymenoclea salsola* T. and G.. and *Ambrosia dumosa* Payne. Oecologia 125:1–10.

Comstock, J. P., and Ehleringer, J. R. 1988. Contrasting photosynthetic behavior in leaves and twigs of *Hymenoclea salsola*, a green-twigged warm desert shrub. Am. J. Bot. 75:1360–1370.

Comstock, J., and Ehleringer, J. 1990. Effect of variations in leaf size on morphology and photosynthetic rate of twigs. Funct. Ecol. 4:209–221.

Comstock, J. P., Cooper, T. A., and Ehleringer, J. R. 1988. Seasonal patterns of canopy development and carbon gain in nineteen warm desert shrub species. Oecologia 75:327–335.

Cordell, S., Goldstein, G., Mueller-Dombois, D., Webb, D., and Vitousek, P. M. 1998. Physiological and morphological variation in *Metrosideros polymorpha*, a dominant Hawaiian tree species, along an altitudinal gradient: The role of phenotypic plasticity. Oecologia 113:188–196.

Cordell, S., Goldstein, G., Meinzer, F. C., and Handley, L. L. 1999. Allocation of nitrogen and carbon in leaves of *Metrosideros polymorpha* regulates carboxylation capacity and $\delta^{13}C$ along an altitudinal gradient. Funct Ecol. 13:811–818.

Dudley, S. A. 1996. Differing selection on plant physiological traits in response to environmental water availability: A test of adaptive hypotheses. Evolution 50:92–102.

Egerton, J. J. G., Banks, J. C. G., Gibson, A., Cunningham, R. B., and Ball, M. C. 2000. Facilitation of seedling establishment: Reduction in irradiance enhances winter growth of *Eucalyptus pauciflora*. Ecology 81:1437–1449.

Ehleringer, J. 1982. The influence of water stress and temperature on leaf pubescence development in *Encelia farinosa*. Am. J. Bot. 69:670–675.

Ehleringer, J. R., and Cook, C. S. 1990. Characteristics of *Encelia* species differing in leaf reflectance and transpiration rate under common garden conditions. Oecologia 82: 484–489.

Ehleringer, J., Mooney, H. A., Gulmon, S. L., and Rundel, P. W. 1981. Parallel evolution of leaf pubescence in *Encelia* in coastal deserts of North and South America. Oecologia 49:38–41.

Ehleringer, J. R., Comstock, J. P., and Cooper, T. A. 1987. Leaf-twig carbon isotope ratio differences in photosynthetic-twig desert shrubs. Oecologia 71:318–320.

Ellsworth, D. S., and Reich, P. B. 1993. Canopy structure and vertical patterns of photosynthesis and related leaf traits in a deciduous forest. Oecologia 96:169–178.

Evans, J. R., and Loreto, F. 2000. Acquisition and diffusion of CO_2 in higher plant leaves. In: Photosynthesis: Physiology and Metabolism. R. C. Leegood, T. D. Sharkey, and S. von Caemmerer (eds.), pp 321–351. Dordrecht, The Netherlands: Kluwer Academic Publishers.

Ezcurra, E., Montana, C., and Arizaga, S. 1991. Architecture, light interception, and distribution of *Larrea* species in the Monte Desert, Argentina. Ecology 72:23–34.

Ezcurra, E., Arizaga, S., Valverde, P. L., Mourelle, C., and Flores-Martinez, A. 1992. Foliole movement and canopy architecture of *Larrea tridentata* DC. Cov. in Mexican deserts. Oecologia 92:83–89.

Gauslaa, Y. 1984. Heat resistance and energy budget in different Scandanavian plants. Holarctic Ecol. 7:1–78.

Geller, G. N., and Nobel, P. S. 1986. Branching patterns of columnar cacti: Influences on PAR interception and CO_2 uptake. Am. J. Bot. 73:1193–1200.

Germino, M. J., and Smith, W. K. 1999. Sky exposure, crown architecture, and low-temperature photoinhibition in conifer seedlings at alpine treeline. Plant Cell Environ. 22:407–415.

Germino, M, J., and Smith, W. K. 2000a. High resistance to low-temperature photoinhibition in two alpine, snowbank species. Physiol. Plant 110:89–95.

Germino, M. J., and Smith, W. K. 2000b. Differences in microsite, plant form, and low-temperature photoinhibition in alpine plants. Arctic Antarctic Alp. Res. 32:388–396.

Germino, M. J., and Smith, W. K. 2001. Relative importance of microhabitat, plant form and photosynthetic physiology to carbon gain in two alpine herbs. Funct. Ecol. 15:243–251.

Gibson, A. C. 1996. Structure-Function Relations of Warm Desert Plants. Berlin: Springer-Verlag.

Gibson, A. C. 1998. Photosynthetic organs of desert plants. BioScience. 48:911–920.

Goldstein, G., Drake, D. R., Melcher, P., Giambelluca, T. W., and Heraux, J. 1996. Photosynthetic gas exchange and temperature-induced damage in seedlings of the tropical alpine species *Argyroxiphium sandwicense*. Oecologia 106:298–307.

Groom, P. K., and Lamont, B. B. 1997. Xerophytic implications of increased sclerophylly: Interactions with water and light in *Hakea psilorrhyncha* seedlings. New Phytologist 136:231–237.

Gutschick, V. P., and Wiegel, F. W. 1988. Optimizing the canopy photosynthetic rate by patterns of investment in specific leaf mass. Am. Nat. 132:67–86.

Hadley, J., Smith, W. K. 1987. Influence of krummholz mat microclimate on needle physiology and survival. Oecologia 73:82–90.

Hultine, K. R., and Marshall, J. D. 2000. Altitude trends in conifer leaf morphology and stable carbon isotope composition. Oecologia 123:32–40.

Jarvis, P. G., and Leverenz, J. W. 1983. Productivity of temperate, deciduous, and evergreen forests. In: Encyclopedia of Plant Physiology, New Series, Vol. 12D. O. L. Lange, P. S. Nobel, C. B. Osmond, and H. Ziegler (eds.), pp. 233–280. Berlin: Springer-Verlag.

Jordan, D. N., and Smith, W. K. 1994. Energy balance analysis of nighttime leaf temperatures and frost formation in a subalpine environment. Agric. For. Meteorol. 71:359–372.

Jordan, D. N., and Smith, W. K. 1995a. Microclimate factors influencing the frequency and duration of growth season frost for subalpine plants. Agric. For. Meteorol. 77:17–30.

Jordan, D. N., and Smith, W. K. 1995b. Radiation frost susceptibility and the association between sky exposure and leaf size. Oecologia 103:43–48.

Kaiser, W. M. 1987a. Effects of water deficit on photosynthetic capacity. Physiol. Plant. 71:142–149.

Kaiser, W. M. 1987b. Methods for studying the mechanisms of water stress effects on

photosynthesis. In: Plant Responses to Stress: Functional Analysis in Mediterranean Ecosystems. J. D. Tenhunen, F. M. Catarino, O. L. Lange, and W. C. Oechel (eds.), pp. 77–93. New York: Berlin Heidelberg: Springer-Verlag.
Körner, C. 1999. Alpine Plant Life: Functional Plant Ecology of High Mountain Systems. Berlin: Springer-Verlag.
Küppers, M. 1989. Ecological significance of aboveground architectural patterns in woody plants: a question of cost-benefit relationships. Trends Ecol. Evol. 4:375–379.
Lamont, B. B., Groom, P. K., and Cowling, R. M. 2002. High leaf mass per area of related species assemblages may reflect low rainfall and carbon isotope discrimination rather than low phosphorous and nitrogen concentrations. Funct. Ecol. 16:403–412.
Leuning, R., and Cremer, K. W. 1998. Leaf temperatures during radiation frost. Part 1. Observations. Agric. For. Meteorol. 42:121–133.
Leuwchner, C. 2000. Are high elevations in tropical mountains arid environments for plants? Ecology 81:1425–1436.
Meziane, D., and Shipley, B. 1999. Interacting determinants of specific leaf area in 22 herbaceous species: effects of irradiance and nutrient availability. Plant Cell Environ. 22:447–459.
Monson, R. K., Smith, S. D., Gehring, J. L., Bowman, W. D., and Szarek, S. R. 1992. Physiological differentiation within an *Encelia farinosa* population along a short topographic gradient in the Sonoran Desert. Funct. Ecol. 6:751–759.
Morgan, J. M. 1984. Osmoregulation and water stress in higher plants. Annu. Rev. Plant Physiol. 35:299–319.
Neufeld, H. S., Meinzer, F. C., Wisdom, C. S., Sharifi, M. R., Rundel, P. W., Neufeld, M. S., Goldring, Y., and Cunningham, G. L. 1988. Canopy architecture of *Larrea tridentata* DC. Cov., a desert shrub: Foliage orientation and direct beam radiation interception. Oecologia 75:54–60.
Niinemets, Ü. 1999. Components of leaf dry mass per area—thickness and density—alter leaf photosynthetic capacity in reverse directions in woody plants. New Phytologist 144: 35–47.
Niinemets, Ü. 2001. Global-scale climatic controls of leaf dry mass per area, density, and thickness in trees and shrubs. Ecology 82:453–469.
Niinemets, Ü., and Kull, O. 1998. Stoichiometry of foliar carbon constituents varies along light gradients in temperate woody canopies: Implications for foliage morphological plasticity. Tree Physiol. 18:467–479.
Niinemets, Ü., and Lukjanova, A. 2003. Needle longevity, shoot growth and branching frequency in relation to site fertility and within-canopy light conditions in *Pinus sylvestris*. Ann. For. Sci. 60:195–208.
Niinemets, Ü., and Tenhunen, J. D. 1997. A model separating leaf structural and physiological effects on carbon gain along light gradients for the shade-tolerant species *Acer saccharum*. Plant Cell Environ. 20:845–866.
Niinemets, Ü., Kull, O., and Tenhunen, J. D. 1999. Variability in leaf morphology and chemical composition as a function of canopy light environment in co-existing trees. Int. J. Plant Sci. 160:837–848.
Niinemets, Ü., Ellsworth, D. S., Lukjanova, A., and Tobias, M. 2001. Site fertility and the morphological and photosynthetic acclimation of *Pinus sylvestris* needles to light. Tree Physiol. 21:1231–1244.
Niinemets, Ü., Cescatti, A., Lukjanova, A., Tobias, M., and Truus, L. 2002a. Modification of light-acclimation of *Pinus sylvestris* shoot architecture by site fertility. Agric. For. Meteorol. 11:121–140

Niinemets, Ü., Portsmuth, A., and Truus, L. 2002b. Leaf structure and photosynthetic characteristics, and biomass allocation to foliage in relation to foliar nitrogen content and tree size in three *Betula* species. Ann. Bot. 89:191–204.

Nilsen, E. T. 1991. The relationship between freezing tolerance and thermotropic leaf movement in five *Rhododendron* species. Oecologia 87:63–71.

Nilsen, E. T., Meinzer, F. C., and Rundel, P. W. 1989. Stem photosynthesis in *Psorothamnus spinosus* smoke tree in the Sonoran Desert of California. Oecologia 79:193–197

Nobel, P. S. 1988. Environmental biology of agaves and cacti. Cambridge: Cambridge University Press.

Oleksyn, J., Modrzynski, J. et al. 1998. Growth and physiology of *Picea abies* populations from elevational transects: Common garden evidence for altitudinal ecotypes and cold adaptation. Funct. Ecol. 12:573–590.

Osmond, C. B., Smith, S. D., Ben, G.-Y., and Sharkey, T. D. 1987. Stem photosynthesis in a desert ephemeral, *Eriogonum inflatum*: Characterization of leaf and stem CO_2 fixation and H_2O vapor exchange under controlled conditions. Oecologia 72:542–549.

Palmroth, S., Stenberg, P., Smolander, S., Voipio, P., and Smolander, H. 2002. Fertilization has little effect on light-interception efficiency of *Picea abies* shoots. Tree Physiol. 22: 1185–1192.

Poorter, L. 1999. Growth responses of fifteen rain forest tree species to a light gradient: The relative importance of morphological and physiological traits. Funct. Ecol. 13:396–410.

Poorter, L., and Werger, M. J. A. 1999. Light environment, sapling architecture, and leaf display in six rain forest tree species. Am. J. Bot. 86:1464–1473.

Rambal, S., Ourcival, J.-M., Joffre, R., Mouillot, F., Nouvellon, Y., Reichstein, M., and Rocheteau, A. 2003. Drought controls over conductance and assimilation of a Mediterranean evergreen ecosystem: Scaling from leaf to canopy. Global Change Biol. 9:1813–1824.

Rasmuson, K. E., Anderson, J. E., and Huntly, N. 1994. Coordination of branch orientation and photosynthetic physiology in the Joshua tree *Yucca brevifolia*.. Great Basin Nat. 54:204–211.

Reich, P. B., Walters, M. B., and Ellsworth, D. S. 1997. From tropics to tundra: Global convergence in plant functioning. Proc. Natl. Acad. Sci. USA 94:13730–13734.

Reich, P. B., Ellsworth, D. S., and Walters, M. B. 1998a. Leaf structure specific leaf area. modulates photosynthesis-nitrogen relations: Evidence from within and across species and functional groups. Funct. Ecol. 12:948–958.

Reich, P. B., Walters, M. B., Ellsworth, D. S. et al. 1998b. Relationships of leaf dark respiration to leaf nitrogen, specific leaf area and leaf life-span: A test across biomes and functional groups. Oecologia 114:471–482.

Roberntz, P. 1999. Effects of long-term CO_2 enrichment and nutrient availability in Norway spruce. I. Phenology and morphology of branches. Trees 13:188–198.

Roderick, M. L., Berry, S. L., and Noble, I. R. 2000. A framework for understanding the relationship between environment and vegetation based on the surface area to volume ratio of leaves. Funct. Ecol. 14:423–437.

Rosati, A., Day, K. R., and DeJong, T. M. 2000. Distribution of leaf mass per unit area and leaf nitrogen concentration determine partitioning of leaf nitrogen within tree canopies. Tree Physiol. 20:271–276.

Sandquist, D. R., Ehleringer, J. R. 1997. Intraspecific variation in leaf pubescence and drought adaptation in *Encelia farinosa* associated with contrasting desert environments. New Phytologist 135:635–644.

Sandquist, D. R., and Ehleringer, J. R. 1998. Intraspecific variation in drought adaptation in brittlebush: Leaf pubescence and timing of leaf loss vary with rainfall. Oecologia 113:162–169.

Smith, W. K., and Brewer, C. A. 1994. The adaptive importance of shoot and crown architecture in conifer trees. Am. Nat. 143:528–532.

Smith, W. K., and Carter, G. A. 1988. Shoot structural effects on needle temperatures and photosynthesis in conifers. Am. J Bot. 75:496–500.

Smith, W. K., and Geller, G. A. 1979. Plant transpiration at high elevations: Theory, field measurements, and comparisons with desert plants. Oecologia 41:109–122.

Smith, W. K., and Nobel, P. S. 1977. Influences of seasonal changes in leaf morphology on water-use efficiency for three desert broadleaf shrubs. Ecology 58:1033–1043.

Smith, W. K., and Nobel, P. S. 1978. Influence of irradiation, soil water potential, and leaf temperature on leaf morphology in a desert broadleaf, *Encelia farinosa* Compositae. Am. J. Bot. 65:429–432.

Smith, S. D., and Osmond, C. B. 1987. Stem photosynthesis in a desert ephemeral, *Eriogonum inflatum*: Morphology, stomatal conductance and water-use efficiency in field populations. Oecologia 72:533–541.

Smith, W. K., Knapp, A. K., Pearson, J. A., Varman, J. H., Yavitt, J. B., and Young, D. R. 1983. Influence of microclimate and growth form on plant temperatures of early spring species in a high-elevation prairie. Am. Midl. Nat. 109:380–389.

Smith, S. D., Didden-Zopfy, B., and Nobel, P. S. 1984. High temperature responses of North American cacti. Ecology 65:643–651.

Smith, S. D., Monson, R. K., Anderson, J. E. 1997a. Physiological ecology of North American desert plants. New York: Springer-Verlag.

Smith, W. K., Vogelmann, T. C., DeLucia, E. H., Bell, D. T., and Shepherd, K. A. 1997b. Leaf form and photosynthesis. BioScience. 47:785–793.

Smith, W. K., Bell, D. T., and Shepherd, K. A. 1998. Associations between leaf structure, orientation, and sunlight exposure in five Western Australian communities. Am. J. Bot. 85:56–63.

Steenbergh, W. F., Lowe, C. H. 1969. Critical factors during the first years of life of the saguaro *Cereus giganteus* at Saguaro National Monument, Arizona. Ecology 50:825–834.

Stenberg, P., Smolander, H., Sprugel, D. G., and Smolander, S. 1998. Shoot structure, light interception, and distribution of nitrogen in an *Abies amabilis* canopy. Tree Physiol. 18:759–767.

Syvertsen, J. P., Lloyd, J., McConchie, C., Kriedemann, P. E., and Farquhar, G. D. 1995. On the relationship between leaf anatomy and CO_2 diffusion through the mesophyll of hypostomatous leaves. Plant Cell Environ. 18:149–157.

Szarek, S. R., and Woodhouse, R. M. 1978. Ecophysiological studies of Sonoran Desert plants. IV. Seasonal photosynthetic capacities of *Acacia greggii* and *Cercidium microphyllum*. Oecologia 37:221–229.

Tardieu, F., and Davies, W. J. 1993. Integration of hydraulic and chemical signaling in the control of stomatal conductance and water status of droughted plants. Plant Cell Environ. 16:341–349.

Ustin, S. L., Jacquemoud, S., and Govaerts, Y. 2001. Simulation of photon transport in a three-dimensional leaf: Implications for photosynthesis. Plant Cell Environ. 24: 1095–1103.

Valladares, F. 1999. Architecture, ecology, and evolution of plant crowns. In: Handbook of Functional Plant Ecology. F. I. Pugnaire and F. Valladares (eds.), pp. 121–194. New York: Marcel Dekker.

Valladares, F., and Pearcy, R. W. 1997. Interactions between water stress, sun-shade acclimation, heat tolerance and photoinhibition in the sclerophyll *Heteromeles arbutifolia*. Plant Cell Environ. 20:25–36.

Valladares, F., and Pearcy, R. W. 1998. The functional ecology of shoot architecture in sun and shade plants of *Heteromeles arbutifolia* M. Roem., a California chaparral shrub. Oecologia 114:1–10.

Walters, M. B., and Reich, P. B. 1999. Low-light carbon balance and shade tolerance in the seedlings of woody plants: Do winter deciduous and broad-leaved evergreen species differ? New Phytologist 143:143–154.

Witkowski, E. T. F., and Lamont, B. B. 1991. Leaf specific mass confounds leaf density and thickness. Oecologia 88:486–493.

Part 7

Overview

11
Summary and Future Perspectives

William K. Smith, Park S. Nobel, William A. Reiners,
Thomas C. Vogelmann, and Christa Chritchley

Although adaptations in the photosynthetic process occur across the hierarchy of botanical organization, evolutionary change by natural selection acts only on the organism, within the framework of the population. However, selective pressure for specific organism traits can be generated at higher levels of organization and complexity due to emerging constraints on resource acquisition (Chapter 1, Fig. 1.1). It is also important to understand that upscale adaptations may provide the selective pressure for downscale adaptations that will be complementary. As demonstrated in the preceding chapters, evidence for adaptations in photosynthesis continue to emerge at higher levels of the structural/spatial hierarchy, and may often be accompanied by corresponding metabolic changes at the cell and chloroplast level. However, these metabolic, biochemical traits may be more highly conserved compared with those governing diversity in form.

Sunlight Capture and Processing

Chapter 2 points to the lack of information about the mechanics of light absorption at the chloroplast level (both photosystems), the control of chloroplast exposure (including motility), and the potentially large impact of internal cell and tissue structure on light capture, including concerns for both too few (shade chloroplasts) and too many (sun chloroplasts) photons. Differences among species at the chloroplast level seem much more conservative than species differences in anatomy, and little is known regarding photosynthetic capacity per unit chlorophyll (cell and tissue level) among species. A key question is the wavelength dependence of quantum yield. Finally, more information is needed on the effects of upscale characteristics such as leaf orientation to direct sunlight and the complex distribution of diffuse and direct sunlight that results on both leaf surfaces.

Chapter 3 probes the potentially strong influence of higher structural/spatial organization on sunlight capture and photosynthetic performance (e.g. linearization of the light response). The constraints of plant height for competitive advantages is reflected in the accompanying decrease in leaf width, allowing for crown/canopy regulation of penumbra (diffuse sunlight) and numbra (direct-beam sunlight). The need for more accurate characterization of sunlight penetration into crowns and canopies, and onto leaf surfaces, is an important hurdle for predicting quantitatively the impacts of leaf, crown, and canopy architecture on photosynthetic carbon gain. These data will enable more accurate estimates of the potential for influencing photosynthetic quantum efficiency further upscale. Improved techniques such as Monte Carlo ray tracing may provide some breakthroughs in this area.

In Chapter 4 an important point is made regarding the processing of absorbed sunlight for photosynthesis. The efficiency in quantum yield appears to always be greater than 85% of its theoretical maximum, a remarkably high value considering all the possible constraints caused by the possibility of too few, as well as too many photons for maximal photosynthetic excitation inside a leaf or individual cell. Also, changes in growth responses leading to structural responses, although on a slower time scale, seem to be in strong concert with adaptations involving photosynthetic metabolism.

CO_2 Capture and Processing

The adaptive venues for CO_2 capture and processing by plants appear to be divided into a less equal dichotomy compared with sunlight. While the processing of harvested CO_2 involves a spectrum of metabolic pathways incorporating C_3, C_4, and CAM pathways, plus capabilities for switching between modes, CO_2 capture alternatives via stomatal pore effects seems less diverse. This is despite the fact that stomatal limitations to CO_2 capture are often considered as dominating the diffusion pathway to carboxylation (although see Ethier and Livingston 2004 for rebuttal). Chapter 5 points out that more research is needed to evaluate the potential role of chloroplast distribution and mobility for increasing the liquid-phase conductance of fixed carbon. A key to quantifying this term could be provided by measurements of chloroplast area most available for absorption, for example, oppressed close to the plasmalemma and cell wall. The role of aquaporins may also be an important component for increasing cellular conductance that needs further study and clarification.

In Chapter 6, it is emphasized that there are few adaptations recognized that could influence directly CO_2 capture at the leaf, branch, crown, canopy, or landscape levels. Yet, effects of aerodynamic properties on ambient CO_2 concentrations could occur at all structural levels, especially for aggregated leaves and branches within crowns, canopies and stands (e.g., heat/mass transfer and source/sink effects). For one example, understory forest herbs with leaf rosettes oppressed tightly to the ground have utilized high concentrations of CO_2 emanating from the soil surface, especially during early morning hours when concentrations were greatest and stomata fully open (W. K. Smith and D. R. Young, unpublished data). Other environmental factors such as sunlight penetration can also be strongly influenced at higher levels of organization, and with potentially major influences on photosynthesis at the organism level (Chapter 3). In contrast, although the Calvin cycle seems to be a highly conserved metabolic pathway among all plant species, variations in starch metabolism and C_4 pathways seem much more diverse (Chapter 7). Also, there appears to a strong association between photosynthetic carbon processing and plant growth via sugar and nitrogen sensing. All of which appear to be tightly coupled to hormone actions, although on different time scales. The importance of CO_2 processing from the leaf to the landscape level involves the capacity and efficiency for allocating fixed carbon to growth processes, without the potentially strong limitations of sink effects (Chapter 8). Furthermore, a strong coupling between photosynthetic production and respiratory support for growth and repair processes will be an important driving force for future changes under a warming global climate with enriched levels of atmospheric CO_2.

Environmental Constraints

Chapters 9 and 10 focus on natural stress factors, as well as the recent appearance of anthropogenic stresses. From the chloroplast to leaf level, stomatal and photosystem II responses are seen as particularly sensitive to these environmental changes. At the leaf level and beyond, a host of factors can influence sunlight capture for photosynthesis, as well as important factors related to water and temperature stress. The problem is one of too much as well as too little sunlight, both of which appear to be major limitations to photosynthetic carbon gain among different species and habitats.

In complex vegetation, as opposed to single cells, and in the heterogeneous environments that characterize most landscapes, the question of too much or too little sunlight cannot be evaluated without accounting for the structure of plants, the aggregate structure of plant assemblages in ecosystems, and in terms of the variation in resource supply and disturbance in the complicated time and space scales of landscapes. Too much sunlight might prevail in the tops of canopies and too little in understories. Too much sunlight may be the dominant photon flux stress for assemblages where vegetation is scanty on xeric ridgetops of rolling landscapes, and too little may be the dominant stress for most of the canopies in well-watered portions of landscapes. The situation might shift in landscape space depending on season, frequency of water supply through precipitation or groundwater supply, and the occurrence of disturbances and subsequent reestablishment and growth of different plant species adapted for dispersal, establishment, growth rates, temporal extent of growth, and longevity. Nutrient supply over landscapes and over time is yet another component of complexity controlling enzyme synthesis, biomass accrual, and photosynthetic area.

Much more information is needed regarding the interactions and possible synergisms of multiple stress factors, the most common situation in the field. While progress is being made at the leaf, shoot, and branch levels, data are sorely needed that evaluate responses at the crown, canopy, stand, and landscape levels to environmental constraints before accurate estimates of community and ecosystem carbon dynamics, now or in the future, can be formulated.

Photosynthesis and Complexity

"Complexity" is defined as a system of two or more parts organized in a complicated way that is difficult to understand, solve, or explain. Photosynthesis is complex, as is any part of nature, but the nature of complexity depends on the scale at which it is examined. Photosynthesis is explained at several hierarchical levels in this book, from molecular to landscape levels. Each of these levels is complex in itself and amenable to asking different kinds of questions (Allen 1998). Interestingly, the relevance of different aspects of photosynthesis changes as the viewpoint moves scales. Central factors at one scale of focus may possibly disappear at others due to offsetting or compensating mechanisms. Thus, the nature and perhaps the degree of complexity is related to the level of detail, or scale, at which an entity is examined. Photosynthesis, along with other natural phenomena can be effectively conceptualized as a hierarchical system, each hierarchical level having its own level of complexity. "Hierarchy theory" is well developed and widely accepted as a conceptual framework for understanding systems of all kinds (Allen and Starr 1982). In these few paragraphs, the nature of a hierarchical approach is sketched out as an intellectual framework that may be useful for organizing all the complexity surrounding photosynthesis.

Hierarchy theory organizes systems of interest into nested systems or subsystems within higher systems. Moving from one level of system to another requires a change of focus to another system of interacting parts. For example, a landscape can be viewed as consisting of patches, patches viewed as consisting of individual plants and animals, plants consist of organ systems like branches, branches consist of leaves, leaves of cells, and so forth. Obviously, there are many ways one can subdivide a system into subsystems depending on the question at hand. For example, in addressing a hydrological question, the landscape could be viewed as consisting of subbasins rather than patches, which are better demarcated by vegetation discontinuities (Allen and Hoekstra 1990). Similarly, a summation of photosynthesis at an annual time step could be interpreted in terms of patches having similar vegetative structure, or as gradients of leaf area.

Each of the nested subsystems can be treated as a self-organizing entity existing within the context of the next larger system. Thus, individual plants have the genetic information for potential photosynthesis and growth, but those functions are constrained by limits imposed by the larger, subtending system through effects on resource availability. Viewed differently, the aggregate performance of plants within a patch, such as annual primary productivity, is an emergent property of that patch, as well as input into the subtending, higher order, landscape system. Theoretically, every layer of a nested, hierarchical system influences the other layers through a system of constraints and performance criteria (see Chapter 1, Fig. 1.1) (O'Neil et al. 1986). This theory also states that temporal variation in input signals, such as varying sunlight, is smoothed and linearized with each successive hierarchical level. Thus, as sunlight varies at the minute, hourly, daily, and seasonal frequency scales, it is perceived at different amplitudes at each level until eventually all of that variability is attenuated into a smooth annual curve at the highest hierarchical level. This theory is extended somewhat in the observation that nonlinear response systems (e.g., light response of photosynthesis) become more linearized as structural, spatial, and temporal complexity increases up the scale of biological complexity (e.g., Ruimy et al. 1996). In general, it is thought that subsystem dynamics also demonstrate slower response times, often with delays and declining output amplitudes as one analyzes performance from small- to large-scale subsystems. Thus, adaptive interactions between the hierarchical levels represented in Chapter 1, Figure 1.1 follow the same interactive constraints involved in subsystem dynamics. Both feedforward and feedback interactions are possible, involving both structural/spatial and metabolic adaptations in photosynthetic performance. Photosynthesis and other natural phenomena, especially evolved, biological phenomena, are very complex and difficult to comprehend. It is useful to formally organize our knowledge in hierarchical levels, knowing of the upscale and downscale effects that one hierarchical level has on the other.

Case Histories of Photosynthetic Scaling

Desert Succulents

Numerous studies have considered photosynthetic adaptation across two levels of structural organization, for example, chloroplasts and leaf level, or leaves and crowns, but rarely more. For illustration, consider the relatively unusual life form of the typical cactus species. A distinguishing feature of most CAM species compared with most C_3 and C_4 species is the rigidity of the shoots and the opaqueness of the photosynthetic organs. Thus, wind or orientational behavior have essentially no influence on the absorption of sunlight by the photosynthetic surfaces of CAM species, which are fixed in space. Relatively little research has been done on photosynthetic reactions of chloroplasts isolated from agaves (also a rigid form) and cacti. Moreover, because the cellular contents of the cells are not accurately known, photosynthetic rates in vitro would not necessarily equal rates in vivo. Thus, scaling from chloroplasts to photosynthetic organs is not readily feasible for CAM plants. A rather unique factor for CAM species is the thickness of the chlorenchyma, which ranges from 1 to 6 mm for agaves and cacti; for example, the chlorenchyma can be 4 mm thick for *Opuntia ficus-indica*, which is native to eastern Mexico but now cultivated in more than 20 countries. The chlorophyll content of *O. ficus-indica* steadily decreases with depth in the chlorenchyma, from 0.32 g m^{-2} in the outer 1 mm-layer to 0.06 g m^{-2} in the fourth 1-mm layer, whereas the activity of PEPCase, the enzyme that initially binds CO_2 in CAM plants, is similar for the outer three 1-mm-thick layers of the chlorenchyma and decreases 60% for the next layer (Nobel et al. 1994). Even though less than 10% of the photosynthetic photon flux (PPF, wavelengths of 400 to 700 nm) apparently penetrates beyond 1 mm, the nocturnal acidity increase and the incorporation of radioactivity from $^{14}CO_2$ is similar for the outer two 1-mm-thick layers, becoming progressively halved in each of the next two layers.

The total chlorophyll content per unit stem area for *O. ficus-indica* is 0.65 g m^{-2} (Nobel et al. 1994), which is somewhat higher than the 0.4 to 0.5 g m^{-2} typical of leaves of C_3 and C_4 species (Nobel 1999), but occurs over a region that generally is at least 10-fold thicker. The relatively low amount of chlorophyll per unit thickness for CAM plants may be a consequence of the relatively large amount of vacuolar space needed to store the malate and other organic acids that accumulate at night. For *O. ficus-indica*, the acidity increase can easily be over 800 mmol m^{-2}, which corresponds to 200 mM when distributed over 4 mm of solid tissue, but would be more than 2 M if distributed over the less than the 0.4-mm leaf thickness representative of C_3 and C_4 species. The steadily decreasing amount of chlorophyll with depth in the chlorenchyma and low volume fraction of intercellular air spaces has complicated any application of the A^{mes}/A approach (Nobel et al. 1975, Nobel 1999) to CAM plants. Indeed, the demarcation between chlorenchyma and water storage parenchyma is not necessarily sharp, as a few cells containing rather pale chloroplasts occur beneath the main chlorenchyma. Thus, uniformity on a cell surface area basis cannot be assumed for the green tissues of agaves and cacti (Garcia de Cortazar and Nobel 1986). Moreover, air spaces often constitute only 2 to 4% of the chlorenchyma volume (Nobel 1988), suggesting that aqueous-phase diffusion is necessary to move substrates and products of CO_2 fixation between cells (Raveh et al. 1998, Nobel, 1999).

How would such autecological data scale to the community level? Actually, the data are at a community level and reflect the wide spacing between plants and minimal interactions with neighboring plants that characterize the sparse vegetation in many deserts. The PPF index indicates a major difference in the approach for studying the photosynthetic responses of certain agaves and cacti to solar irradiation compared with C_3 and C_4 species. In particular, the rigid shoot surface can be divided into a set of areas whose orientations allow calculation hour-by-hour of the solar irradiation, taking into consideration the sun's trajectory at different latitudes and seasons (Nobel 1988). An important consideration is the fact that PPF does not penetrate through the opaque photosynthetic organs. Thus, the upper side of an agave leaf or the south-facing side of a cactus stem may be well positioned to receive PPF, whereas the leaf underside or the north-facing side of the stem witnesses a low-PPF environment. Early modeling attempts were done with a mature *A. deserti* with 60 leaves, each leaf divided into three subsurfaces for the upper surface and six for the lower surface, or 540 subsurfaces in total (Woodhouse et al. 1980). To provide more architectural realism, the orientations of 27 cladodes on two plants of *O. ficus-indica* were determined in the field; each cladode face was divided into 145 subsurfaces on each side for a total of 7830 subsurfaces (Garcia et al., 1985). Because the cladodes were fixed in three-dimensional space, intercladode shading could be calculated using a ray-tracing technique. Moreover, the spacing between plants could readily be changed in the model, allowing for simulation of effects of stem area index (SAI, total area of both sides of the stems per unit ground area) in daily net CO_2 uptake and hence productivity. An even more ambitious architectural model was done for a mature agave, whose 160 leaves occurred radially in a spiral, each leaf occurring at a 137° angle from the previously unfolded leaf (Nobel and Garcia de Cortazar 1987). Each leaf was divided into 2200 subsurfaces for a total of 355,200 subsurfaces for the plant, each with a specific orientation and location in space. A total of 200,000 parallel solar subbeams were then directed at the "plant" and solar irradiation calculated hourly, taking into consideration self-shading within a plant and shading by an adjacent identical plant, whose spacing could be varied in the model to vary the leaf area index. For predicting net CO_2 uptake by agaves and cacti in the field, scaling from measurements on small parts of the shoot in the laboratory are less of a problem than knowing the actual environmental conditions in the field.

Conifer Trees

Another successful, as well as distinct, life form in the plant kingdom is the characteristic conifer tree, including the typical bottle-brush arrangement of needle-like leaves on individual stems, a layered branch architecture within the crown, and the distinguishable spire-shaped crowns within the stand canopy (Smith and Brewer 1989). As discussed in Chapter 1, the life form of conifer tree species provide an extant example of early adaptations to high sunlight regimes in terrestrial plants (Smith et al. 1997a, Field et al. 2004). There appear to be important adaptive qualities associated with each level of this structural hierarchy extending from the leaf to the shoot, branch, and whole-crown configuration. The small dimension and more circular cross-sectional geometry of a conifer needle (Jordan and Smith 1993), the extreme clustering of needles on sun shoots in contrast to the more planar arrangements on shade shoots, the layered branch architecture of the crown, and the spacing and tapered crowns of individual trees result in apparent advantages that involve a variety of functional relationships directly related to effects on needle temperature and sunlight absorption (Carter and Smith 1989, Smith and Brewer 1990). The cylindrical shape and more-circular cross section of individual needles may be a solution to high irradiance and more xeric regimes at higher elevations, as well as the pervasive threat of low-temperature photoinhibition (Smith et al. 1998). This circular, cross-sectional geometry also invokes radial diffusion instead of planar, and minimizes the need for photon focusing, propagation, and entrapment that may be fundamental in angiosperm leaves (e.g., sun/shade leaves). The same advantage can be ascribed to CO_2 absorption in the mesophyll where CO_2-concentrating mechanisms (e.g., increase in the degree of amphistomy in thicker, sun-type laminar leaves) are no longer necessary. At the shoot level, the highly clustered arrangement of needles also insures boundary-layer warming without significant loss of sunlight interception, maximizing photosynthesis (Smith and Carter 1989). However, this tight clustering with little mutual shading among leaves is possible only with needle-like leaves. In addition, the strength of a cylindrical, sclerophyllous needle is also required to survive severe mechanical damage that would occur during the high winds and snow-loading characteristic of winter. Finally, crown architecture (branch layering) generates greater light penetration to branches lower in the canopy (especially at higher latitudes and lower sun angles), while this same conical crown configuration also reduces mechanical stress from high winds and snow loading (Smith and Brewer 1990). By allowing greater sunlight penetration into the crown where older leaves occur more distally on greater foliated lengths, branch layering at the crown level also complements the longer leaf longevity (evergreen foliage) of conifer branches lower in the crown and canopy. This combination of form and function is necessary for a more efficient investment of leaf biomass in species with such abbreviated growth seasons. Differences in the relative adaptive advantages of each of these structural/spatial features at the needle, shoot, branch, and crown level have not been compared quantitatively for any conifer species. Yet, the idea that new adaptive properties emerge as the level of structural organization increases is readily apparent, as is the idea that subsequent adaptive changes at lower structural levels can occur in response to newly emerged, adaptive properties. Thus, despite the realization that fundamental properties of adaptation occur across the organizational complexity of the conifer tree life form, the quantitative importance of each to photosynthesis is unknown. Interestingly, cellular adjustments to low-temperature photoinhibition of photosynthesis appeared to be substantially less in treeline conifer species than alpine, herbaceous species (Germino and Smith 2000). Thus, adaptations due to differences in plant form and structure across the organizational hierarchy of conifer tree species may dominate, as well as dictate, photosynthetic adaptations at the cell level and below. Regardless, the apparent coupling of both structural and metabolic adaptations that complement one another is a recurring theme in the photosynthetic process of conifer trees. Of course, many other adaptive qualities may be found in the conifer tree form (e.g., evergreen habit, freeze tolerance, seed cones, etc.). A similar, dramatic evolution to needle-like leaves has occurred in the dominant chaparral genus, *Adenostoma*, in response to increasing summer drought conditions following the Late Pliocene (D. D. Ackerly, unpublished).

Adaptive Integration Across the Organizational Hierarchy

It is apparent from the previous chapters that more complex levels of organization (e.g., trees versus leaves; increased plant size and height) may generate new adaptive alternatives, and these same emergent benefits contribute to subsequent adaptive changes at lower organizational levels, for example, the common occurrence of shade and sun leaves inside the shade-generating crowns of larger shrubs and trees. Thus, adaptive properties at higher levels of organization may also drive compensatory, as well as complementary, changes at lower levels. The close association of differences in leaf anatomy and morphology according to the amount of incident sunlight during development is well documented for many plant species. However, such basic relationships as the evolutionary coupling of leaf structure with leaf orientation to direct sunlight and photosynthetic performance is much less documented (Jordan and Smith 1995, Smith et al. 1998, Field et al. 2004). In general, very little is known about the comparative effectiveness of adaptation on one structural level compared with another, or the functional interaction of complementary adaptations across the broad hierarchy of plant structural and spatial complexity. The relative occurrence and coupling of adaptive advantages due to structural versus metabolic modifications are also poorly understood (e.g., Anten and Hirose 2003).

Rarely do studies traverse multiple levels of organization and complexity when evaluating the comparative importance of each level to a fundamental physiological process such as photosynthesis. Thus, it is particularly difficult to evaluate past evolutionary changes in photosynthetic capabilities among different species, or predict future changes under current scenarios of climate change. For example, at what structural/organizational levels are the most important adaptations influencing photosynthesis for extant species today, for example, carbon pathways for photosynthesis (Chapter 7) or sun/shade leaf structure (Chapters 3 and 10)? How do past and future changes at one organizational level scale functionally to other levels, either from the top-down, or from the bottom-up? These questions are fundamental to any comprehensive understanding of the photosynthetic process and its functional importance to the critical carbon exchange dynamics of the earth ecosystem.

Adaptive Potential

The possibility of adaptations at different structural/spatial levels, within the broad spectrum of botanical organization and complexity, presents a rather daunting challenge for understanding past evolutionary changes and, certainly, future changes associated with the rapid, unprecedented anthropogenic change now occurring on a global scale (see review by Ackerly and Monson 2003). In terms of the photosynthetic process, which types of adaptive modes will be most likely to occurr? For example, are adaptations in metabolic physiology, ranging from stomatal function to light and CO_2 processing by mesophyll cells, more likely to occur than adaptations related to structure, such as those that influence light capture (e.g., leaf orientation and arrangement)? To answer this question requires an understanding of the relative effectiveness of each of these broad categories of adaptation, and the potential for selection to act on specific traits within each structural/spatial level of the organizational hierarchy.

In regard to the above, the evolutionary patterns observed within island systems points to an interesting hypothesis. During relatively short evolutionary periods since their formation, islands show a remarkable diversity in form and structure within the same array of descendents (common ancestry), but a highly conserved chloroplast genome (Clegg et al. 1995, Field et al. 2004, Remington and Purugganan 2004). Although, comprehensive studies are lacking, a revealing investigation might be to compare photosynthetic characteristics at the metabolic level with those due to differences in form and structure among island species (e.g., Galpagos finches; Hawaiian silverswords, honey creepers, and *Drosophila*). Among a recently evolved lineage, would differences in the metabolic processes of photosynthesis be greater than changes in photosynthetic performance? Within the plant kingdom, in general, which of these broad categories of adaptive change have had the most important impacts on photosynthetic performance? As one example that illustrates the meaning of this question, three different metabolic pathways have been recognized (C_3, C_4, and CAM), along with distinct intermediates (see Schulze and Caldwell 1995, Sage 2001, Ackerly and Monson 2003 for reviews). Corresponding changes in structure have also been identified (e.g., Kranz anatomy) that are associated characteristically with these carbon pathways, and may also be necessary for their adaptive function. In several of the few studies designed to evaluate gas exchange properties in plants, the ratio of photosynthetic carbon gain to transpiration (water-use efficiency) and a smaller leaf size have been associated with greater survivability (Ehleringer 1993; Dudley 1996a,b). However, accelerated growth (rapid photosynthesis) and phenology was more strongly linked than water-use efficiency to increased survival in a desert shrub (Donovan and Ehleringer 1994). Thus, the relative importance of structural or life history traits versus metabolic traits remains virtually unknown.

In addition to adaptations related to metabolic carbon pathway, there are also ubiquitous patterns in plant form and structure that are characteristic of vascular terrestrial plants. One example is the prevalence of leaves with contrasting sun/shade structure found commonly on the same plant, different plants of the same species, and among different species. Some time ago, the recognition that leaf internal anatomy may have an important quantitative influence on photosynthesis measured for sun versus shade leaves, beyond changes measured at the cell level (Nobel et al. 1975, Björkman 1981). Photosynthetic CO_2 uptake, expressed on a leaf-area basis, increased proportionally with mesophyll cell surface area per unit leaf surface area due to greater leaf thickness and more cells, or a similar leaf thickness and more, smaller cells. By simply stacking more mesophyll cells beneath the same unit area of leaf surface, the higher photosynthesis of sun leaves was achieved, without major changes in cellular light/CO_2 capture and processing within individual cells (Nobel et al. 1987, Börjkman 1981). This finding was one of the first to recognize the potential importance of structural changes alone for modifying photosynthetic performance on a leaf area basis, and without altering the photosynthetic capacity of individual cells or photosynthesis per unit leaf mass. This same phenomenon also accounts for the positive correlation between a higher specific leaf mass (1/specific leaf area) and photosynthesis commonly reported (Reich et al. 2003). However, an inherent assumption that has persisted without testing is that CO_2 uptake does, in fact, remain constant when expressed on a leaf volume, or biomass, basis (Niinemets 1999). This comparison—photosynthesis per leaf area versus per leaf biomass—is a fundamental, quantitative test of the evolution of physiological versus structural adaptation. Photosynthesis per unit leaf area is not a measure of investment efficiency (cost/benefit ratio), but an efficiency based solely on sunlight and CO_2 capture per unit leaf area. Additional arguments were made that leaf morphology and anatomy could also act to increase CO_2 uptake per unit biomass by maximizing the overlap of light and CO_2 gradients inside leaves (Smith et al. 1997). Both the distribution of photons and CO_2 inside the leaf appear to be influenced by the strong interaction between leaf orientation and the occurrence of different cell types (e.g., palisade and spongy mesophyll). In addition, leaf orientation, morphology, and internal anatomy may also interact to enhance photosynthetic performance, and are coupled functionally according to differences in habitat and leaf/plant form that generate specific levels of sunlight interception by the leaf (Smith et al. 1998). Extending these earlier finding to the more recent focus on the measurement of chloroplast surface area could provide another venue for quantifying adaptive effects on CO_2 capture at the cell and internal leaf level (see Chapter 2). Thus, these combined effects at the cell and whole-leaf level may have a large impact on photosynthetic performance due primarily to structural changes, independent of changes in the metabolic biochemistry of individual cells and chloroplasts. Similar evaluations of crown, canopy, stand, and landscape effects on light and CO_2 harvesting, as well as processing, are needed before a comprehensive understanding of the relative importance and coupling of adaptive properties at each organizational level will be possible. Evidence seems to be accumulating in support of a greater adaptive importance for developmental control of functional diversity in plant form, and even spatial patterns, compared with the importance of changes in photosynthetic metabolism associated with chloroplast biochemistry (also see Meinzer 2003). Moreover, it has been suggested that " . . . the evolution of coordinated suites of traits, and adaptive differentiation among populations may often consist of divergence in suites of traits," similar to the examples given above (Geber and Griffin 2003)

References

Ackerly, D. D., and Monson, R. K. (eds.) 2003a. Evolution of Functional Traits in Plants. Int. J. Plant Sci. (Suppl.) 164.

Allen, T. F. H. 1998. The landscape "level" is dead: Persuading the family to take it off the respirator. In: Ecological Scale: Theory and Application. D. L. Peterson and V.T. Parker (eds.), pp. 35–54. New York: Columbia University Press;

Allen, C. D., and Breshears, D. D. 1998. Drought-induced shift of a forest-woodland ecotone: rapid landscape response to climate variation. Proc. Nat. Acad. Sci. 95: 14839–14842

Allen, T. F. H., and Hoekstra, T. W. 1990. The confusion between scale-defined levels and conventional levels of organization in ecology. J Veget. Sci. 1:5–12.

Allen, T. F. H., and Starr, T. B. 1982. Hierarchy Perspectives for Ecological Complexity. Chicago: University of Chicago Press.

Anten, N. P. R., and Hirose, T. 2003. Shoot structure, leaf physiology, and daily carbon gain of plant species in a tall grass prairie. Ecology 84:955–968.

Björkman, O. 1981. Responses to different quantum flux densities. In: Encyclopedia of

Plant Physiology, New Series, Vol. 12A. Physiological Plant Ecology. O. L. Lange, P. S. Nobel, C. B. Osmond, and H. Ziegler (eds.), pp. 57–107. Berlin: Springer-Verlag.

Donovan, L. A., and Ehleringer, J. R. 1994. Carbon isotope descrimination, water use efficiency, growth and mortality in natural shrub populations. Oecologia 100:347–354.

Dudley, S. A. 1996a. Differing selection on plant physiological traits in response to environmental water availability: A test of adaptive hypothesis. Evolution 50:92–102.

Dudley, S. A. 1996b. The response to differing selection on plant physiological traits: Evidence for local adaptation. Evolution 50:103–110.

Ehleringer, J. R. 1993. Carbon and water relations in desert plants: An isotopic perspective. In: Stable Isotopes and Plant Carbon-Water Relations. J. R. Ehleringer, A. E. Hall, and G. D. Farquhar (eds.), pp. 155–172. San Deigo: Academic Press.

Ethier, G. J., and Livingston, N. J. 2004. On the need to incorporate sensitivity to CO_2 transfer conductance into the Farquar-von Caemmerer-Berry leaf photosynthesis model. Plant Cell Environ. 27:137–153.

Field, T. S., Arens, N. C., and Dawson, T. E. 2003. The ancestral ecology of angiosperms: Emerging perspectives from extant basal lineages. In: Evolution of Functional Traits in Plants. D. A. Ackerly and R. C. Monson (eds.), pp. 129–142. Int. J. Plant Sci. (Suppl.) 164.

Garcia de Cortazar, V., and Nobel, P.S. 1986. Modeling of PAR interception and productivity of a prickly pear cactus, *Opuntia ficus-indica* L., at various spacings. Agron. J. 78:80–85.

Geber, M. A., and Griffen, L. R. 2003. Inheritance and natural selection on functional traits. In Evolution of Functional Traits in Plants, eds. D. A. Ackerly and R. C. Monson, pp. S129–S142. Int. J. Pl. Sci. (Supplement) 164, Chicago: University of Chicago Press.

Germino, M. J., and Smith, W. K. 1999. Sky exposure, crown architecture, and low-temperature photoinhibition in conifer seedlings at alpine treeline. Plant Cell Environ. 22:407–415.

Germino, M. J., and Smith, W. K.. 2000. High resistance to low-temperature photoinhibition in two alpine, snowbank species. Physiologia Plantarum 110:89–95.

Givnish, T. J., Montgomery, R. A., and Goldstein, G. 2004. Adaptive radiation of photosynthetic physiology in Hawaiian lobeliads: light regimes, static light responses, and uhde-plant compensation points. Am. J. Bot. 91:228–246.

Jordan, D. N., and Smith, W. K. 1993. Simulated influence of leaf geometry on sunlight interception and photosynthesis in conifer needles. Tree Physiol. 13:29–39.

Jordan, D. N., and Smith, W. K.. 1995. Radiation frost susceptibility and the association between sky exposure and leaf size. Oecologia 103:43–48.

Meinzer, F. C. 2003. Functional convergence in plant responses to the environment. Oecalogia 134:1–11.

Niinemets, Ü. 1999. Components of leaf dry mass per area thickness and density alter photosynthetic capacity in reverse directions in woody plants. New Phytol. 144:35–47.

Nobel, P. S. 1986. Form and orientation in relation to PAR interception by cacti and agaves. In: On the Economy of Plant Form and Function. T. J. Givnish (ed.), pp. 83–103. Cambridge: Cambridge University Press.

Nobel, P. S. 1988. Environmental Biology of Agaves and Cacti. New York: Cambridge University Press.

Nobel, P. S. 1999. Physiochemical and Environmental Plant Physiology, 2nd edition. San Diego: Academic Press.

Nobel, P. S., and Garcia de Cortazar, V. 1987. Interception of photosynthetically active radiation and predicted productivity for *Agave* rosettes. Photosynthetica 21:261–272.

Nobel, P. S., and Hartsock, T. L. 1986. Temperature, water, and PAR influences on predicted and measured productivity of *Agave deserti* at various elevations. Oecologia 68:181–185.

Nobel, P. S., and Loik, M. E. 1999. Form and function of cacti. In: Ecology of Sonoran Desert Plants and Plant Communities. R. H. Robichaux (ed.), pp. 143–163. Tucson: University of Arizona Press.

Nobel, P. S., Zaragoza, L. J., and Smith W. K. 1975. Relation between mesophyll surface area, photosynthetic rate, and illumination level during development for leaves of *Plectranthus parviflorus* Hanckel. Plant Physiol. 55:1067–1070.

Nobel, P. S., Cui, M., and Israel, A. A. 1994. Light, chlorophyll, carboxylase activity and CO_2 fixation at various depths in the chlorenchyma of *Opuntia ficus-indica* (L.) Miller under current and elevated CO_2. New Phytologist 128:315–322.

Raveh, E., Wang, N., and Nobel, P. S. 1998. Gas exchange and metabolite fluctuations in green and yellow bands of variegated leaves of the monocotyledonous CAM species *Agave americana*. Physiol. Plantarum 103: 99–106.

Reich, P. B., Wright, I. J., Cavender-Bares, J., Craine, J. M., Oleksyn, J., Westoby, M., and Walters, M. B. 2003. The evolution of plant functional variation: Traits, spectra, and strategies. In: Evolution of Functional Traits in Plants. Int. J. Plant Sci. (Suppl.) 164:1–6.

Nobel, P. S. 1999. Physicochemical and Environmental Plant Physiology, 2nd ed. San Deigo: Academic Press.

O'Neill, R. V., DeAngelis D. L., Waide J. B, and Allen, T. F. H. 1986. A Hierarchical Concept of Ecosystems. Princeton: Princeton University Press.

Ruimy, A. L., Kergoat, Field, C. B., and Saugier, .B. 1996. The use of CO_2 flux measurements in models of the global terrestrial carbon budget. Global Change Biol. 2:287–296.

Sage, R. F. 2001. Environmental and evolutionary preconditions for the origin and diversification of the C_4 photosynthetic syndrome. Plant Biol. 3:202–213

Schulze, E.-D., and Caldwell, M. M. 1995. Ecophysiology of Photosynthesis. New York: Springer-Verlag.

Smith, W. K., and Brewer, C. A. 1994. The adaptive importance of shoot and crown architecture in conifer trees. Am. Nat. 143:528–532.

Smith, W. K., Vogelmann, T. C., Bell, D. T., DeLucia, E. H., and Shepherd, K. A. 1997. Leaf form and photosynthesis. BioScience 47:785–793.

Smith, W. K., Bell, D. T., and Shepherd, K. A. 1998. Associations between leaf orientation, structure and sunlight exposure in five western Australian communities. Am. J. Bot. 85:56–63.

Woodhouse, R. M., Williams, J. G., and Nobel, P. S. 1980. Leaf orientation, radiation interception, and nocturnal acidity increases by the CAM plant *Agave deserti* (Agavaceae). Am. J. Bot. 67:1179–1185.

Index

Ecological Studies

Volumes published since 1995

Volume 111
Peatland Forestry: Ecology and Principles (1995)
E. Paavilainen and J. Päivänen

Volume 112
Tropical Forests: Management and Ecology (1995)
A.E. Lugo and C. Lowe (Eds.)

Volume 113
Arctic and Alpine Biodiversity: Patterns, Causes and Ecosystem Consequences (1995)
F.S. Chapin III and C. Körner (Eds.)

Volume 114
Crassulacean Acid Metabolism: Biochemistry, Ecophysiology and Evolution (1995)
K. Winter and J.A.C. Smith (Eds.)

Volume 115
Islands: Biological Diversity and Ecosystem Function (1995)
P.M. Vitousek, H. Andersen, and L. Loope (Eds.)

Volume 116
High-Latitude Rainforests and Associate Ecosystems of the West Coast of the Americas: Climate, Hydrology, Ecology and Conservation (1995)
R.G. Lawford, P.B. Alaback, and E.R. Fuentes (Eds.)

Volume 117
Anticipated Effects of a Changing Global Environment on Mediterranean-Type Ecosystems (1995)
J.M. Moreno and W.C. Oechel (Eds.)

Volume 118
Impact of Air Pollutants on Southern Pine Forests (1995)
S. Fox and R.A. Mickler (Eds.)

Volume 119
Freshwaters of Alaska: Ecological Synthesis (1997)
A.M. Milner and M.W. Oswood (Eds.)

Volume 120
Landscape Function and Disturbance in Arctic Tundra (1996)
J.F. Reynolds and J.D. Tenhunen (Eds.)

Volume 121
Biodiversity and Savanna Ecosystem Processes: A Global Perspective (1996)
O.T. Solbrig, E. Medina, and J.F. Silva (Eds.)

Volume 122
Biodiversity and Ecosystem Processes in Tropical Forests (1996)
G.H. Orians, R. Dirzo, and J.H. Cushman (Eds.)

Volume 123
Marine Benthic Vegetation: Recent Changes and the Effects of Eutrophication (1996)
W. Schramm and P.H. Nienhuis (Eds.)

Volume 124
Global Change and Arctic Terrestrial Ecosystems (1996)
W.C. Oechel (Ed.)

Volume 125
Ecology and Conservation of Great Plains Vertebrates (1997)
F.L. Knopf and F.B. Samson (Eds.)

Volume 126
The Central Amazon Floodplain: Ecology of a Pulsing System (1997)
W.J. Junk (Ed.)

Volume 127
Forest Design and Ozone: A Comparison of Controlled Chamber and Field Experiments (1997)
H. Sanderman, A.R. Wellburn, and R.L. Heath (Eds.)

Volume 128
The Productivity and Sustainability of Southern Forest Ecosystems in a Changing Environment (1998)
R.A. Mickler and S. Fox (Eds.)

Volume 129
Pelagic Nutrient Cycles (1997)
T. Andersen

Volume 130
Vertical Food Web Interactions (1997)
K. Dettner, G. Bauer, and W. Völkl (Eds.)

Volume 131
The Structuring Role of Submerged Macrophytes in Lakes (1998)
E. Jeppesen, M. Søndergaard, M. Søndergaard, and K. Christoffersen (Eds.)

Volume 132
Vegetation of the Tropical Pacific Islands (1998)
D. Mueller-Dombois and F.R. Fosberg

Volume 133
Aquatic Humic Substances (1998)
D.O. Hessen and L.J. Tranvik

Volume 134
Oxidant Air Pollution Impacts in the Montane Forests of Southern California: A Case Study of the San Bernardino Mountains (1999)
P.R. Miller and J.R. McBride (Eds.)

Volume 135
Predation in Vertebrate Communities: The Bialowieza Primeval Forest as a Case Study (1998)
B. Jedrzejewska and W. Jedrzejewski

Volume 136
Landscape Disturbance and Biodiversity in Mediterranean-Type Ecosystems (1998)
P.W. Rundel, G. Montenegro, and F.M. Jaksic (Eds.)

Volume 137
Ecology of Mediterranean Evergreen Oak Forests (1999)
F. Roda, J. Retana, C.A. Gracia, and J. Bellot (Eds.)

Volume 138
Fire, Climate Change, and Carbon Cycling in the Boreal Forest (2000)
E.S. Kasischke and B.J. Stocks (Eds.)

Volume 139
Responses of Northern U.S. Forests to Environmental Change (2000)
R.A. Mickler, R.A. Birdsey, and J. Hom (Eds.)

Volume 140
Rainforest Ecosystems of East Kalimantan: El Nino, Drought, Fire and Human Impacts (2000)
E. Guhardja, et al. (Eds.)

Volume 141
Activity Patterns in Small Mammals: An Ecological Approach (2000)
S. Halle and N.-C. Stenseth (Eds.)

Volume 142
Carbon and Nitrogen Cycling in European Forest Ecosystems (2000)
E.-D. Schulze (Ed.)

Volume 143
Global Climate Change and Human Impacts on Forest Ecosystems: Postglacial Development, Present Situation and Future Trends in Central Europe (2001)
P. Joachim and U. Bernhard

Volume 144
Coastal Marine Ecosystems of Latin America (2001)
U. Seeliger and B. Kjerfve (Eds.)

Volume 145
Ecology and Evolution of the Freshwater Mussels Unionoida (2001)
B. Gerhard and W. Klaus (Eds.)

Volume 146
Inselbergs: Biotic Diversity of Isolated Rock Outcrops in Tropical and Temperate Regions (2001)
S. Porembski and W. Barthlott (Eds.)

Volume 147
Ecosystem Approaches to Landscape Management in Central Europe (2001)
J.D. Tenhunen, R. Lenz, and R. Hantschel (Eds.)

Volume 148
A Systems Analysis of the Baltic Sea (2001)
F.V. Wulff, L.A. Rahm, and P. Larsson (Eds.)

Volume 149
Banded Vegetation Patterning in Arid and Semiarid Environments: Ecological Processes and Consequences for Management (2001)
D.J. Tongway, C. Valentin, and J. Seghieri (Eds.)

Volume 150
Biological Soil Crusts: Structure, Function and Management (2001)
J. Belnap and O.L. Lange (Eds.)

Volume 151
Ecological Comparisons of Sedimentary Shores (2001)
K. Reise (Ed.)

Volume 152
Future Scenarios of Global Biodiversity (2001)
F.S. Chapin, III, O.E. Sala, and E. Huber-Sannwald (Eds.)

Volume 177
A History of Atmospheric CO_2 and Its Effects on Plants, Animals, and Ecosystems (2005)
J.R. Ehleringer, T.E. Cerling, and M.D. Dearing (Eds.)

Volume 178
Photosynthetic Adaptation: Chloroplast to Landscape (2004) W.K. Smith, T.C. Voqelmann, and Christa Critchley (Eds.)